Energy Systems in Electrical Engineering

Series Editor

Muhammad H. Rashid, Florida Polytechnic University, Lakeland, USA

More information about this series at http://www.springer.com/series/13509

Rohit Dhiman · Rajeevan Chandel
Editors

Nanoscale VLSI

Devices, Circuits and Applications

Springer

Editors
Rohit Dhiman
Department of Electronics and Communication Engineering
National Institute of Technology Hamirpur (HP)
Hamirpur, Himachal Pradesh, India

Rajeevan Chandel
Department of Electronics and Communication Engineering
National Institute of Technology Hamirpur (HP)
Hamirpur, Himachal Pradesh, India

ISSN 2199-8582 ISSN 2199-8590 (electronic)
Energy Systems in Electrical Engineering
ISBN 978-981-15-7936-3 ISBN 978-981-15-7937-0 (eBook)
https://doi.org/10.1007/978-981-15-7937-0

This Springer imprint is published by the registered company Springer Nature Singapore Pte Ltd.
The registered company address is: 152 Beach Road, #21-01/04 Gateway East, Singapore 189721, Singapore

To my revered Parents, for their affection and untiring efforts in my upbringing. Also dedicated to my wife, Anjali, and our loving daughter, Shipra Dhiman, for their precious time and patience.

—Rohit Dhiman

My loving family, Prof. Ashwani Kumar Chandel, our dear son, Ayush Chandel, and very affectionate daughter-in-law, Nilanshi. My respected mother, mother in law & teachers and dear students who have always been a source of my continued efforts for academic excellence.

—Rajeevan Chandel

Preface

The perpetual scaling of complementary metal-oxide semiconductor (CMOS) technology has resulted in significant performance improvements in very large-scale integration (VLSI) circuit design techniques and system architectures. According to ITRS, Intel's next-generation *i8* billion transistor processors have set out to achieve an industry leading performance of the order of GHz. This trend is expected to continue in future also, but will require breakthroughs in the design of nanoscale VLSI and post-CMOS technologies, generally known as nanoelectronics. With the development of novel materials and nanoscale devices, research is being directed to gain better physical insights of the parameters that influence the device, circuit, and system characteristics. This book titled, Nanoscale VLSI: Devices, Circuits and Applications, is written by such researchers in the respective areas of nanoelectronic devices, integrated circuits (ICs), nanomagnetic computation, and other relevant areas. The 15 chapters of the book are classified under four parts that cover modeling, simulation, and applications of electronic, magnetic, and compound semiconductors in the nanoscale VLSI devices, circuits, and systems. This comprehensive volume eloquently presents the design methodologies for ultra-low-power VLSI design, potential post-CMOS devices, and circuits and their applications from the architectural and system perspective. The book shall serve as an invaluable reference book for the graduate students, Ph.D./M.S./M.Tech. scholars, researchers, and practicing engineers working in the frontier areas of nanoscale VLSI design of devices, circuits, systems, and their applications.

The first part of the book addresses the importance of low voltage and low power in current IC design. Chapter 1: Low-Voltage Analog Integrated Circuit Design, of the book deals with detailed insights about the low-voltage design techniques for analog ICs. In the modern ultra-low-power analog CMOS design, sub-threshold current is utilized as the driving current. Chapter 2: Design Methodology for Ultra-Low-Power CMOS Analog Circuits for ELF-SLF Applications, explores a systematic design methodology based on the inversion coefficients for the design of an operational transconductance amplifier, suitable for extreme low-frequency (ELF) regime. Chapter 3: Orthogonally Controllable Versatile Quadrature Oscillator for Low-Voltage Applications, introduces a dual-mode quadrature oscillator circuit comprising of a single fully differential current conveyor with three grounded resistors and two grounded capacitors. Practical realization of the proposed quadrature oscillator using commercially available ICs has also been illustrated. It is essential to retain power and energy efficiency in ICs over a wide range of load current and voltage. Chapter 4: Design Techniques for Low-Power Integrated Circuits, discusses pulse width modulation, pulse frequency modulation, and pulse skip modulation that result in reduced power dissipation and improved energy efficiency in power ICs.

With the advancements in CMOS scaling, the power density constraint puts a limit on the number of transistors that can be simultaneously switched on. Therefore, to exploit full benefits of scaling, novel post-CMOS devices are extensively investigated and covered in the second part of the book. The application of bilayer (BL) graphene nanoribbon (GNR)-based tunnel field-effect transistor (TFET) for its potential applications in post-CMOS electronics is provided in Chap. 5: Bilayer Graphene Nanoribbon Tunnel FET for Low-Power Nanoscale IC Design. It also covers a comprehensive description of BL-GNR TFET as a potential alternative to monolayer GNR TFETs due to its high ON-state current and low sub-threshold swing. The incorporation of compound semiconductors like InAs, InGaAs, InSb, SiGe that have outstanding carrier transport properties has opened up new vistas to the device designers with faster and better device performance. The impressive potential of SiGe source/drain Si-nanotube junctionless FET for reduced short-channel effects and its threshold voltage behavior is explored in Chap. 6: A Threshold Voltage Model for SiGe Source/Drain Silicon-Nanotube-Based Junctionless Field-Effect Transistor. In Chap. 7, the architecture and electrical performance of III–V nanoscale quantum well FETs for high performance, low-power solid-state IC technology is presented. FinFET technology has seen a major increase in the adoption for use within ICs with faster and better performance. However, process variability is detrimental and can affect the performance of FinFET. Chapter 8: FinFET—A Beginning of Non-planer Transistor Era, elucidates the influence of work function fluctuations on various FinFET-based logic circuits.

The third part of the book provides the possibilities of IC design with some emerging technologies and addresses the challenges that still need to be addressed. Recently, gallium nitride (GaN) has gained tremendous attention due to its high band gap energy and high electric breakdown voltage. This part of the book provides physical insights about the GaN technology and its design space exploration in Chap. 9, Gallium Nitride: Emerging Future Technology for Low-Power Nanoscale IC Design. Voltage-controlled oscillator (VCO) plays a significant role in the realization of phase-locked loop, radio frequency ICs, analog-to-digital converter, and other circuits. In Chap. 10: A Low-Power Hybrid VS CNTFET-CMOS Ring Voltage-Controlled Oscillator using Current Starved Power Switching Technology, a ring VCO is designed using a virtual source carbon nanotube (CNT) FET, CMOS, and current starved power switching technology. In today's sophisticated nano-era and densely packed IC designs, on-chip interconnects determine the overall performance of VLSI circuits and systems. Novel optical interconnects promise many attractive features which make these the most prominent interconnect technology in future silicon-on-insulator chips. The readers can explore some research aspects of optical interconnects in Chap. 11: Chip-Level Optical Interconnect in Electro-optics Platform, of the book. Accurate analytical modeling and simulation of graphene FET for the realization of VLSI circuits is reported in Chap. 12: Emerging Graphene FETs for Next-Generation IC Design.

Over the years, a rapid growth has been witnessed in semiconductor industry because of the huge demand for system level designs. System level designs are prominently used for the various applications such as high-performance computing, control system, telecommunications, image and video processing, consumer electronics, and others. To address this, a holistic approach from the architectural and system perspective is required and is addressed in the last part of the book. The adverse effects of More than Moore and emerging demands of computing on the edge devices necessitate a significant improvement in the energy and area-efficient rebooting computing architecture design. Chapter 13: Power and Area-Efficient Architectural Design Methodology for Nanomagnetic Computation, discusses nanoscale architecture design and its implementation using nanomagnets. This chapter also throws light on graphene-based on-chip clocking interconnect replacing the traditional copper. The VLSI design process of digital signal processing (DSP) hardware is dependent on high-level synthesis framework that comprises of design space exploration of power and area-delay tradeoff and is addressed in Chap. 14: Design Space Exploration of DSP Hardware using Adaptive PSO and Bacterial Foraging for Power/Area-Delay Tradeoff. The part concludes with Chap. 15: Register-Transfer-Level Design for Application-Specific Integrated Circuits.

This book is a unique coverage of topics covering recent advancements in the field of post-CMOS IC design, modeling and simulation, and other potential research areas for the efficient design exploration of low voltage, low power, VLSI devices, circuits, and their applications at the system level.

Hamirpur, India

Rohit Dhiman
Rajeevan Chandel

Acknowledgements We would like to express our heartfelt gratitude to the authors of the individual chapters who have devoted their significant time and contributed their expertise to shape the book. We express our sincere thanks to all the authors for their excellent insights, as we are sure that this edited volume will be a useful text to many readers interested in nanoscale VLSI devices, circuits, and their applications. We are grateful to the Editorial team of the Springer for its tremendous support through the stages of preparation and finally bringing out this book as an excellent academic treasure to its readers.

Contents

About the Editors

Rohit Dhiman received his B.Tech. in Electronics and Communication Engineering from HP University Shimla, India in 2007. He did his M.Tech. Degree in VLSI Design from National Institute of Technology (NIT) Hamirpur in 2009. He was awarded Ph.D. Degree from NIT Hamirpur in 2014. Presently Dr. Rohit Dhiman is working as an Assistant Professor in Electronics and Communication Engineering Department at NIT Hamirpur and is the author/co-author of Research Publications in International Journals and Conference proceedings of repute. He has also edited/ authored three books published by the IET and Springer. He has been awarded with the *Young Scientist Award* from the Science and Engineering Research Board, Department of Science and Technology GoI, New Delhi. He is recipient of the prestigious *Young Faculty Research Fellowship* of the Ministry of Electronics and Information Technology (MeitY), Government of India and has two sponsored research projects to his credit. His major research interest is in device and circuit modelling for low power nanoscale IC design.

Rajeevan Chandel received B.E. Degree in E&CE from Thapar Institute of Engineering and Technology (now Thapar University), Patiala, India in 1990. She is a double gold medalist of Himachal Pradesh University, Shimla in Pre-University and Pre-Engineering in 1985 and 1986 respectively. She did her M.Tech. Degree in Integrated Electronics and Circuits, from Indian Institute of Technology (IIT) Delhi in 1997. She was awarded Ph.D. Degree from IIT Roorkee, India in 2005. Dr. Chandel joined Department of E&CE, NIT Hamirpur as Lecturer in 1990, where presently she is working as Professor. Dr. Chandel has been Head of the E&CE Department twice and was formerly Dean (Research & Consultancy) at NIT Hamirpur. She has over 150 research papers in peer reviewed International Journals of repute and Conferences. She has six sponsored projects to her credit from Government of India. Currently, she is the Chief Investigator of the prestigious SMDP-C2SD project of MeitY, New Delhi. She has also edited/authored three books of IET and Springer. Her research interests are electronics circuit modelling and low power VLSI design. She is a Fellow of IETE (I) and life member of ISTE (I) and member of IEEE, ISSS and VSI.

Part I
Low Voltage and Low Power VLSI Design

Chapter 1
Low-Voltage Analog Integrated Circuit Design

Deepika Gupta

Abstract In this chapter, we review the challenges and effective design techniques for ultra-low-power analog integrated circuits. With the miniaturization, having low-power low-voltage mixed signal IC is essential to maintain the electric field in the device. This constraint presents bottleneck for the researchers to design robust analog circuits. Specifically, the low value of supply voltage with small technology influences many specifications of analog IC, e.g., power supply rejection, dynamic range and immunity to noise, etc. In addition, it also affects the ability of the MOS transistor to be operated in the strong inversion region. Note that with the technology reduction, power supply V_{DD} is reducing but the threshold voltage V_T is not decreasing proportionally to maintain low leakage current. However, this process reduces the overdrive voltage and limits the staking of transistors. In this case, the transistor can be made to work in weak inversion to work and reduce the power consumption. Further, reduction in V_{DD} to achieve low-power consumption causes many other circuit-related issues such as PVT variations, degradation of dynamic range, mismatching in circuits element and differential paths. There have been many design methods developed for the ultra-low-power analog ICs. In this chapter, we will discuss some of the design techniques to reduce the power consumption in analog ICs. In addition, we will also discuss the basic building blocks of analog circuits with discussed design techniques.

Keywords Low-voltage analog circuits · Subthreshold circuit · Bulk-driven circuit · Dynamic threshold MOS transistor · Floating gate MOS transistor

---

D. Gupta (✉)
Department of Electronics and Communication Engineering,
International Institute of Information Technology Naya Raipur, Raipur, India
e-mail: deepika@iiitnr.edu.in

R. Dhiman and R. Chandel (eds.), *Nanoscale VLSI*,
Energy Systems in Electrical Engineering,
https://doi.org/10.1007/978-981-15-7937-0_1

1.1 Introduction

Owing to the rapid miniaturization of electronic circuits, the low-power consumption has become essential in the battery operated systems like wearable health monitoring devices, cellphones, tablets, etc. In these handheld devices, low-power consumption is crucial to maintain the high battery life. Low-power consumption even becomes more crucial for small systems such as Internet of things (IoT) and smart cards, which require batteryless operation. To reduce the power consumption for a electronic system, active cooling can be used. However, they are bulky, may create noise in circuit and are expensive (Svensson 2015). In addition, other logic families such BiCMOS can be used for better performance of electronic circuits with low voltage but at the cost of manufacturing expenses (Svensson 2015). Also, some cutting-edge alternatives such as solar power, fuel cells and RF power may be used to have low-power application in integrated circuits (Zimmer et al. 1989). However, the low-voltage levels of these power sources must be handled very carefully for the proper operations of an integrated circuit (IC). Therefore, there is an increasing demand for the development of sophisticated design techniques to improve the low-power performance of IC (Zhiyuan et al. 2017).

Further, we know that any electronic system is made from both analog and digital component. So, the power needs for both must be understood well to optimize the power of whole system. Importantly, power-aware designs for digital systems are very well known these days. In 1990, researchers Chandrakasan et al. (1992) and Liu and Svensson (1993) started the analysis of power in digital circuits. Due to those efforts, the power consumption for a digital circuit can be summarized in a single equation

$$P = \frac{1}{2}\alpha f C_{\mathrm{L}} V_{\mathrm{DD}}^2$$

where α is the probability for output to change its logic in one clock cycle or the switching activity, C_{L} is the switched capacitance, f is the clock frequency, and V_{DD} is the supply voltage. However, finding such single equation for power consumption of analog IC is quite difficult. Indeed, some researchers have reported the power analysis for the analog circuits, but it is not as systematic as digital circuits (Vittoz 1980, 1990; Bult 2000; Annema et al. 2005). Recently, few researchers have also started focusing on power conscious design for wireless networks (Abidi et al. 2000; Nilsson and Svensson 2014).

Nowadays, analog IC design techniques are required to aim at attaining higher speed with large dynamic range. It is to note that the circuit capacitances contributed from intrinsic capacitance and interconnect parasitic capacitance strongly influence the speed and bandwidth of analog circuits. With scaling, interconnect capacitance starts dominating over the other and simply reducing the size of transistor will not proportionally improve the circuit bandwidth and speed (Rajput and Jamuar 2002). This occurs due to aggressive transistor size scaling as compared to interconnect. Therefore, scaling in analog design may provide higher packing density but there is no significant advantage for speed optimization.

The speed, bandwidth and dynamic range for analog circuits depend largely on the power consumption. For any analog IC design, the power consumption has three main components:

- Continuous charging and discharging of circuit capacitances result in the dynamic power consumption.
- Biasing currents for MOS transistor cause static power consumption in circuit.
- Power consumption due to the flow of current when both PMOS and NMOS transistors are in the ON state is the short-circuit power consumption.

The net power consumption in a circuit can be given as the sum of the power consumptions due to reasons mentioned above. Note that for analog circuits, similar to the digital counterpart, the total power consumption is directly proportional to the supply voltage (Svensson and Wikner 2010). Therefore, to minimize net power consumed by the analog circuit, the obvious way to reduce it would be to operate the circuits at low supply voltages. Also, the reduction of parasitic capacitances can also help in minimization of power consumption in analog circuits.

1.2 Challenges in Low-Voltage Design

Working with low voltage can be a general solution for low-power operation of analog circuits. However, the low-voltage levels available at input of an analog circuit degrade the performance metrics such as bandwidth, voltage swing and dynamic range. In this section, we will review some challenges faced while designing of the low-voltage analog circuits.

1.2.1 Supply Voltage Scaling

To match with the current scaling scenario, the supply voltage available for analog circuit has been scaled down aggressively. Importantly, due to the lower supply voltage at scaled technology, the available voltage headroom for the MOS transistor operation has reduced. Note that the lower voltage headroom for MOS transistor degrades the voltage swing and the dynamic range for analog circuits. Further, the scaling of technology also leads to the higher density on chip, increasing the power dissipation per unit area on substrate. As the substrate is able to handle only specific amount of heat, the power consumption with increased density has to reduce for maintaining the proper operation.

Scaling of the MOS transistor threshold voltage can be adapted to mitigate the effect of low-voltage levels and to increase the voltage swing in scaled analog circuits. However, the reduction of threshold voltage of MOS transistor is quite less significant over the years compared to the scaling scenario of supply voltage, as shown in Fig. 1.1 (Zhao and Cao 2006). From figure, it is clear that the supply voltage is

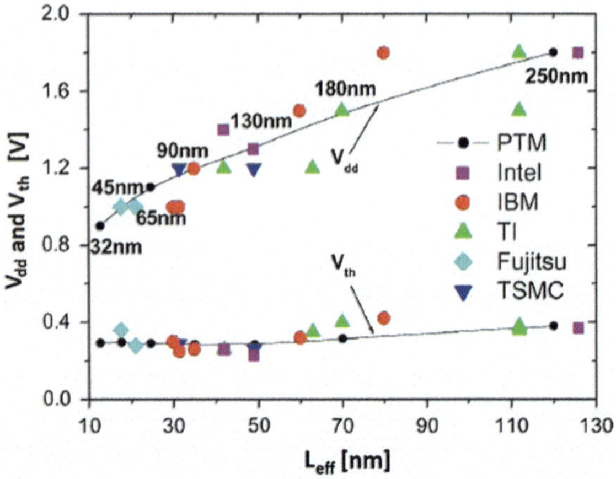

Fig. 1.1 Threshold voltage (V_T) and supply voltage (V_{dd}) scaling trend versus effective channel length (L_{eff}) Zhao and Cao (2006)

scaling very rapidly compared to the threshold voltage. This results in poor ON/OFF characteristics of MOS transistor with degraded voltage swing in analog circuits. Apart from this, due to scaling, the ICs are allowed to operate at high frequency with lower-power consumption. However, for analog circuits, some specific amount of current may be needed to maintain the performance of the circuit with scaling of supply voltage. This may result to an insignificant increment in power consumption. So, one can say, direct reduction of supply voltage and scaling may not reduce the power consumption proportionally in analog circuits. Therefore, some innovative design techniques must be used to overcome the supply voltage scaling limitation and to reduce power consumption in analog IC design.

1.2.2 Transistor Inversion

The cutting-edge fabrication processes pose numerous limitations to low-voltage analog IC design. Importantly, it causes severe constraints on the level of inversion in a MOS transistor. Generally, a MOS transistor is considered to have zero current, hence no channel inversion when its turn on voltage V_{GS} is less than the V_T of the MOS transistor. On the other hand, current flows through the MOS transistor for $V_{GS} > V_T$ and channel is said to be strongly inverted (Sedra and Smith 2011). However, in real transistor, such a sudden transition from OFF state to ON is not possible. Therefore, all the transistors have some smaller amount of channel inversion even for $V_{GS} < V_T$. This region of inversion is called as the weak region of inversion of MOS transistor (Ueno et al. 2009). In strong inversion region, the conducting channel of MOS device is fully formed above V_T, whereas in weak inversion region it is only available partly below V_T. Note that a MOS transistor shows different behavior in strong and weak inversion region. For strong inversion region, the transistor exhibits a square law characteristics, whereas exponential characteristics are observed in weak inversion (Swanson and Meindl 1972).

In conventional analog IC with higher technology nodes, transistor operates in strong inversion region. However, the modern CMOS technology needs significantly lower supply voltage to minimize the high electric field within the small devices. As the reduction of threshold voltage is quite unreasonable compared to the supply voltage scaling for scaled devices (Zhao and Cao 2006), this results in voltage headroom reduction and limits the available turn on voltage for MOS transistor. So, it gets difficult to operate the MOS transistor in strong inversion region with the small input voltage. Furthermore, low-voltage operation with scaled technology also restricts the number of stacked transistors in the circuit which need to be operated in saturation region or strong inversion region (Keane et al. 2008). Here, the stacked transistors in analog circuits need the minimum supply voltage to increase in order to achieve strong inversion. However, for low-voltage operation, the transistors can be operated in weak inversion region of MOS transistor operation at the cost of bandwidth (BW) and area, as will be explained later in this chapter.

1.2.3 Device Models

The optimized low-voltage analog IC design requires minimum consumption of power and/or silicon area while meeting all performance specification such as speed, bandwidth and dynamic range. The unconventional inversion modes of MOS transistor can be explored to achieve this goal in scaled technologies. While operating MOS transistor extremely into the weak inversion and strong inversion, often it will not provide a good trade-off between frequency response, power consumption and silicon area for analog IC design. So to operate the MOS transistor/circuit over unconventional inversion region of MOS operation, one needs innovative design techniques and new accurate simulation models of MOS devices. The MOS transistor model should be able provide single equation solution over the all available inversion regions. The charge-based EKV model can be a very suitable example of a MOS simulation model to be used in all inversion regions of transistor operations (Enz et al. 2018). In EKV model, the smallest number of core parameters is needed for the accurate behavioral modeling of transistor. Particularly, charge-based EKV model is beneficial for the analysis of analog circuits because it allows the analysis with simple calculations over different inversion regions. Hence, developing new device simulation models specific for analog circuit design is crucial.

1.2.4 PVT Fluctuations

Owing to the miniaturization of semiconductor devices, it is very difficult to make sure the fixed geometrical dimensions, various doping profiles, thickness of dielectric region during fabrication of MOS transistor. Due to these physical structural fluctuations of MOS transistor, various electrical parameters of analog IC such as parasitic of chip interconnects and threshold voltage get affected (Onabajosilva and Martinez 2012; Chang et al. 2017). As a result, the performance of circuit deteriorates. In addition, aging also affects the performance of analog IC. Also, aging in circuits results in a long-term slow variations in device performance, causing an organized error. Further, these effects are also described by negative bias temperature instability (NBTI) (Schroder 2007).

These days designed ICs have to deal with industrial real-time problems and take requirements into account. Hence, the analog ICs are also needed to be able to handle temperature variations according to the industry standards in a very wide range. When these ICs are used with such wide range of temperature, some electrical MOS transistor/circuit parameter may get affected (Hosticka et al. 1985). These variations may result in another systematic error in the performance of analog ICs. These variations must be handled very carefully while designing low-voltage analog ICs.

Furthermore, as we know that with low-voltage circuit design, scaling of threshold voltage is not significant compared to the supply voltage. Therefore, very small voltage headroom is available with advanced ICs. However, this voltage headroom problem becomes even more worse with the temperature and other physical variations. In addition, the drain current (I_D) of the MOS transistor shows the extreme temperature sensitivity with the supply voltage scaling (Wolpert and AmpaDdu 2012). So, for MOS transistor, as the voltage scales, the temperature sensitivity of drain current increases. Importantly, for low-voltage IC design, temperature sensitivity of I_D increases significantly if the supply voltage scales below 500 mV.

1.2.5 Dynamic Range

These days, the fabrication technology available with miniaturization also limits the dynamic voltage range of analog ICs, hence affecting the signal-to-noise ratio (SNR). Here, the dynamic range is defined as the ration between maximum supply voltage swing and noise signal voltage. Importantly, for analog ICs, the noise signal voltage is limited by the thermal noise. The behaviors of thermal noise are inverse of the bias current. Hence, the expression of dynamic range of analog IC can be given as

$$\mathrm{DR} = \frac{(V_{\mathrm{DD}} - 2V_{\mathrm{sat}})^2}{\alpha/I}$$

Here, $V_{\mathrm{DD}} - 2V_{\mathrm{sat}}$ shows the full signal swing, I represent the bias current and α is the constant. Therefore, the dynamic range of any analog circuit is the function of bias current (Baschirotto et al. 2009).

Hence, it can be inferred that rail-to-rail operation and differential representation of signals can be used to improve dynamic range of analog IC at the cost of circuit complexity and high-voltage operation.

1.2.6 Mismatching

Differential path and circuit element layout mismatching strongly affect the proper operation of high-performance analog ICs. Therefore, any deviation, either random or systematic creating mismatch, is crucial for the production and reliability of the circuit. Interestingly, the input offset voltage in operational amplifier is an example of mismatching consequences in analog circuits. One can note that input offset voltage is an important parameter while designing an amplifier and affects its other AC/DC specification. Moreover, the performance degradation of analog circuits even becomes poorer for low supply voltages. Many critical parameters of amplifier get degraded due to both mismatch and supply restriction. Therefore, suitable layout techniques should be used to avoid effects of mismatch in analog ICs.

Owing to the different circuit requirement and to overcome the design challenges mentioned above, low-voltage analog IC requires completely different design techniques from the high-voltage counterpart. This generates a requirement for some innovative design techniques to be adapted for the low-voltage analog circuits. One such technique can be the use of current levels for the operation of analog circuits. This current mode technique furnishes a better replacement for low-voltage high-performance analog circuit design. In this case, voltage levels existing at different nodes become irrelevant. In the next section of this chapter, we will review some design techniques for low-voltage analog ICs.

1.3 Low-Voltage Design Techniques

As discussed earlier, while designing a low-voltage circuit, crucial parameters to consider are the noise voltage level and the reduction in threshold voltage of transistor. Better noise immunity can be achieved with the MOS transistor of high threshold voltage but at the cost of voltage headroom. Here, lower threshold voltage of transistor may result in higher-voltage headroom but poorer noise immunity, hence lower SNR. These days, the scaling of threshold voltage is limited to the noise floor level. Any reduction below this may introduce sufficient noises in circuit operation. Here, to overcome the limitation of threshold voltage scaling with valid noise performance, efficient design techniques are needed for low-voltage operation of analog ICs. Now, we will review some of the low-voltage design techniques available in the literature for analog circuits.

1.3.1 Subthreshold Circuits

As already discussed, the operating region of MOS transistor is important to decide various parameters of analog IC design. We know that operation of transistor in weak inversion region allows the designer to work with low supply voltages, hence low-power consumption in analog IC design. Whereas, the strong inversion region of MOS transistor operation can make circuit to work with good frequency response. We know that for a MOS transistor when V_{GS} is greater than V_T (i.e., strong inversion region), drain current flows and the MOS transistor is said to be ON. The considered MOS transistor model with all terminal voltages is shown in Fig. 1.2.

The drain current in strong inversion region can be given as (Shah 1964)

$$I_{DS} = (K' W/L)[(V_{GS} - V_T) - V_{DS}/2]V_{DS}.$$

Further, according to above expression V_{GS} less than V_T, no current flows through the MOS transistor and it is considered to be OFF. However, in reality, a very small amount of current flows through the MOS transistor for V_{GS} less than V_T due to weak channel inversion. The operating region of MOS transistor with weak channel inversion is called as the subthreshold region of operation. In subthreshold region of MOS transistor operation, the current is exponentially proportional to the applied voltage (Ueno et al. 2009; Shah 1964; Geiger et al. 1990) and is given as follows:

$$I_{DS} = \frac{2K_{\prime}W}{L}\left(\frac{nkT}{qe}\right)^2 \exp\frac{q(V_{GS} - V_{T,nmos})}{nkT}$$

Here, n represents the subthreshold slope factor. Its value typically lies between 1.2 and 2. Further, q, $V_{T,nmos}$, k, T represent electronic charge, threshold voltage

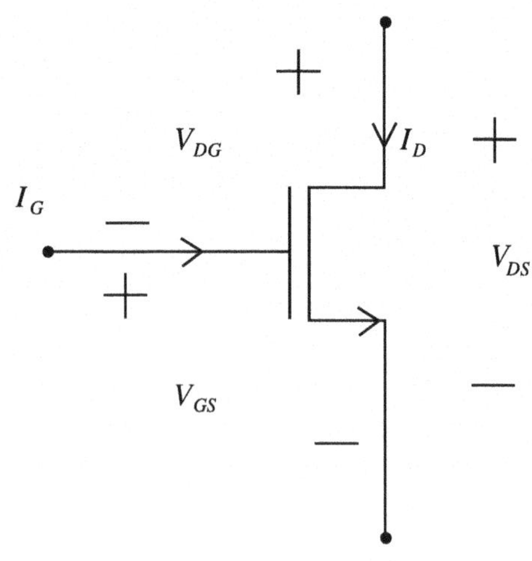

Fig. 1.2 MOS transistor model

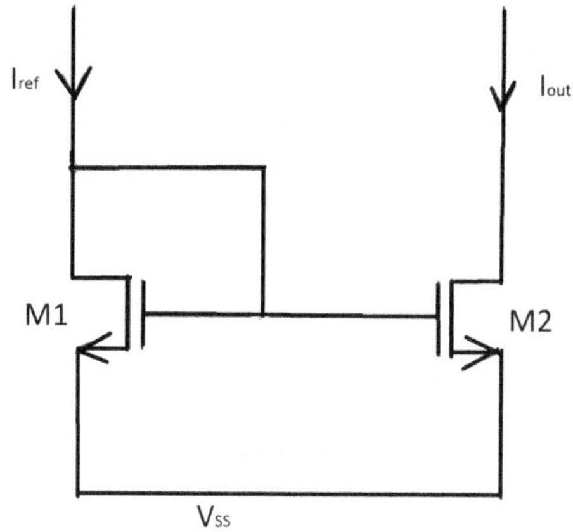

Fig. 1.3 Basic current mirror circuit with MOS transistor operating in subthreshold region

of considered NMOS transistor, Boltzmann constant and temperature, respectively. Note that the MOS transistor has lower saturation voltage (approximately 100 mV) in subthreshold region. Hence, large voltage swing can be achieved at low supply voltages. Specifically, this technique for low-voltage analog design is very effective to achieve proper operation of cascaded MOS transistors.

Figure 1.3 shows the design of current mirror circuit with subthreshold MOS transistors. This seems as a conventional current mirror, however the MOS transistors are being operated in subthreshold region. Hence, one will obtain the similar characteristics as with the conventional current mirror. Nevertheless, the value required for the input voltage will be small compared to the conventional counterpart to get the analog IC working. In addition, the input voltage to the circuit can further be reduced using other low-voltage design technique, discussed later in this chapter.

Other than the advantages of MOS transistors in subthreshold region in analog IC, it suffers from poor frequency response and linearity for $V_{DS} < 3V_{th}$. Here, V_{th} represents the thermal voltage and is equal to $\frac{KT}{q}$. Additionally, the leakage current due to reverse-biased drain and source with substrate is not negligible as compared to the drain current.

1.3.2 *Bulk-Driven MOS Transistor*

A. Guzinsky first introduced bulk-driven (BD) MOS transistors as active element (Guzinski et al. 1987) for the input differential pair. Many other works have been published depicting the advantages of BD technique on the performance of low-voltage operation of MOS transistor (Guzinski et al. 1987; Blalock et al. 1998; Sinencio and Andreou 1999). The primary purpose was to get low transconductance (g_m) to achieve improved linearity of MOS amplifier. In general, for MOS transistor to process any data, some current must flow through the drain terminal of transistor.

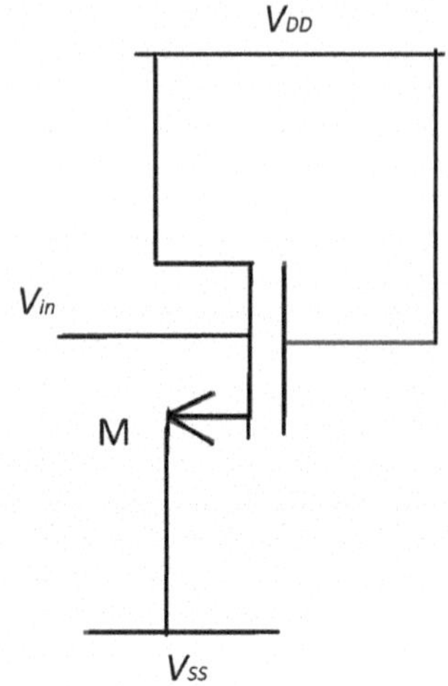

Fig. 1.4 Bulk-driven MOS transistor model

In conventional general operation of MOS transistor, this current is obtained when the applied bias at the gate terminal of MOS transistor becomes greater than the threshold voltage of the transistor. However, to reduce the required input voltage to turn ON the MOS transistor (or to avoid necessity of higher-voltage headroom) in BD technique, the MOS transistor is biased in saturation mode so that it can flow a continuous current and the input is applied to the bulk contact, as shown in Fig. 1.4. Clearly, it is understood that the drain current (I_D) of a traditionally connected MOS transistor is governed by gate to source voltage V_{GS}; whereas in BD approach, it is controlled by bulk to source voltage V_{BS}.

A close observation of BD MOS transistor suggests its resemblance with junction field-effect transistor (JFET). Here, the bulk contact plays the role of the gate terminal of virtual JFET and controls the current of MOS transistor. Hence, one can understand that with BD approach, MOS transistor works as a depletion transistor. So, it can also work with positive, negative and zero bias voltages. The advantages of the bulk-driven approach can be summarized as follows:

- The depletion characteristics of BD MOS transistor significantly minimize the need of overcoming the threshold voltage of the transistor. This increases the voltage headroom and improves the low-voltage performance of MOS transistor.
- This allows the MOS transistor to be used with lower supply voltages.
- BD approach is suitable to be used with current CMOS technology.

However, the BD approach forces the MOS transistor to have an isolated bulk terminal; hence, fabrication becomes complex. Other drawbacks of the BD approach can be listed as follows:

- The transconductance (g_m) of the MOS transistor with BD technique is quite smaller than the g_m of conventional counterpart. This affects the bandwidth offered by the IC employing BD MOS transistors. The bandwidths of BD and conventional MOS transistors are related as follows:

$$f_{T,BD} = \frac{n}{3.8} f_{T,conventional}$$

Here, n is the ration of body transconductance (g_{mb}) and the transconductance (g_m). As the g_{mb} of BD transistor can be 3–4 times lower than its g_m (generally n has value from 0.2 to 0.4), hence BD transistor has poorer frequency response than the conventional.

- Bulk-driven MOS transistors have higher input capacitance in comparison with the conventional MOS counterpart.
- The BD approach depends significantly on process technology. Therefore, the P well process can only result in the fabrication of N-channel BD MOS transistor.
- BD MOS transistors are highly prone to the latch up effect.

The example of BD MOS transistor-based analog circuit is shown in Fig. 1.5. It is a differential amplifier having their body terminal tied to the input voltage (Guzinski et al. 1987; Blalock et al. 1998; Rajput and Jamuar 2001; Sinencio and Andreou 1999). The gate terminal of transistors M1 and M2 is connected to a fixed supply voltage to ensure their working in saturation region. Note that the range of common mode input voltage V_{CM} of a conventional differential amplifier is limited due to the need to high threshold voltage.

So, when the BD MOS transistors are used, rail-to-rail input common mode volt age range can be obtained due to its operation in saturation mode. Hence, sufficient voltage headroom can be obtained for the operation of differential amplifier with the supply voltage as low as 0.6 V. Also, BD MOS transistors in differential amplifier

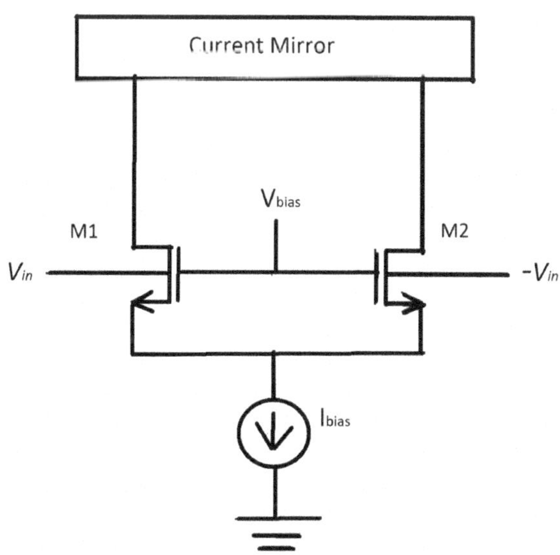

Fig. 1.5 Bulk-driven differential amplifier model

allow the circuit to achieve linear transconductance with respect to the differential inputs. Therefore, BD MOS transistor aids for the design of low-voltage analog IC.

1.3.3 Self-cascode Structure

With the scaling of technology, the output resistance of the MOS transistor is also reducing. One must note that high output resistance of MOS transistor is needed to achieve high gain in analog circuits. Hence, short channel MOS transistors are not able to provide sufficient large gain in analog IC design. The cascode structure can be general solution to have high gain with the scaled technologies (Sinencio and Andreou 1999). However, due to the biasing structure used in cascode, this method results in decreased output voltage swing. Hence, these structures cannot be used in low-voltage design techniques. To overcome this limitation and to achieve both high gain and voltage swing, self-cascode arrangement of two transistor can used, as shown in Fig. 1.6.

In self-cascode structure, the two transistors M1 and M2 can be considered as a single equivalent transistor with same gate voltage V_G. As the gate biasing is same, hence the structure is called as the self-cascode structure. Here, with self-cascode structure, the effective channel length of the equivalent MOS transistor is comparatively large than the M1 or M2, minimizing the effect of channel length modulation of MOS transistor on analog IC performance. Here, the lower transistor M1 acts as the resistor whose value depends on the input voltage. Hence, the self-cascode structure increases the output resistance and consequently gain. It also reduces the effect of

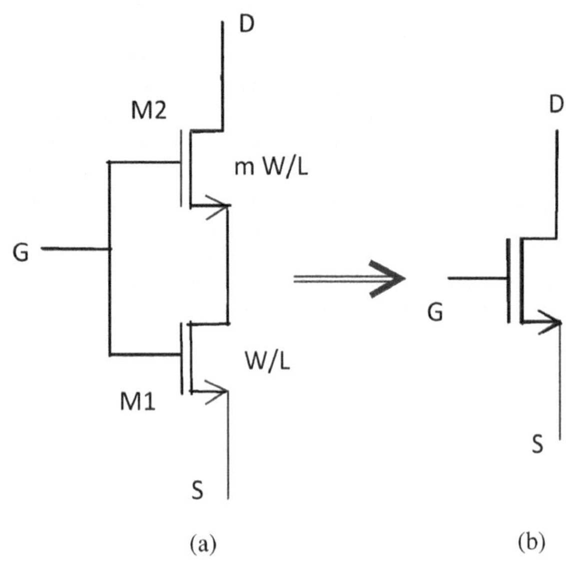

Fig. 1.6 **a** Self-cascode structure. **b** Equivalent structure

Miller's capacitance on the transistors gates. This approach has a strong application in the low-voltage analog circuit design. It enables the low-voltage analog circuits to operate with larger-voltage headroom (Bhardwaj and Rajput 2009; Baek et al. 2013).

If both the transistors, i.e., M1 and M2 have same aspect ratio, transistor M1 operates in linear region and M2 operates in saturation region. Hence, the equivalent transistor will not achieve the desired operation. The optimized performance of analog circuit with self-cascode structure is obtained when the W/L ratio of MOS transistor M2 is kept larger compared to the M1. In this case, the equivalent transistor will be completely in the saturation region. As the saturation voltages for transistor M1 and M2 are already small, there is no appreciable change in the drain to source saturation voltage of equivalent transistor and individual transistor M1 and M2. In self-cascode arrangement, the saturation voltage for equivalent transistor can be given as $V_{\mathrm{DSAT}} = V_{\mathrm{DSATM1}} + V_{\mathrm{DSATM2}}$. Hence, the self-cascode structure does not need high compliance voltage at output nodes.

Further, the transconductance of equivalent transistor can be given as $\frac{g_{\mathrm{m2}}}{m}$, where g_{m2} is the transconductance of MOS transistor M2. Here, m can be given as $\frac{(W/L)_{\mathrm{M2}}}{(W/L)_{\mathrm{M1}}}$. The output resistance of the equivalent MOS transistor is observed proportional to the parameter m. Moreover, note that the self-cascode structure operates at a very lower-power supply compared to the regular cascode structure. So the main advantage of self-cascode structure is to provide low-voltage operation with high gain. Moreover, the analog circuits design with self-cascode approach can be found in various literature (Baek et al. 2013; Xu et al. 2016).

1.3.4 Level Shifter Approach

The voltage level shifting is a technique to operate analog circuits with low input voltages. Importantly, a MOS transistor can be either operating in saturation region of operation or in the subthreshold region (Ismail and Fiez 1994; Rajput and Jamuar 2001; Johns and Martin 1997). Generally, this technique uses resistors in the circuit to shift the input common mode voltage to the region of operation of input differential amplifier (Carillo et al. 2000). Figure 1.7 is representing simple current mirror circuit based on the level shifter approach (Rajput and Jamuar 2002).

We can recall that the input current in a conventional current mirror circuit is given as $K' \frac{W}{2L}(V_{\mathrm{GS1}} - V_{\mathrm{T}})^2$. Here, parameter k', W, L, V_{GS}, V_{th} have their conventional meaning. Hence, the input voltage for conventional CM must be greater than the V_{T} voltage of the input MOS transistor. However, from Fig. 1.7, one can observe that the input voltage for a CM based on voltage level shifter approach is $V_{\mathrm{GS1}} - V_{\mathrm{GS3}}$. Therefore, with voltage level shifter approach, input voltage restriction to be greater than V_{T} can be relaxed. Consequently, level shifter approach is beneficial to design low-voltage analog IC. Also, this approach allows analog IC to have higher bandwidth at low voltage. Apart from this, rail-to-rail operation can be obtained both at input and output with level shifter approach.

Fig. 1.7 Simple current mirror with voltage level shifter approach

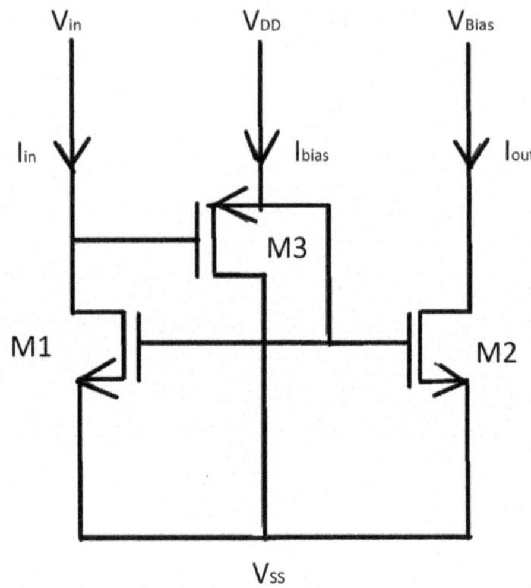

Offset current (I_{offset}) in the output MOS transistor M2 for no input current is the main drawback of voltage level shifting approach. The effect of I_{offset} becomes significant on the circuit performance when the input current I_{in} is of the same order as the I_{offset}. For such circuits, the range of operation is decided by the value of I_{offset}. In addition, voltage level shifter approach uses high number of MOS transistor for low-voltage analog IC design. As a result, this approach increases the power dissipation in the circuit.

1.3.5 Floating Gate MOS Transistor

Floating gate MOS transistors are popular for their application in digital circuit as memory element. Also, sometimes another dielectric layer known as charge trap layer can be used in place of FG as digital memory element (Gupta and Vishvakarma 2016). Further, these FG MOS transistors are becoming popular in analog IC design as adaptive circuit element and as capacitive-biased analog memory element. The structure of floating gate MOS transistor is very similar to the conventional transistor. The only difference lies in the polysilicon/metallic isolated floating gate (FG) between the main control gate and the conductive channel. Here, the main control gate and floating gate are not connected physically. However, they are connected electrically due to the capacitively coupled structure at gate.

Further, the FG MOS transistor has this ability to tune its threshold voltage according to the circuit needs; hence, these devices are also getting accepted by community for low-voltage analog circuit design. Dynamic threshold voltage reduction with FG MOS transistor allows it to be used with low-voltage supply. Many structures have been proposed for this purpose (Villegas and Barnes 2003; Wang et al. 2006; Yan

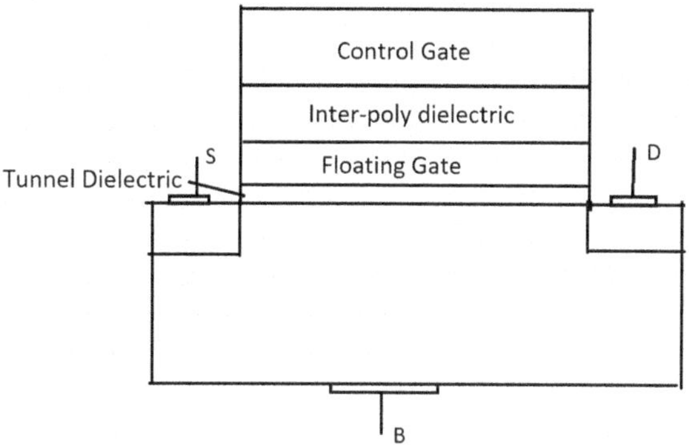

Fig. 1.8 Floating gate MOS transistor structure. Here, source (S), drain (D) and bulk (B) are contacts to MOS transistor

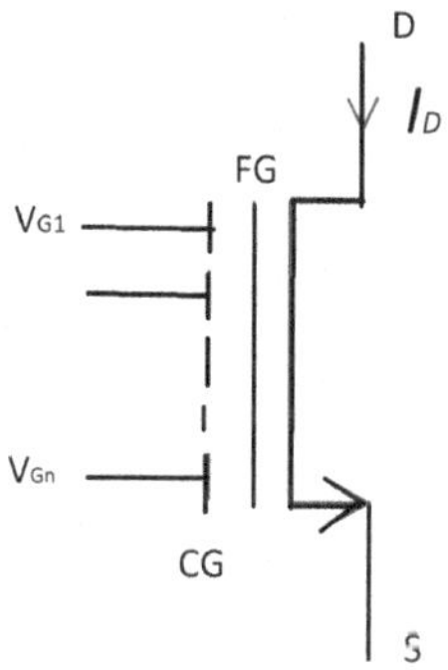

Fig. 1.9 Multi-input floating gate MOS transistor

and Sanchez-Sinencio 2000; Cunha et al. 1998). One such structure is shown in Fig. 1.8. Here, a high voltage is applied to the control gate of the MOS transistor. This voltage causes a high electric field at tunnel dielectric which attracts the channel electron to the control gate. These electrons while traveling to the control gate get trapped at the floating gate. Note that the amount of electrons at floating gate decides to the threshold voltage of the FG MOS transistor. Specifically, amount of charge at floating gate can be changed by several ways such as ultraviolet radiations, hot electrons and Fowler–Nordheim tunneling. Further, the discharging of FG MOS transistor is quite difficult due to the dielectric between FG and channel/control gate of the transistor. Note that FG MOS transistor can retain this FG charge for several years with variation as low as 2% on room temperature.

Furthermore, a FG is considered to have no accumulated charge for low-voltage application. Therefore, a multi-input FG MOS transistor as shown in Fig. 1.9 is used by the researchers for low-voltage analog IC design with FG MOS transistor (Mehrvarz et al. 1996). Here, the control gates array is formed over a single FG polysilicon layer. All the control gates are given voltages (i.e., V_{G1}, $V_{G2} \ldots V_{Gn}$) such that the total charge at the floating gate is conserved. Hence, the FG MOS transistor can be used with low-voltage supply voltage with dynamic threshold voltage characteristics.

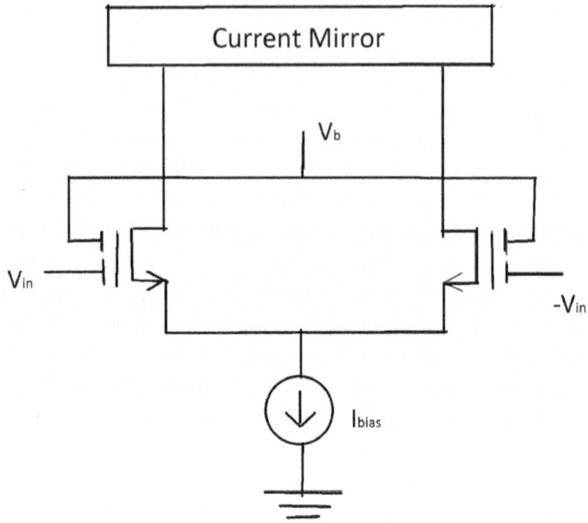

Fig. 1.10 Design of simple current mirror circuit using floating gate MOS transistor

Figure 1.10 shows the two input FG MOS transistor (Rajput and Jamuar 2002) in simple current mirror circuit. Here, a DC voltage is applied to the one gate, and the signal is applied to the another gate. For this multi-input FG MOS transistor, V_{T} adjust automatically for $V_{\mathrm{T,new}}$. The value of $V_{\mathrm{T,new}}$ is given as

$$V_{\mathrm{T,new}} = \frac{V_{\mathrm{T}} - V_{\mathrm{b}} K_1}{K_2}$$

where $K_1 = \frac{C_{\mathrm{G1}}}{C_{\mathrm{Total}}}$ and $K_2 = \frac{C_{\mathrm{G2}}}{C_{\mathrm{Total}}}$. Here, C_{G1} and C_{G2} are the capacitance between floating gate and control gate for two inputs and C_{Total} is the total capacitance between (i) control gates and floating gates (ii) floating gate and drain terminal (iii) floating gate and source terminal (iv) floating gate and bulk terminal (Rajput and Jamuar 2002).

Importantly, by careful selection of fixed DC voltage V_{b}, K_1 and K_2, the $V_{\mathrm{T,new}}$ will be less than V_{T}. Therefore, smaller new threshold voltage can be obtained for MOS transistor. In addition, the overall transconductance of multi-input FG structure results in smaller transconductance from the single input counterpart as $g_{\mathrm{m,overall}} = K_2 g_{\mathrm{m}}$. Here, $g_{\mathrm{m,overall}}$ is the transconductance of multi-input FG MOS transistor and g_{m} is the transconductance for single input counterpart. Clearly, $g_{\mathrm{m,overall}}$ has reduced from g_{m} by a factor of K_2. Moreover, the FG MOS transistor also presents smaller output impedance and smaller output conductance than the MOS transistor working at the same biasing condition. Therefore, one can conclude that the multi-input FG transistors can be used to design low-power analog electronic design.

However, as FG MOS transistor results in only low output resistance, hence only low gain circuits are possible with this technique. Also, the fabrication of additional gate results in increased fabrication cost and sets up complexity compared to the conventional technology.

1.3.6 Dynamic Threshold Voltage MOS Transistor

With the scaling trends of current CMOS design, the dynamic threshold (DT) technique was proposed. In fact, DT MOS transistor technique for low-voltage analog IC design is derived from the BD MOS transistor (Mehrvarz et al. 1996; Assaderaghi et al. 1994). The only difference lies in the biasing condition. In DT MOS transistor, gate and bulk electrodes are tied together and the biasing is applied dynamically. Schematic of DT MOS transistor is shown in Fig. 1.11. As the gate and bulk terminal are tied together, hence there is no need to operate the MOS transistor above the cut-in voltage (i.e., 0.7 V) of P-N junction between source/drain and substrate. This results in dynamic reduction of threshold voltage of MOS transistor. Hence, the DT MOS transistors can be used in analog IC design to operate at low supply voltage. Clearly, for DT MOS transistor, potential at any point in the conductive channel is governed by both gate and bulk voltage. As a result, high overall transconductance $g_m + g_{mb}$ is achieved with this technique.

Furthermore, the main difference between DT and BD technique lies in the input capacitance and maximum transit frequency. The maximum transit frequency for DT MOS transistor can be given as follows:

$$f_{t(DT)} = \frac{g_m + g_{mb}}{2\pi(C_{GS} + C_{BD} + C_{GS} + C_{BS})}$$

Here, g_m, g_{mb}, C_{GD}, C_{BD}, $C_{GS} and C_{BS}$ are parameters related to the DT MOS transistors. Here, one can note that as the transconductance of DT MOS transistor is larger than the conventional counterpart; hence, the former provides higher transit frequency as compared to the later. Hence, DT MOS transistor not only allows circuit to work at low voltage but also it provides better frequency response.

Furthermore, Fig. 1.12 is representing a simple current mirror circuit with DT MOS technique. Clearly, the gates and bulks of two transistors are tied together. Here, the voltage between the bulk/gate and source terminal controls both the transistors together dynamically (Metaj et al. 2017).

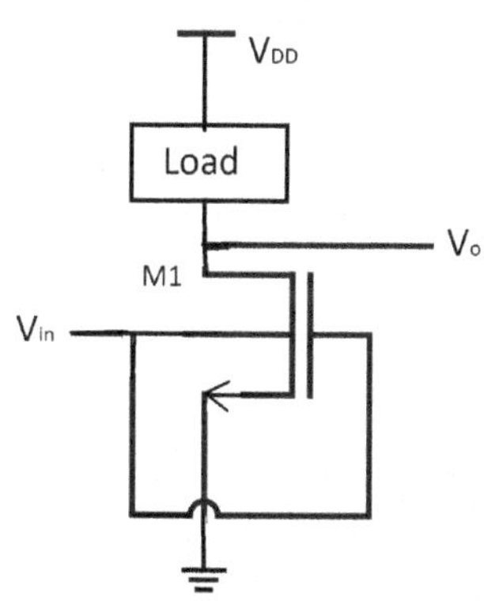

Fig. 1.11 Dynamic threshold MOS transistor

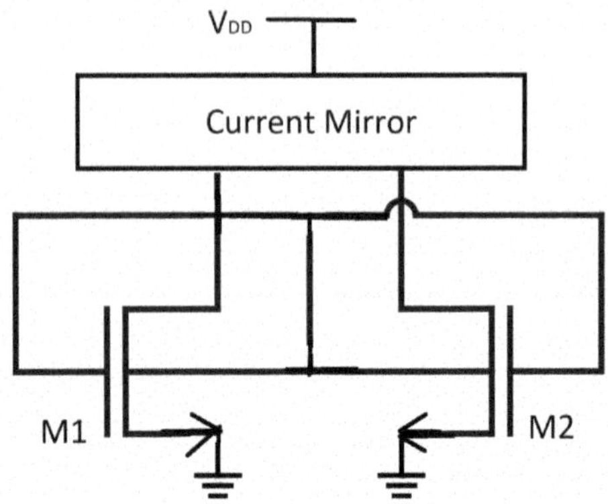

Fig. 1.12 Simple current mirror with dynamic threshold voltage approach

1.3.7 Low-Voltage Analog Cells

Any analog IC can be thought as a collection of many sub-circuits. These sub-circuits can be called as analog cells. In fact, the properties possessed by these analog cells largely determine the overall performance of an analog IC. Hence, if these circuits can be designed to operate at a low voltage, then the analog IC consisting these analog cells would automatically operate at low voltage. This technique has been used in the design of various low-voltage analog circuits (Sanchez-Sinencio 2000; Rajput and Jamuar 2001). For an instance, a low-voltage current mirror circuit can be used to design a low-voltage analog circuit (Yan and Sanchez-Sinencio 2000; Sanchez-Sinencio 2000).

1.4 Conclusion

In this chapter, the analysis of analog circuits is presented at low supply voltage. This chapter also covers different issues that may affect the designing of analog circuit at low voltage such as supply voltage scaling and transistor inversion modes. Also, some techniques are discussed to overcome these design issues.

References

Abidi A, Pottie G (2000) Kaiser W (200) Power-conscious design of wireless circuits and systems. Proc IEEE 88(10):1528–1545

Annema AJ, Nauta B, van Langevelde R, Tuinhout H (2005) Analog circuits in ultradeep-submicron CMOS. IEEE J Solid-State Circuits 40(1):132–143

Assaderaghi F, Sinitsky D, Parke S, Bokor J, Ko PK, Hu C (1994) A dynamic threshold voltage MOSFET (DT-MOS) for ultra-low voltage operation. In: IEDM Technical Digest, pp 809–812

Baek KJ, Gim JM, Kim HS, Na KY, Kim NS, Kim YS (2013) Analogue circuit design methodology using self-cascode structures. Electron Lett 49(9)

Baschirotto A, Chironi V, Cocciolo G, DAmico S, De Matteis M, Delizia P (2009) Low power analog design in scaled technologies. In: Proceedings of topical workshop electron particle physics. France, pp 103–110

Bhardwaj K, Rajput SS (2009) 1.5V high performance OP AMP using self cascode structure. IEEE Student Conference on Research and Development (SCOReD), https://doi.org/10.1109/SCORED.2009.5443054, 16-18 Nov.2009

Blalock BJ, Allen PE, Rincon-Mora GAR (1998) Designing 1-V Op amps using standard digital CMOSTechnology. IEEE Trans Circuits Syst II 45(7):769–780

Bult K (2000) Analog design in deep sub-micron CMOS. In: Proceeding 26th European solid-state circuits conference (ESSCIRC), pp 126–132

Carillo JFD, Ausin JL, Torelli G et al (2000) 1-V rail-to-rail operational amplifiers in standard CMOS technology. IEEE J Solid-State Circuits 35(1):33–44. https://doi.org/10.1109/4.818918

Chandrakasan A, Sheng S, Brodersen R (1992) Low power CMOS digital design. IEEE J Solid-State Circuits 27(4):473–484

Chang J, Chen Y, Chan W et al (2017) A 7nm 256Mb SRAM in high-k metal-gate finFET technology with write-assist circuitry for low-VMIN applications. In: Proceedings of the IEEE international solid-state circuits conference, pp 206–208

Cunha AIA, Schneider MC, Montoro CG (1998) An MOS transistor model for analog circuit design. IEEE J Solid-State Circuits 33(10):1510–1519. https://doi.org/10.1109/4.720397

Enz C, Chicco F, Pezzotta A (2017) Nanoscale MOSFET modeling-part 1: the simplified EKV model for the design of low-power analog circuits. IEEE Solid-State Circuits Magazine 9(3):26–35

Geiger RL, Allen PE, Strader NR (1990) VLSI Design Techniques for Analog and Digital Circuits. Mc-Graw Hill, New York

Gupta D, Vishvakarma SK (2016) Improved short channel characteristics with long data retention time in extremely short channel NAND flash device. IEEE Trans. Electron Dev 63(2):668–674

Guzinski A, Bialko M, Matheau JC (1987) Bodydriven differential amplifier for application continuousime active-C filter. Eur Conf Circuit Theo Des

Hosticka BJ, Dalsab KG, Krey D, Zimmer G (1985) Behavior of analog MOS integrated circuit at high temperatures. IEEE J Solid-State Circuits sc-20(4):871–874

Ismail M, Fiez T (1994) Analog VLSI signal and information processing. McGraw-Hill, New York

Johns DA, Martin K (1997) Analog integrated circuit design. Wiley, New York

Keane J, Eom H, Kim T-H, Sapatnekar S, Kim C (2008) Stack sizing for optimal current drivability in subthreshold circuits. IEEE Trans Very Large Scale Integr (VLSI) Syst 16(5):598–602

Liu D, Svensson C (1993) Trading speed for low power by choice of supply and threshold voltages. IEEE J Solid-State Circuits 28(1):10–17

Mehrvarz HR, Kwok CY (1996) A novel multi-input floating-gate MOS four-quadrant analog multiplier. IEEE J Solid-State Circuits 31(8):1123–1131

Metaj R, Stopjakova V, Arbet D (2017) Design techniques for low voltage analog integrated circuits. J Electr Eng 68(4):245–255

Nilsson E, Svensson C (2014) Power consumption of integrated low power receivers. IEEE J Emerging Sel Top Circuits and Systems 4(3):273–283

Onabajosilva M, Martinez J (2012) Analog circuit design for process variation-resilient systems-on-a-chip. Springer Science & Business Media,

Rajput SS, Jamuar SS (2001) Design techniques for low voltage analog circuit structures. NSM 2001/IEEE, Malaysia, November 2001

Rajput SS, Jamuar SS (2001) Low voltage high performance ccii for analog signal processing applications. ISIC2001, Singapore

Rajput SS, Jamuar SS (2001) Low voltage, low power high performance current mirror for portable analogue and mixed mode applications. Proceedings of IEE circuits devices and systems 148(5):273–278

Rajput SS, Jamuar SS (2002) Low Voltage Analog CIRCuit Design Techniques. IEEE Circuits and Syst Mag 2(1):24–42

Sanchez-Sinencio E (2000) Low voltage analog circuit design techniques. IEEE Dallas CAS workshop

Schroder DK (2007) Negative bias temperature instability: What do we understand? Microelectron Reliab 47(6):841–852

Sedra AS, Smith KC (2011) Microelectronic CIRCuits. Oxford Univ. Press, Oxford, U.K

Shah CT (1964) Characterization of the metal oxide semiconductor transistors. IEEE Trans Electron Dev 11(7):324–345

Sinencio ES, Andreou AG (ed) Low voltage/low power integrated circuits and systems-low-voltage mixed-signal circuits. IEEE Press (1999). Online ISBN 9780470545065

Svensson C (2015) Towards power centric analog design. IEEE Circuits Syst Magz 15(3):44–51

Svensson C, Wikner J (2010) Power consumption of analog circuits: a tutoria. Analog Integr Circuits Sig Proc 65(2):171–184

Swanson RM, Meindl JD (1972) Ion-implanted complementary MOS transistors in low-voltage circuits. IEEE J. Solid-State Circuits Sc-7(2):146–153

Ueno K, Hirose T, Asai T, Amemiya Y (2009) A 300 nW, 15 ppm/-C, 20 ppm/V CMOS Voltage Reference Circuit Consisting of Subthreshold MOSFETs. IEEE J Solid-State Circuits 44(7):2047–2054

Villegas ER, Barnes H (2003) Solution to trapped charge in FGMOS transistors. Electronics Lett 39(19):1416–1417

Vittoz E (1980) Micropower IC. Proc IEEE Eur Solid-State Circuits Conf 2:174–189

Vittoz E (1990) Future of analog in the VLSI environment. In: Proceeding IEEE international symposium circuits and systems, vol 2, pp 1372–1375

Wang A, Calhoun BH, Chandrakasan AP (2006) Sub-threshold design for ultra low-power systems, 1st edn. Springer Science+Business Media, New York (USA). ISBN 0-387-33515-3

Wolpert D, AmpaDdu P (2012) Managing temperature effects in nanoscale adaptive systems. Springer, New York. ISBN 978-1-4614-0748-5

Xu D, Liu L, Xu S (2016) High DC gain self-cascode structure of OTA design with bandwidth enhancement 52(9):740–742

Yan S, Sanchez-Sinencio E (2000) Low voltage analog circuit design techniques: a tutorial. In: Proceedings of the IEICE transactions on fundamentals of electronics, communications and computer sciences, vol 83, No 2, pp 179–196. ISSN: 0916-8508

Zhao W, Cao Y (2006) New generation of predictive technology model for sub-45 nm early design exploration. IEEE Trans Electron Devices 53(11):2816–2823

Zhiyuan C, Law MK, Mak PI, Martins RP (2017) A single-chip solar energy harvesting IC using integrated photodiodes for biomedical implant applications 11(1):44–53

Zimmer G, Esser W, Fichtel J, Hosticka B, Rothermel A, Schardein W (1989) BiCMOS: technology and circuit design 20(1-2):59–75

Chapter 2
Design Methodology for Ultra-Low-Power CMOS Analog Circuits for ELF-SLF Applications

Soumya Pandit

Abstract For extreme low-frequency (ELF) and super low-frequency (SLF) applications like biomedical applications (brain wave signal processing and brain–computer interface circuits), seismic signal processing applications, submarine communication applications, ultra-low-power dissipation of the electronic circuits is the most essential criterion. With the scaling of CMOS technology in the nanoscale, the contribution of leakage power becomes very significant compared to any other sources of power dissipation like switching power, bias power, etc. Subthreshold leakage current is an important component of all sources of leakage current. In modern design methodology for ultra-low-power analog circuits, this component of leakage current has been made use of for design purpose. The physics of the MOS transistor in the subthreshold region or weak inversion region is different from that when the transistor operates in the strong inversion region. Therefore, a good understanding of this physics is important for ultra-low-power design. Compact models play significant role in modern design methodologies. This chapter briefly discusses compact model for MOS transistor operating in the weak inversion region. Inversion coefficient-based design methodology for ultra-low-power analog circuits is discussed in detail. Implementation of the design methodology is then exemplified by a complete design of operational transconductance amplifier, operating in the extreme low-frequency region. Application areas of the design methodology are also discussed.

Keywords Extreme low frequency (ELF) · Drain-induced barrier lowering (DIBL) · Operational transconductance amplifier (OTA) · Subthreshold current · Super low frequency (SLF) · Transconductance

---

S. Pandit (✉)
Centre of Advanced Study, Institute of Radio Physics and Electronics, University of Calcutta, Kolkata, India
e-mail: sprpe@caluniv.ac.in

R. Dhiman and R. Chandel (eds.), *Nanoscale VLSI*,
Energy Systems in Electrical Engineering,
https://doi.org/10.1007/978-981-15-7937-0_2

2.1 Introduction

The fundamental idea behind the evolution of a transistor device stems from the concept of controlled switch, where the device operates in two states: OFF state and ON state. In the ON state, the device acts as short circuit by forming a direct path between the input and the output terminals. On the other hand, in the OFF state, the device acts as open circuit, where there does not exist any direct path between the input and the output terminals. For a metal–oxide–semiconductor (MOS) transistor, the two terminals through which the current flows are referred to as source and drain terminals. On the other hand, the controlling terminal is referred to as gate terminal (Pandit 2015). The transfer of resistance from very low (short-circuit condition) to very high (open-circuit condition) is achieved by applying voltage signal to the gate terminal. The controllability of the gate terminal over the flow of current, therefore, plays very significant role in this transfer process. The power dissipation during the OFF state is therefore very small, owing to some non-ideal leakage current. This physics of MOS transistor appears to be fine, when the geometry of the transistor is large. However, with scaling down of the feature size of the transistor, the gate terminal looses its control over the current flow (Pandit 2018). Even in the OFF state, significant amount of current starts flowing between the source and the drain terminals. In other words, leakage current becomes an important concern to scaled down MOS transistor, and the behavior of a MOS transistor becomes more like a resistor compared to a switch. Therefore, in scaled technology, the design of integrated circuits (ICs) becomes a challenging task. It requires good understanding of the physics of MOS transistor in the OFF state, the various sources of leakage current components, necessary modeling of those.

2.2 Physics of MOS Transistor Operating in the Weak Inversion Mode/OFF State

2.2.1 Concept of OFF State of a MOS Transistor

The concept of OFF current of a MOS transistor can be explained with reference to Fig. 2.1, which shows the typical variation of drain current (I_D), measured in logarithmic scale, of a MOS transistor versus the gate-to-source voltage (V_{GS}). The transistor OFF current is measured when the gate voltage is zero. From Fig. 2.1, the OFF currents are measured to be 5.91 and 0.96 pA at $V_{DS} = 1.8$ V and 50 mV, respectively. Thus, we see that the OFF current depends upon the magnitude of the drain voltage applied to the transistor. Apart from this, the OFF current of a MOS transistor also depends upon several other factors such as threshold voltage, physical dimensions of the channel, doping profile of the channel, depth of the source/drain junction and thickness of the gate oxide. The conduction current that flows between the drain and the source terminals of a MOS transistor, when the gate voltage is below

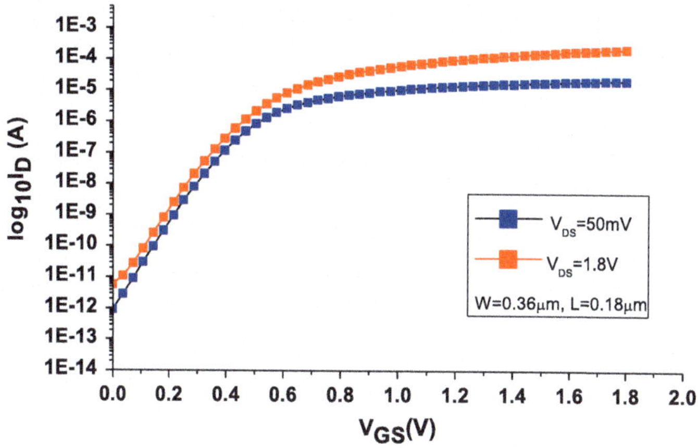

Fig. 2.1 The I_D versus V_{GS} characteristics of a MOS transistor for $L = 0.18\,\mu\text{m}$, $W = 0.36\,\mu\text{m}$

the threshold voltage is referred to as subthreshold leakage current. Another major component of leakage current is the reverse diode leakage currents at the transistor drain. Apart from these two major components, there are several other sources of leakage current in a nanoscale MOS transistor. These are tunneling current into and through the gate oxide, leakage current due to injection of hot carriers from substrate to gate oxide, gate-induced drain leakage current and punch through current. A comprehensive overview of various sources of leakage current in a nanoscale MOS transistor is provided in Roy et al. (2003). In this section, we discuss the subthreshold leakage current which critically affects the OFF current of scaled integrated circuits. The subthreshold leakage current is often referred to as weak inversion current, especially by the analog designers.

2.2.2 *Subthreshold Leakage Current/Weak Inversion Current*

When a MOS transistor operates in a condition where the effective gate voltage $(V_{GS} - V_T)$ is quite low, i.e., $(V_{GS} - V_T <\approx -2nU_T)$, $n \approx 1.4$ and $U_T \approx 26\,\text{mV}$ at room temperature, the inversion charge is much less than the depletion charge and the flow of drain current is primarily due to diffusion of minority carriers (Pandit 2013). The weak inversion mode is defined by the condition $\Phi_F \leq \psi_s \leq 2\Phi_F$, where $\psi_s = 2\Phi_F = 2\frac{k_B T}{q}\ln\left(\frac{N_A}{n_i}\right)$ is the surface potential at strong inversion and $N_A = N_{\text{sub}}$ is the uniform substrate concentration. Under this condition, the behavior of a MOS transistor is similar to that of a bipolar transistor, where source terminal acts as emitter, substrate acts as base, and drain acts as the collector terminal.

In VLSI circuit simulation, compact models play very significant role. Compact models of a circuit element are simple mathematical description of the behavior of that circuit element that are used for computer-aided analysis and design (Saha 2016). The two compact models most widely used in semiconductor industries are Berkeley short channel IGFET (BSIM) model and Enz-Krummenacher-Vittoz (EKV) model. The weak inversion/subthreshold operation of a MOS transistor is very effectively modeled in both of the compact models. The weak inversion drain current is modeled as follows Saha (2016).

$$I_{\mathrm{DS}} = \mu_n C'_{\mathrm{ox}} \frac{W}{L} (n-1) U_T^2 \exp\left(\frac{V_{\mathrm{GS}} - V_T}{nU_T}\right) \left[1 - \exp\left(-\frac{V_{\mathrm{DS}}}{U_T}\right)\right] \tag{2.1}$$

where $U_T = \frac{k_{\mathrm{B}}T}{q}$ is the thermal voltage, $n = 1 + \frac{C_{\mathrm{dm}}}{C'_{\mathrm{ox}}}$ is the subthreshold swing factor, μ_n is the surface mobility of the electrons, C'_{ox} is the oxide capacitance per unit area, W and L are the effective channel width and length of the MOS transistor, respectively, and V_T is the threshold voltage. The weak inversion drain current, as approximated from EKV model, using source as the reference terminal is given by Enz and Vittoz (2006)

$$I_{\mathrm{DS}} = 2n\mu_n C'_{\mathrm{ox}} U_T^2 \left(\frac{W}{L}\right) \exp\left(\frac{V_{\mathrm{GS}} - V_T}{nU_T}\right) \tag{2.2}$$

While writing (2.2), we assume that the drain–source voltage V_{DS} exceeds its saturation value, V_{DSat}. In the weak inversion mode, the saturation value of drain–source voltage is $\approx 4U_T$.

The transconductance of a MOS transistor operating in weak inversion mode is defined as

$$g_m(WI) = \frac{I_{\mathrm{DS}}}{nU_T} \tag{2.3}$$

The transconductance efficiency is, therefore, written as

$$\frac{g_m}{I_{\mathrm{DS}}}(WI) = \frac{1}{nU_T} \tag{2.4}$$

It may be noted that the transconductance and the transconductance efficiency of a MOS transistor operating in the weak inversion mode neither depend upon the geometry of the transistor device, nor on the process parameters.

2.2.2.1 Drain-Induced Barrier Lowering

As noted from the preceding discussion, the drain current in the weak inversion mode exponentially depends upon the overdrive voltage. Therefore, with the reduction of threshold voltage, by some means, the weak inversion drain current changes significantly. The weak inversion drain current increases with the application of the high drain bias, through the reduction of threshold voltage. The dependence of threshold voltage of a MOS transistor on the drain bias is referred to as drain-induced barrier lowering.

When the transistor is in the OFF state, a potential barrier (in the p-type region) prevents the electrons to flow from the source to the drain, for an n-channel MOS transistor. With the application of gate-to-source voltage, this barrier reduces and eases the conduction of the electrons. For long channel MOS transistor, such barrier lowering is controlled by the potential applied on the gate terminal, and the drain bias does not have any role over it. However, for a short channel MOS transistor, the drain and the source fields penetrate deeply into the middle of the channel, which lowers the potential barrier between the source and the drain. The result of this is that even at lower gate voltage, the carriers can overcome the barrier between the source and the channel. In other words, the threshold voltage of a short channel MOS transistor reduces from its long channel value. With application of high drain bias to a short channel MOS transistor, the barrier height is further lowered, resulting in further decrease of the threshold voltage. This phenomenon is called drain-induced barrier lowering Pandit et al. (2014).

The threshold voltage model for a short channel MOS transistor is modeled as

$$V_T = V_{T0} - \Delta V_T \tag{2.5}$$

where V_{T0} is the long channel threshold voltage and ΔV_T is the amount of the reduction of threshold voltage due to short channel effect and ΔV_T is given by Saha (2016), Pandit et al. (2014)

$$\Delta V_T = \theta_T(L)\left[2\left(\psi_{\mathrm{bi}} - \psi_s\right) + V_{\mathrm{DS}}\right] \tag{2.6}$$

Here $\theta_T(L)$ is the short channel effect coefficient depending on the channel length and is given by

$$\theta_T(L) = \frac{1}{2\cosh\left(\frac{L}{l_t}\right) - 2} \tag{2.7}$$

2.3 Theoretical Formulation of the Design Methodology

2.3.1 All Region Drain Current Model of a MOS Transistor

In the strong inversion mode, the drain current for an n-channel MOS transistor (NMOS) is proportional to the square of the effective gate voltage $(V_{\mathrm{GS}} - V_{\mathrm{Tn}})$ and is written as Enz and Vittoz (2006)

$$I_{\mathrm{DS}} = \frac{1}{2}\left(\frac{\mu_n C'_{\mathrm{ox}}}{n}\right)\left(\frac{W}{L}\right)\left(V_{\mathrm{GS}} - V_{\mathrm{Tn}}\right)^2 \tag{2.8}$$

The strong inversion mode occurs at high effective gate voltage $(V_{\mathrm{GS}} - V_{\mathrm{Tn}}) >$ 225 mV. An unified expression for the drain current interpolated from weak through strong inversion is written as

$$I_{\mathrm{DS}} = 2n\mu C'_{\mathrm{ox}} U_T^2 \left(\frac{W}{L}\right) \left[\ln\left(1 + e^{\frac{V_{\mathrm{GS}} - V_{\mathrm{Tn}}}{2nU_T}}\right)\right]^2 \tag{2.9}$$

$$= 2n\mu C'_{\mathrm{ox}} U_T^2 \left(\frac{W}{L}\right) \left[\ln\left(1 + e^{\nu}\right)\right]^2 \tag{2.10}$$

The effective gate voltage is normalized to $2nU_T$ and is represented by the factor ν. It may be noted that the velocity saturation component is omitted here. This is based on the assumption that for ultra-low-power operation, the supply voltage should also be very small and the drift velocity would not therefore saturate with the applied electric field. Small values of ν characterize the weak inversion mode. Also $\ln(1 + e^{\nu}) \approx e^{\nu}$. Therefore, we arrive at the weak inversion drain current expression. On the other hand, large values of ν characterize the strong inversion mode. Further, $\ln(1 + e^{\nu}) \approx \nu$ and we get back the expression for the drain current operating in the strong inversion mode.

2.3.2 *Inversion Coefficient Definition*

The inversion coefficient (IC) factor provides a numerical identity factor characterizing the inversion status of a MOS transistor. For IC < 0.1, the transistor operates in the weak inversion mode, for $0.1 <$ IC < 10, the transistor operates in the moderate inversion mode, and for IC > 10, the transistor operates in the strong inversion mode.

The transition current is defined as Binkley (2008)

$$I_S = 2n\mu C'_{\mathrm{OX}} U_T^2 \frac{W}{L} = I_0 . \frac{W}{L} \tag{2.11}$$

The traditional inversion coefficient is defined as Binkley (2008)

$$\mathrm{IC} = \frac{I_D}{I_S} = \left[\ln\left(1 + e^{\frac{V_{\mathrm{GS}} - V_{\mathrm{Tn}}}{2nU_T}}\right)\right]^2 = \left[\ln\left(1 + e^{\nu}\right)\right]^2 \tag{2.12}$$

The above expression simply becomes $\nu > 1, \mathrm{IC} = \nu^2$ for large values of ν and $\nu < 1, \mathrm{IC} = e^{2\nu}$. This is a fundamental relationship, and some numerical values and design ideas points may be derived as follows.

1. At the threshold voltage V_{Tn}, the effective gate voltage is zero and the value of the inversion coefficient $IC \approx 0.5$. In the weak inversion region, ν is negative, and for $\nu = -2$, we get $IC \approx 0.01$. This corresponds to effective gate voltage approximately equal to -145 mV. This is often used for analog circuit designs with very low supply voltages.
2. The value of IC comes out to be 10 for $\nu = 3.12$. This corresponds to effective gate voltage equal to approximately 0.22 V. This is often used by the designers to ensure that the transistor operates in the strong inversion mode.
3. The transition current expression as defined here involves two parameters. The first one is $\mu C'_{\text{ox}} W/L$ which is written as $K'W/L$. It may be noted that any error in good estimation of the value of K' leads to inaccuracy in determining the value of W/L. Therefore, good estimation of the value of K' is essential. The second parameter is $2nU_T$, which is about 72 mV at room temperature.

The most important small signal parameter of a MOS transistor is transconductance. For low-power circuit design, the transconductance parameter is estimated to be

$$g_m \approx \frac{I_D}{nU_T} \cdot \frac{2}{1 + \sqrt{1 + 4.\text{IC}}} \tag{2.13}$$

2.3.3 *Sizing Methodology*

The sizing relationship is given by the following simple equation Binkley (2008).

$$\text{IC} = \frac{I_D}{I_0 \cdot \frac{W}{L}} \tag{2.14}$$

where I_0 is referred to as the technology current. In the design method, as adopted in the present chapter, the channel width is determined as follows

- Fix up the drain current I_D passing through a transistor. This depends on the desired specifications.
- Fix up the inversion coefficient (IC), depending upon the mode of operation, i.e., weak inversion or strong inversion.
- The technology current I_0 is constant, depending upon the operating temperature. If not mentioned explicitly, room temperature may be assumed.
- Fix up the value of the channel length, considering the gate area and hence device capacitances and other performance parameters, such as noise.
- Compute the value of W from (2.14).

2.4 Implementation of the Design Methodology

2.4.1 Design Example

In this section, the implementation of the design methodology is illustrated through the design of an operational transconductance amplifier (OTA) circuit. We select a doublet input OTA circuit which consists of two input differential pairs for its operation instead of one pair in conventional OTAs. The circuit diagram for a doublet input OTA circuit is shown in Fig. 2.2. The primary target application for this circuit is extreme low frequency, ultra-low-power analog systems. The selection of the mode of operation of the transistors depends upon this. For ultra-low-power requirements, the current flowing through a transistor should be very small, typically few nano amperes or less. This dictates the transistor to work in the weak inversion mode. This is further supported by the fact that the frequency of operation is low. The circuit is to be designed using 180 nm technology node of Semiconductor Laboratory, India. The design specifications of the circuit are tabulated in Table 2.1.

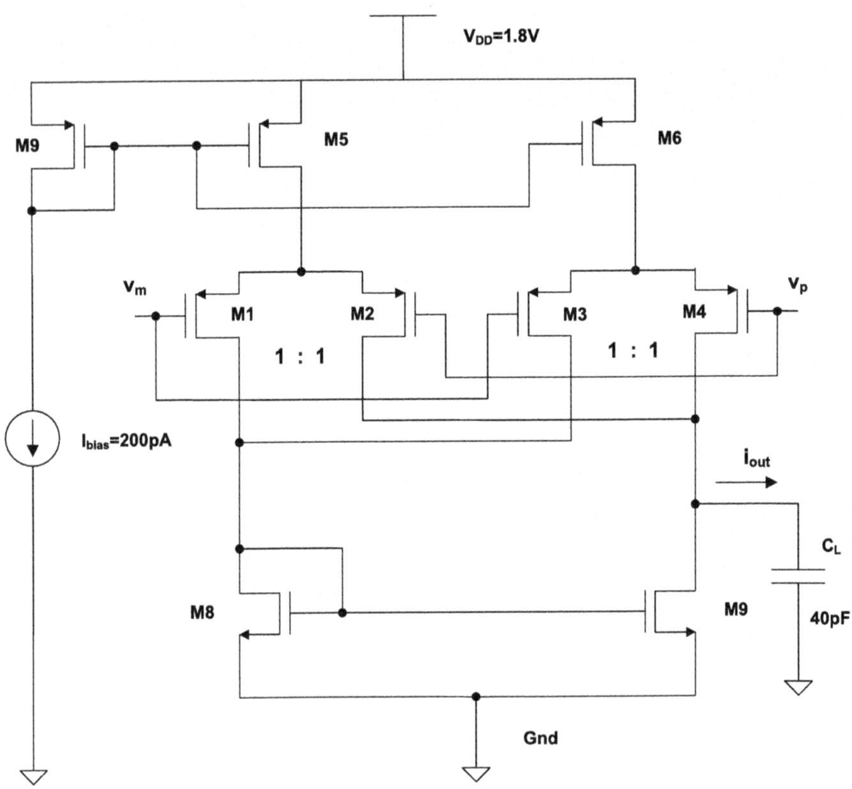

Fig. 2.2 Doublet input OTA circuit

Table 2.1 Design specifications

Parameter	Value
Supply voltage	1.8 V
Bias current	200 pA
Load capacitance	40 pF
Channel length	5 μm
Technology current I_0 NMOS	0.428 μA
Technology current I_0 PMOS	0.1 μA
Threshold voltage V_T NMOS	440 mV
Threshold voltage V_T PMOS	−450 mV

Table 2.2 Calculation of size ratio of each transistor

Devices	IC	Drain current (pA)	W/L
M1, M2, M3, M4	0.001	100	5/5
M5, M6, M7	0.001	200	10/5
M8, M9	0.0009	200	2.5/5

The current flowing through each of the transistors and the corresponding (W/L) ratios as calculated following the sizing methodology as discussed earlier are shown in Table 2.2.

2.4.2 *Design Analysis*

2.4.2.1 Tranconductance of the Complete Circuit

We perform certain simple calculations for detail analysis of the circuit. The small signal analysis of the OTA circuit is shown in Fig. 2.3. From Fig. 2.3, we write

$$i_{\text{out}} = (i_{d4} + i_{d2}) - (i_{d3} + i_{d1}) \tag{2.15}$$

Since each transistor carries equal DC current, their transconductance values are same. Let this be g_m. Now we write

$$i_{\text{out}} = g_m \cdot \left(v_{\text{sg4}} - v_{\text{sg3}} -\right) + g_m \cdot \left(v_{\text{sg2}} - v_{\text{sg1}} -\right) \tag{2.16}$$

Now we see that

$$\left(v_{\text{sg4}} - v_{\text{sg3}} -\right) = \left(v_{\text{sg2}} - v_{\text{sg1}} -\right) = v_p - v_m = v_{\text{id}} \tag{2.17}$$

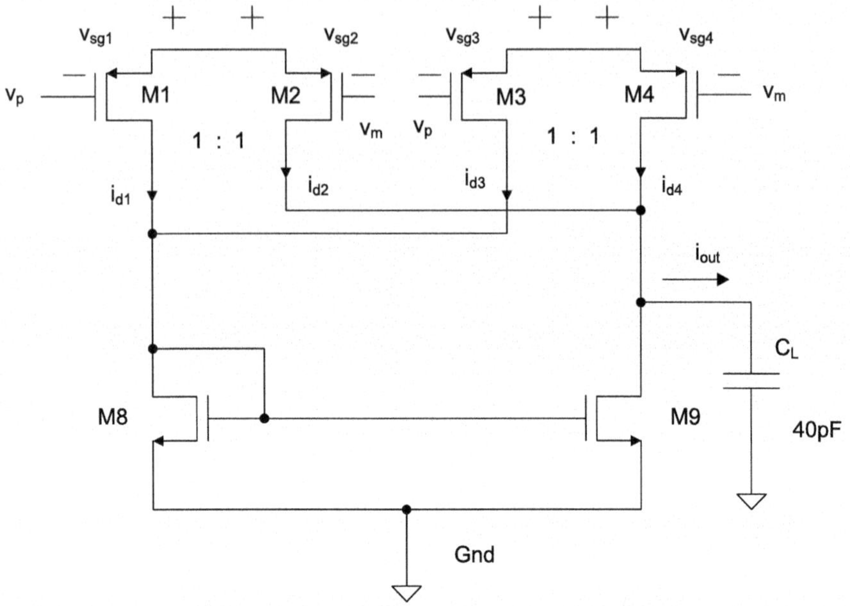

Fig. 2.3 Small signal analysis of the circuit

Thus, we write that the transconductance of the complete OTA circuit is

$$G_{\text{mOTA}} = \frac{i_{\text{out}}}{v_{\text{id}}} = 2g_m \tag{2.18}$$

As mentioned earlier, in the weak inversion mode $g_m = I_D/(nU_T)$. This comes out to be 2.75 nS and hence $G_{\text{mOTA}} = 5.5\,\text{nS}$.

2.4.2.2 Input Common Mode Range

For NMOS transistor, the device operates in the saturation mode under weak inversion condition, if the drain-to-source voltage $V_{\text{DS}} \geq 3U_T$ and for PMOS transistor, the device operates in the saturation mode, under weak inversion condition, if the drain-to-source voltage $V_{\text{SD}} \geq 3U_T$. From Fig. 2.2, we find that to keep the transistor M5 operate in the saturation mode,

$$V_{\text{DD}} - (V_{\text{CM}} + V_{\text{SG1}}) \geq 3U_T \tag{2.19}$$

$$V_{\text{CM}} \leq V_{\text{DD}} - V_{\text{SG1}} - 3U_T \tag{2.20}$$

$$(V_{\text{CM}})_{\text{max}} = V_{\text{DD}} - V_{\text{SG1}} - 3U_T \tag{2.21}$$

Now $|V_{\text{Tp}}| = 450\,\text{mV}$ and $I_{\text{SD1}} = 100\,\text{pA}$. Now in order to find out V_{SG1}, we use (2.1) for PMOS transistor and compute $V_{\text{SG1}} = 198.56\,\text{mV}$. Therefore, $(V_{\text{CM}})_{\text{max}}$ comes out to be 1524.44 mV. Now in order to find out the minimum common mode voltage, we proceed as follows. To keep M1 operating in the saturation mode,

$$V_{\text{SD1}} \geq 3U_T \tag{2.22}$$

$$(V_{\text{CM}} + V_{\text{SG1}}) - V_{\text{GS8}} \geq 3U_T \tag{2.23}$$

$$V_{\text{CM}} \geq V_{\text{GS8}} - V_{\text{SG1}} + 3U_T \tag{2.24}$$

$$(V_{\text{CM}})_{\text{min}} = V_{\text{GS8}} - V_{\text{SG1}} + 3U_T \tag{2.25}$$

Now we have $I_{\text{D8}} = 200\,\text{pA}$ and $W/L_8 = 0.5$, thus using (2.1), we compute $V_{\text{GS8}} = 161\,\text{mV}$. Therefore, $(V_{\text{CM}})_{\text{min}} = 40.44\,\text{mV}$.

2.4.2.3 Unity Gain Frequency

For the analysis of unity gain frequency, we consider the model as shown in Fig. 2.4. From this , we write the following

$$v_{\text{out}} = i_{\text{out}} \cdot \left(\frac{1}{sC_L}\right) \tag{2.26}$$

$$v_{\text{id}} = v_p - v_m = v_{\text{in}} \tag{2.27}$$

$$v_{\text{out}} = G_{\text{mOTA}} . v_{\text{in}} \cdot \left(\frac{1}{j\omega C_L}\right) \tag{2.28}$$

At unity gain frequency, $\left|\frac{v_{\text{out}}}{v_{\text{in}}}\right| = 1$. Solving for ω and hence unity gain frequency, we finally write

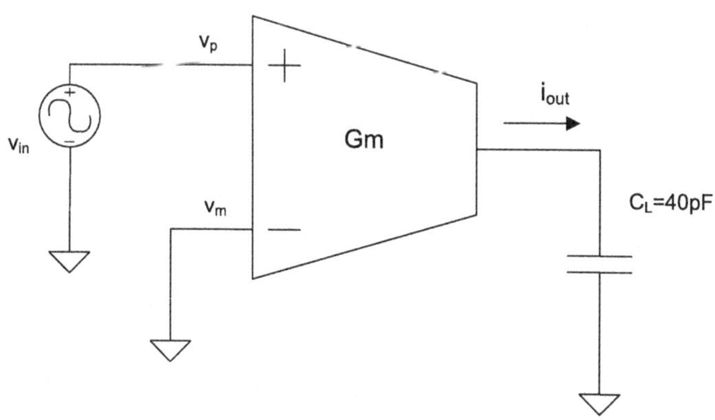

Fig. 2.4 Model for the analysis of unity gain frequency

$$f_u = \frac{G_{\mathrm{mOTA}}}{2\pi C_L} \tag{2.29}$$

Substituting the appropriate values, we get $f_u = 21.8\,\mathrm{Hz}$. Thus, this circuit is applicable for extreme low-frequency applications.

2.4.2.4 DC Gain, 3-DB Frequency and Gain–Bandwidth Product

In order to obtain the Bode plot parameters, we use the small signal model of the OTA where we incorporate the finite output resistance as well. This is shown in Fig. 2.5. We write the following

$$v_{\mathrm{out}} = (G_{\mathrm{mOTA}}.v_{\mathrm{id}}) \cdot \left(R_{\mathrm{out}}||\frac{1}{sC_L}\right) \tag{2.30}$$

$$\frac{v_{\mathrm{out}}}{v_{\mathrm{in}}} = \frac{G_{\mathrm{mOTA}}.R_{\mathrm{out}}}{1 + sC_L.R_{\mathrm{out}}} \tag{2.31}$$

Comparing this with the transfer function of a first-order system, we see that the DC gain and 3-dB frequency are

$$A_0 = G_{\mathrm{mOTA}}.R_{\mathrm{out}} \tag{2.32}$$

$$\omega_{3\mathrm{dB}} = \frac{1}{R_{\mathrm{out}}.C_L} \tag{2.33}$$

The gain–bandwidth product is thus constant and is written as

$$\mathrm{GBW} = \frac{G_{\mathrm{mOTA}}}{2\pi C_L} = f_u \tag{2.34}$$

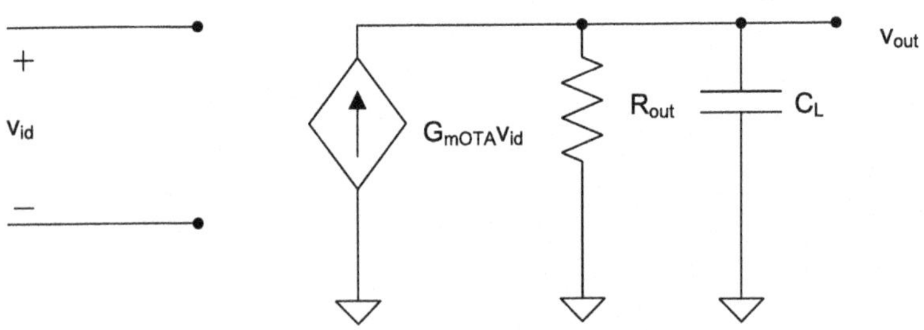

Fig. 2.5 Model for computing the gain–bandwidth product and 3-dB frequency

2.4.2.5 Slew Rate

The slew rate factor is defined as the maximum rate of change of output voltage with time. In the case of the OTA, the rate of change of output voltage will be dependent on the capacitor at the output node. The maximum rate of change of voltage occurs when maximum current flows through the capacitor. To obtain the maximum current, we apply a high enough voltage at the non-inverting terminal which causes the entire bias current (200 pA) to switch to that branch. Since there are two input pairs, the maximum current that will flow through the capacitor is 400 pA. The slew rate is defined as

$$\mathrm{SR} = \frac{I_C}{C_L} \tag{2.35}$$

Substituting values, the slew rate comes out to be 10 V/s. This shows that the circuit is applicable for extreme low frequency.

2.4.3 Simulation Results

The design is simulated using SCL, 180 nm technology node through Cadence framework using BSIM model.

2.4.3.1 Transient Simulation

The setup for transient simulation of the design is shown in Fig. 2.6. As mentioned earlier, we apply a sinusoidal signal of 10 Hz, which lies in the extreme low-frequency range. The transient simulation results showing the variations of the output current of the complete circuit, i.e., i_{out} with time are shown in Fig. 2.7. We find that the output current is sinusoidal in nature which concludes that the circuit is linear. In order to find out G_{mOTA}, we do a parametric sweep of the amplitude of input voltage (v_{id}) from −100 to +100 mV and record the values of amplitude of output current (i_{out}). Then a graph of i_{out} vs v_{id} (amplitude) is plotted and G_{mOTA} is evaluated from the slope of the curve at low values of v_{id} (−50 to +50 mV). This is shown in Fig. 2.8. The value of G_{mOTA} as calculated analytically is 5.5 nS, whereas that coming out from simulation results is 5.428 nS.

2.4.3.2 AC Simulation

AC analysis is performed to obtain several important parameters such as gain margin, phase margin, 3-dB frequency, unity gain frequency and DC gain. The simulation setup is shown in Fig. 2.9. The gain response plot as well as the parameters extracted are shown in Fig. 2.10.

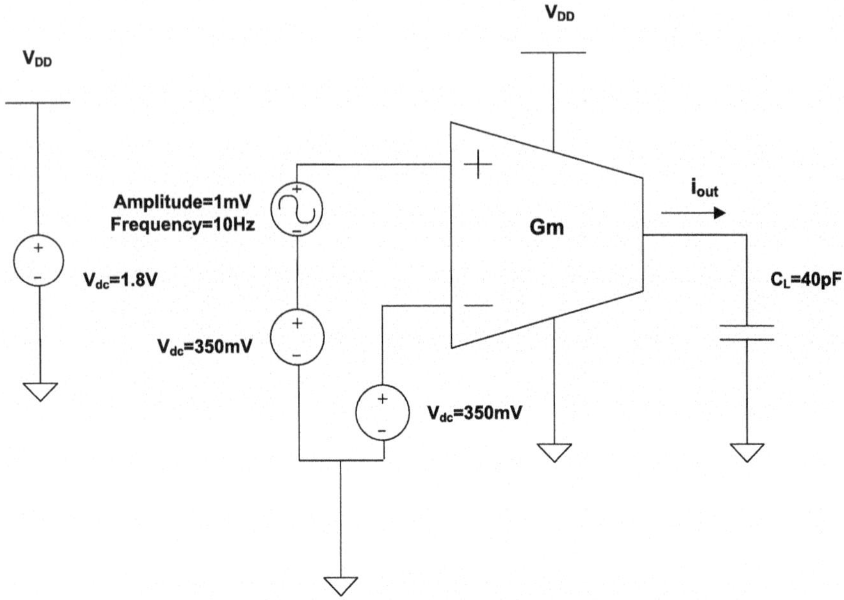

Fig. 2.6 Setup for transient simulation

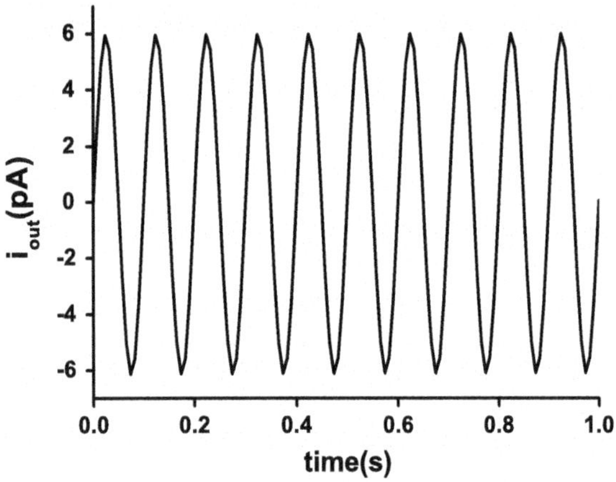

Fig. 2.7 Variation of the output current i_{out} with time

2.4.3.3 ICMR Simulation

Figure 2.11 shows the setup for measuring the input common mode range (ICMR). The OTA is connected in a negative feedback loop. A DC voltage V_{CM}, connected to the non-inverting terminal, is swept from 0 to 1.8 V. The values of V_{CM} for which a constant bias current flows through the biasing transistor, e.g., M6 determine the ICMR. The ICMR simulation graph is shown in Fig. 2.12, and the results are indicated within the graph.

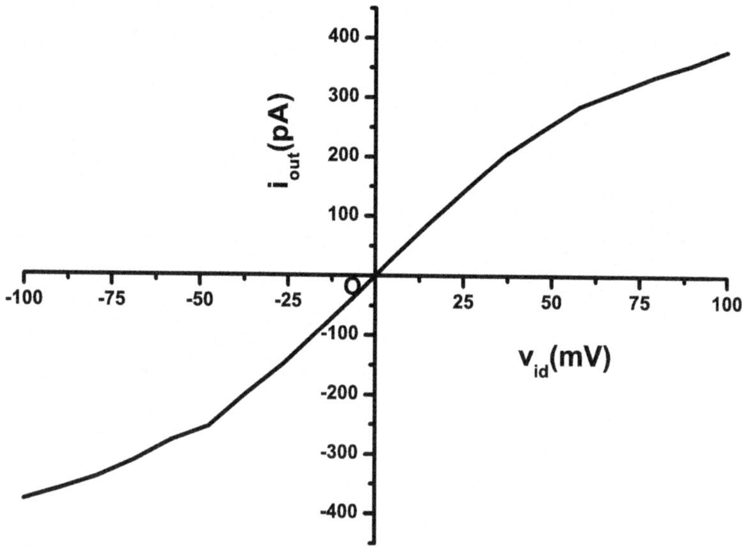

Fig. 2.8 Variation of the output current i_{out} with differential voltage v_{id}

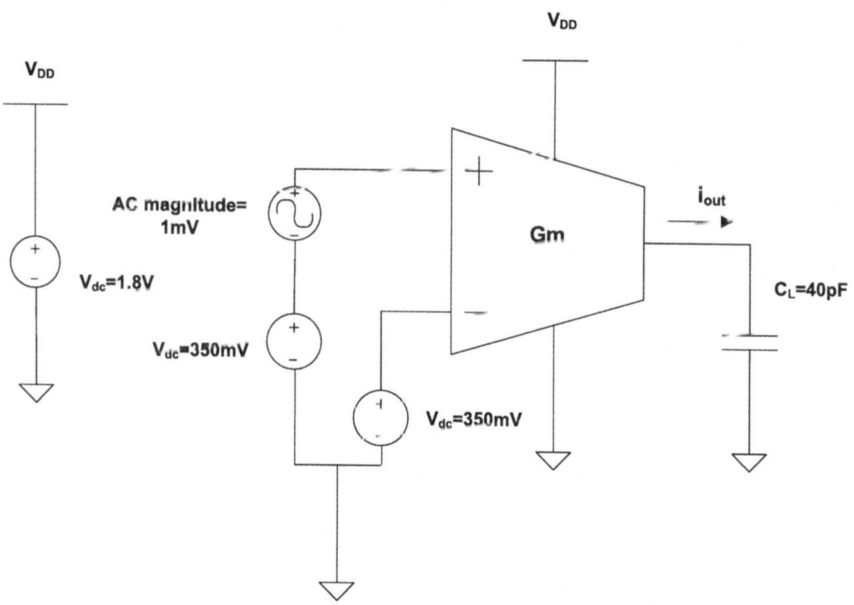

Fig. 2.9 Setup for AC simulation

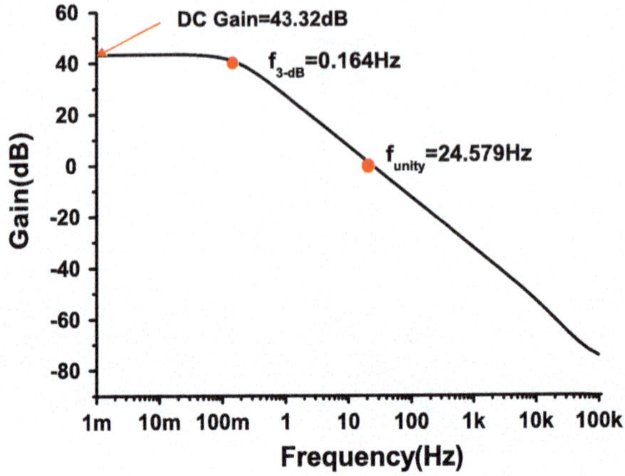

Fig. 2.10 Gain response plot

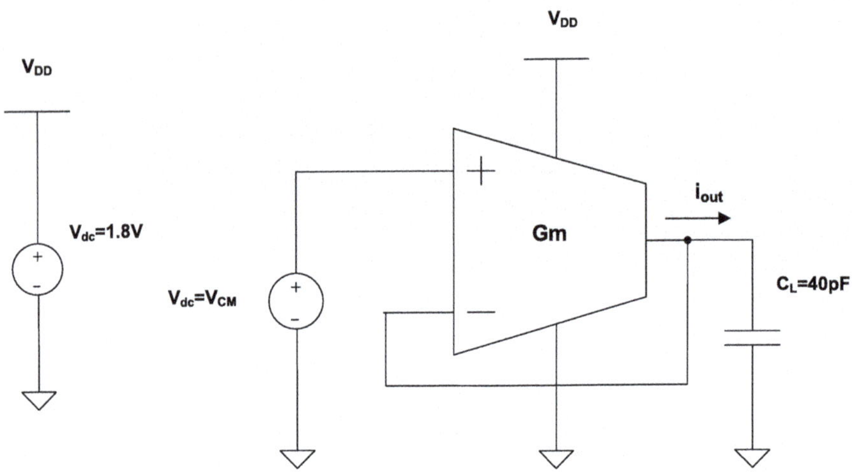

Fig. 2.11 Setup for ICMR simulation

2.4.3.4 PSRR Simulation

Figure 2.13 shows the setup for measuring the power supply rejection ratio (PSRR). The non-inverting terminal is shorted to ground, and the OTA is connected in a negative feedback loop. The AC source connected in series with the supply will produce the output voltage which will help us estimate noise voltages introduced at the output by ripples in the power supply. The positive PSRR versus frequency plot is shown in Fig. 2.14, and the results are indicated in the graph.

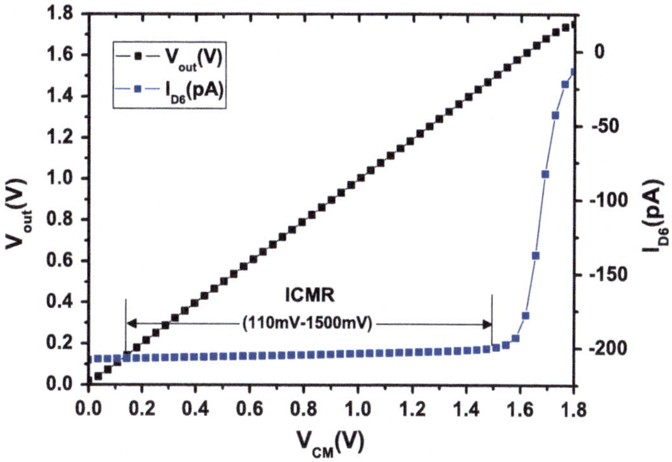

Fig. 2.12 V_{out} versus V_{CM} and I_{D6} versus V_{CM} for OTA

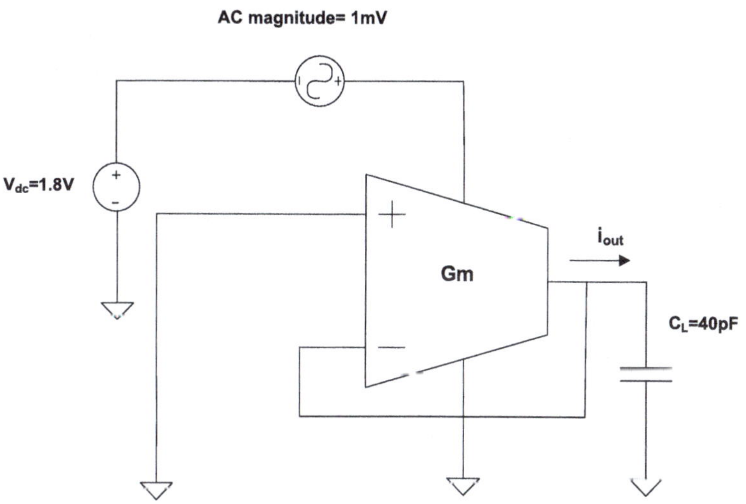

Fig. 2.13 Setup for PSRR simulation

2.4.3.5 Slew Rate Simulation

Figure 2.15 shows the setup for measuring slew rate. The unity gain configuration will force the OTA to track the input voltage but due to the finite slew rate the rising and falling slopes will be finite and different from that of the input. The slopes of the output voltage will give us the value of the slew rate of the OTA. The graph for measuring the slew rate is shown in Fig. 2.16, and the value is indicated in the graph.

The values of all important parameters as extracted from the simulation results are summarized in Table 2.3.

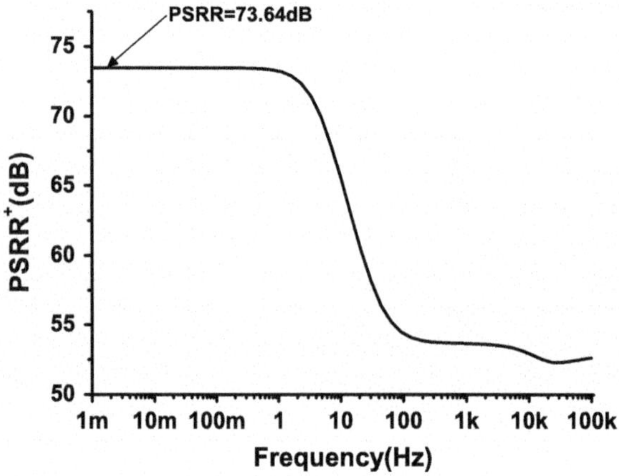

Fig. 2.14 PSRR+ versus frequency for OTA

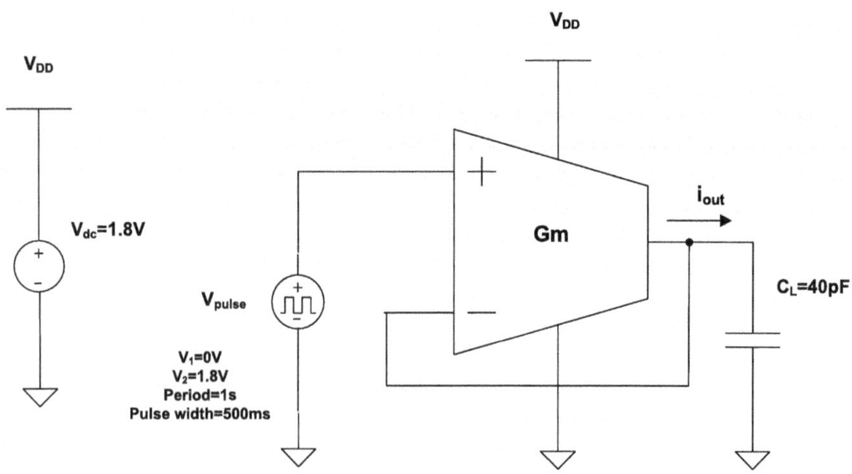

Fig. 2.15 Setup for slew rate simulation

2.5 Application of the Design Methodology

The frequency band 3–30 Hz with wavelength 10^8–10^7 m is identified to be the extreme low-frequency (ELF) band by the International Telecommunications Union (What are the spectrum band designators 2020). The super low-frequency (SLF) band ranges from 30–300 Hz with wavelength from 10^7–10^6 m (What are the spectrum band designators 2020). The radio waves within this band may be generated by lightning and other natural fluctuations in the magnetic field of the earth. In recent literature, it has been reported that most electrical activity in vertebrates and invertebrates occurs at ELF band, with characteristic maxima below 50 Hz Price et al.

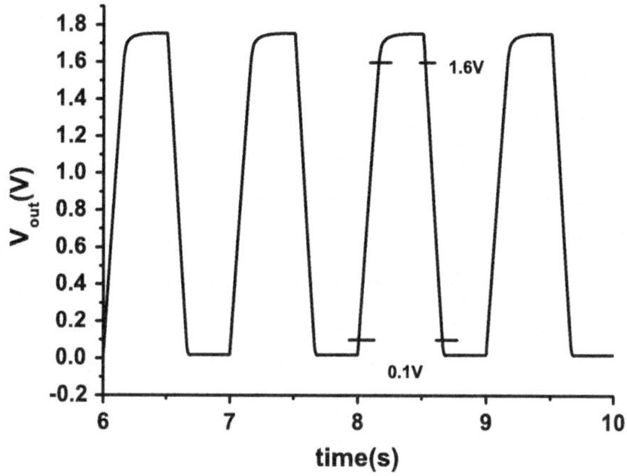

Fig. 2.16 V_{out} versus time for OTA

Table 2.3 Summary of the simulation results

S. No.	Parameters	Simulation values
1	i_{out}	6 pA
2	G_{mOTA}	5.428 nS
3	DC gain	43.32 dB
4	UGB	24.579 Hz
5	Gain Margin	71.598 dB
6	Phase Margin	90.31°
7	ICMR	110–1500 mV
8	PSRR	73.64 dB
9	Slew rate	10.714 V/s
10	Power consumption	1.12 nW

(2020). For human being, the majority of the electrical activities occur in a frequency range below 50 Hz. The brain waves like delta brain wave (0.5–3 Hz), theta waves (3–8 Hz), alpha wave (8–12 Hz), beta wave (12–38 Hz) and gamma wave (38–42 Hz) represent different states of our brain functions starting from that in deepest meditation and dreamless sleep to highly conscious state carrying simultaneous processing of information from different brain areas (Price et al. 2020). Electroencephalography (EEG) is an efficient technique to acquire brain signals which corresponds to various states from the scalp surface area. The EEG spikes have a bandwidth of 0.5 Hz-1 kHz. For acquisitions of EEG signals, wearable devices are preferred because of greater comfort and continuous monitoring (Tohidi et al. 2019). Analog front end circuits are major components for such devices. Ultra-low-power consumption is the primary requirement for analog circuits present in such wearable devices (Karimi-Bidhendi 2017). MOS transistors are used in the weak inversion mode. In this mode, the bandwidth is not high. However, this is not of a major problem, since brain signals lie in the ELF-SLF band. The design methodology, as presented in this chapter, will thus serve as an effective methodology for the design of analog circuits operating in the ELF-SLF band.

2.6 Summary

This chapter presents a systematic design methodology based on inversion coefficient for the design of ultra-low-power CMOS analog circuits. For ELF-SLF applications, the frequency of operation is very low, and power dissipation is the most important challenge. Therefore, the MOS transistors may operate in the weak inversion mode. The sizing technique within this methodology is very simple. The selection of bias current and the channel length play important roles in the design process. The other important parameter to select is value of the inversion coefficient, depending on whether the transistor is to be operated in the weak inversion or moderate inversion or strong inversion mode. The design methodology is implemented through the design of an operational transconductance amplifier circuit, meant for ELF applications.

Acknowledgements The author thanks the SMDP-C2SD project of the University of Calcutta, sponsored by MeitY, Govt. of India, for providing the necessary simulation resources which have been made use of for the work carried. The author expresses his deep gratitude to his student, Shri Rishov Aditya, B.Tech student of Netaji Subhash Engineering College, for carrying out many simulation experiments on this topic as part of his B. Tech project work. He further acknowledges the support provided by Dr. S. Sarkhel of the same institute for her support.

References

Binkley DM (2008) Trade offs and Optimization in Analog CMOS Design. Wiley

Enz CC, Vittoz EA (2006) Charge-based MOS Transistor modeling: the EKV model for low-power and RF IC design. Wiley

Karimi-Bidhendi A et al (2017) CMOS Ultralow power brain signal acquisition front-ends: design and human testing. IEEE Trans Biomed Circuits Syst 11(5):1111–1122

Pandit S (2013) MOSFET characterization for VLSI circuit simulation. In: Sarkar C (ed) Technology computer aided design. CRC Press, Boca Raton. https://doi.org/10.1201/9781315216454

Pandit S (2018) Nanoscale silicon MOS Transistors. In: Roy S, Ghosh C, Sarkar C (eds) Nanotechnology. CRC Press, Boca Raton. https://doi.org/10.1201/9781315116730

Pandit S (2015) Nanoscale MOSFET: MOS transistor as basic building block. In: Sengupta A, Sarkar C (eds) Introduction to Nano. Engineering Materials, Springer, Berlin, Heidelberg

Pandit S, Mandal C, Patra A (2014) Nano-scale CMOS analog circuits. CRC Press, Boca Raton. https://doi.org/10.1201/9781315216102

Price C, Williams E, Elhalel G et al (2020) Natural ELF fields in the atmosphere and in living organisms. Int J Biometeorol. https://doi.org/10.1007/s00484-020-01864-6

Roy K, Mukhopadhyay S, Mahmoodi-Meimand H (2003) Leakage current mechanisms and leakage reduction techniques in deep-submicrometer CMOS circuits. Proc IEEE 91(2):305–327

Saha SK (2016) Compact models for integrated circuit design. CRC Press, Boca Raton. https://doi.org/10.1201/b19117

Tohidi M, Kargaard Madsen J, Moradi F (2019) Low-power high-input-impedance eeg signal acquisition SoC with fully integrated IA and signal-specific adc for wearable applications. IEEE Trans Biomed Circuits Syst 13(6):1437–1450

What are the spectrum band designators and bandwidths? https://www.nasa.gov/directorates/heo/scan/communications/outreach/funfacts/txt_band_designators.html

Chapter 3
Orthogonally Controllable VQO for Low-Voltage Applications

Bhartendu Chaturvedi, Jitendra Mohan, and Atul Kumar

Abstract A versatile quadrature oscillator (VQO) circuit is introduced in this chapter. The circuit comprises a fully differential second-generation current conveyor (FDCCII), three resistors and two capacitors, all of which are grounded. The proposed circuit is versatile as it simultaneously delivers the voltage-mode and current-mode outputs. The oscillator circuit is benefitted with appropriately suited modern integrated circuit (IC) technology attributes such as: availability of two quadrature voltages and two quadrature currents simultaneously, orthogonal controllability of oscillation frequency as well as condition of oscillation (CO), low power consumption, low total harmonic distortion (THD), good sensitivity performance, and use of all grounded components. The proposed oscillator structure operates at ± 0.9 V and hence suitable for low-voltage applications. Effects of device non-idealities and parasitic on the performance of the proposed oscillator are further analyzed. Validation of theoretical aspects of the proposed oscillator circuit is done by carrying out HSPICE simulations using 0.18 µm TSMC CMOS parameters. Furthermore, to exploit the practicality of the proposed VQO, results of experimental verification performed by connecting discrete passive components and commercially available ICs (AD844) on a breadboard are also included.

Keywords Analog circuit design · Oscillator circuit · Orthogonal control · Versatile circuit

---

B. Chaturvedi (✉) · J. Mohan · A. Kumar
Department of Electronics and Communication Engineering, Jaypee Institute of Information Technology, Noida, Uttar Pradesh 201304, India
e-mail: bhartendu.prof@gmail.com

J. Mohan
e-mail: jitendramv2000@rediffmail.com

A. Kumar
e-mail: atul.nit304@gmail.com

R. Dhiman and R. Chandel (eds.), *Nanoscale VLSI*,
Energy Systems in Electrical Engineering,
https://doi.org/10.1007/978-981-15-7937-0_3

3.1 Introduction

Sinusoidal quadrature oscillators have become important cells as these oscillators are widely used in a variety of applications in communication and instrumentation systems. Single-sideband modulation and quadrature mixers are some of the key applications in communication domain, and selective voltmeters and vector generators are the salient applications in instrumentation domain. Consequently, numerous circuits of quadrature oscillator realized using varities of active elements are already described in the literature. These realizations can be categorized as: voltage-mode (Abaci and Yuce 2017; Maheshwari and Chaturvedi 2012; Sotner et al. 2015; Chaturvedi et al. 2019a; Yucel and Yuce 2014; Maheshwari 2007; Minaei and Yuce 2010), current-mode (Maheshwari and Chaturvedi 2011; Minaei and Ibrahim 2005; Keskin and Biolek 2006; Kumar and Chaturvedi 2016, 2018a; Maheshwari 2003; Prommee and Khateb 2014) and versatile-mode (availability of current and voltage signal(s) simultaneously) (Kumar and Chaturvedi 2017, 2018b; Mohan et al. 2016; Chaturvedi and Mohan 2015; Maheshwari 2004, 2008, 2014; Yuce 2017; Maheshwari and Khan 2007; Chaturvedi and Kumar 2019). The oscillation frequency of any oscillator circuit must be controllable independent of condition of oscillation (CO), so that any changes in later one do not affect the former. Similarly, CO must be controllable independent of oscillation frequency. Therefore, orthogonal control of these two parameters is always a desired attribute of any oscillator design. However, this feature is missing in many of the earlier reported circuits (Abaci and Yuce 2017; Maheshwari and Chaturvedi 2011, 2012; Chaturvedi et al. 2019a; Maheshwari 2007, 2008, 2014; Minaei and Yuce 2010; Kumar and Chaturvedi 2016, 2018a, b; Prommee and Khateb 2014; Mohan et al. 2016; Yuce 2017). Moreover, the grounded passive components are preferable from integration point of view. From the literature review, it has been observed that circuits presented in (Abaci and Yuce 2017; Maheshwari and Chaturvedi 2012; Yucel and Yuce 2014; Minaei and Yuce 2010; Minaei and Ibrahim 2005; Keskin and Biolek 2006; Kumar and Chaturvedi 2016, 2017, 2018a, b; Mohan et al. 2016; Yuce 2017; Maheshwari and Khan 2007; Maheshwari 2004) use the floating passive components.

In this chapter, a VQO circuit consists of single fully differential second-generation current conveyor (FDCCII), and five passive components are introduced. All the passive components used in the realization of the proposed quadrature oscillator circuit are grounded. The oscillation frequency is independently adjustable with the help of resistor without influencing the CO. Similarly, CO can also be adjusted via another resistor without disturbing oscillation frequency. A possible practical realization of the proposed circuit using ICs (AD844) and discrete passive components is also shown.

3.2 Proposed Quadrature Oscillator Circuit

The circuit of the proposed versatile quadrature oscillator employs one FDCCII, three grounded resistors and two grounded capacitors, as depicted in Fig. 3.1. FDCCII has proved its versatility as current-mode active element by featuring in numerous signal processing applications (Mohan et al. 2016, 2020; El-Adawy et al. 2000; Chaturvedi et al. 2018, 2019b; Kumar et al. 2017). The terminal characteristics of FDCCII (El-Adawy et al. 2000) which is used to realize the proposed circuit are given in Eq. 3.1.

$$\begin{bmatrix} I_{Y1} \\ I_{Y2} \\ I_{Y3} \\ I_{Y4} \\ V_{X+} \\ V_{X-} \\ I_{Z+} \\ I_{-Z+} \\ I_{-Z-} \end{bmatrix} = \begin{bmatrix} 0 & 0 & 0 & 0 & 0 & 0 \\ 0 & 0 & 0 & 0 & 0 & 0 \\ 0 & 0 & 0 & 0 & 0 & 0 \\ 0 & 0 & 0 & 0 & 0 & 0 \\ 0 & 0 & 1 & -1 & 1 & 0 \\ 0 & 0 & -1 & 1 & 0 & 1 \\ 1 & 0 & 0 & 0 & 0 & 0 \\ -1 & 0 & 0 & 0 & 0 & 0 \\ 0 & -1 & 0 & 0 & 0 & 0 \end{bmatrix} \begin{bmatrix} I_{X+} \\ I_{X-} \\ V_{Y1} \\ V_{Y2} \\ V_{Y3} \\ V_{Y4} \end{bmatrix} \tag{3.1}$$

The analysis of the circuit gives the following characteristic equation.

$$s^2 + \frac{(R_3 - R_1)}{R_1 R_3 C_1} s + \frac{1}{R_2 R_3 C_1 C_2} = 0 \tag{3.2}$$

From Eq. 3.2, oscillation frequency, f_0 and CO are found as

$$f_0 = \frac{1}{2\pi} \sqrt{\frac{1}{R_2 R_3 C_1 C_2}} \tag{3.3}$$

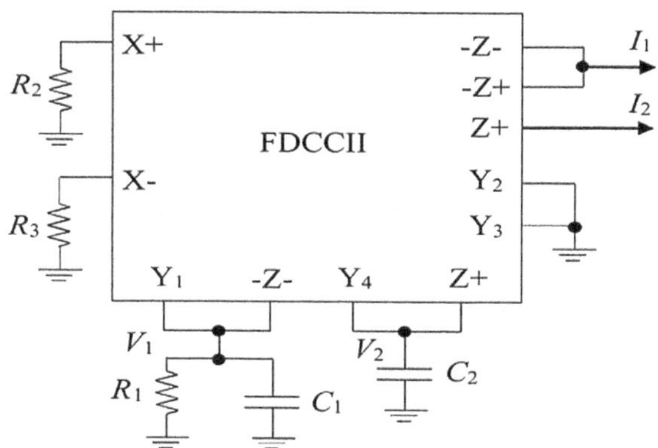

Fig. 3.1 Proposed versatile quadrature oscillator

$$\mathrm{CO}: R_1 \geq R_3 \tag{3.4}$$

It is to be observed from Eqs. 3.3 to 3.4 that tunability of f_0 can be achieved through R_2 without disturbing the CO. Similarly, with the help of R_1, independent control of CO can be achieved without altering the oscillation frequency. The relationships between the voltage outputs and current outputs of the proposed oscillator can be expressed as:

$$V_1 = jK_1V_2 \tag{3.5}$$

$$I_1 = jK_2I_2 \tag{3.6}$$

where, $K_1 = \omega R_2C_2$ and $K_2 = \omega R_2C_1$. Equations 3.5–3.6 reveal that V_1 and V_2; I_1 and I_2 are in quadrature relationship, respectively.

3.2.1 *Non-ideal Aspects*

The port relations of FDCCII for non-ideal scenario are expressed in matrix form as follows.

$$\begin{bmatrix} I_{Y1} \\ I_{Y2} \\ I_{Y3} \\ I_{Y4} \\ V_{X+} \\ V_{X-} \\ I_{Z+} \\ I_{-Z+} \\ I_{-Z-} \end{bmatrix} = \begin{bmatrix} 0 & 0 & 0 & 0 & 0 & 0 \\ 0 & 0 & 0 & 0 & 0 & 0 \\ 0 & 0 & 0 & 0 & 0 & 0 \\ 0 & 0 & 0 & 0 & 0 & 0 \\ 0 & 0 & \beta_1 & -\beta_2 & \beta_3 & 0 \\ 0 & 0 & -\beta_4 & \beta_5 & 0 & \beta_6 \\ \alpha_1 & 0 & 0 & 0 & 0 & 0 \\ -\alpha_2 & 0 & 0 & 0 & 0 & 0 \\ 0 & -\alpha_3 & 0 & 0 & 0 & 0 \end{bmatrix} \begin{bmatrix} I_{X+} \\ I_{X-} \\ V_{Y1} \\ V_{Y2} \\ V_{Y3} \\ V_{Y4} \end{bmatrix} \tag{3.7}$$

In Eq. 3.7, α_1, α_2, α_3, β_1, β_2, β_3, β_4, β_5 and β_6 are the non-ideal transfer gains. The proposed quadrature oscillator has been reanalyzed using Eq. 3.7, and the characteristic equation thus obtained is modified as follows.

$$s^2 + \frac{(R_3 - \alpha_3\beta_4R_1)}{R_1R_3C_1}s + \frac{\alpha_1\alpha_3\beta_1\beta_6}{R_2R_3C_1C_2} = 0 \tag{3.8}$$

The modified f_0 and CO are now given in Eqs. 3.9–3.10.

$$f_0 = \frac{1}{2\pi}\sqrt{\frac{\alpha_1\alpha_3\beta_1\beta_6}{R_2R_3C_1C_2}} \tag{3.9}$$

$$\mathrm{CO}: \alpha_3\beta_4 R_1 \geq R_3 \tag{3.10}$$

The active and passive sensitivities of f_0 are found as follows:

$$S^{f_0}_{\alpha_1,\alpha_3,\beta_1,\beta_6} = -S^{f_0}_{R_2,R_3,C_1,C_2} = \frac{1}{2}, \quad S^{f_0}_{\beta_4} = S^{f_0}_{R_1} = 0 \tag{3.11}$$

Sensitivity figures as ≤ 1 are always considered acceptable; thus, Eq. 3.11 signifies that the proposed quadrature oscillator exhibits good sensitivity performance.

3.2.2 *Parasitic Considerations*

Parasitic associated with various FDCCII ports are described by the following expressions: $R_{Y1}//(1/(sC_{Y1}))$ at Y_1 terminal, $R_{Y2}//(1/(sC_{Y2}))$ at Y_2 terminal, $R_{Y3}//(1/(sC_{Y3}))$ at Y_3 terminal and $R_{Y4}//(1/(sC_{Y4}))$ at Y_4 terminal; the small resistances R_{X+} and R_{X-} appear at $X+$ and X− terminals, respectively, and combinations of $R_{Z+}//(1/(sC_{Z+}))$, $R_{-Z+}//(1/(sC_{-Z+}))$ and $R_{-Z-}//(1/(sC_{-Z-}))$ appear at $Z+$, $-Z+$ and $-Z-$ terminals, respectively. By taking the parasitic of FDCCII into consideration, the proposed circuit of versatile quadrature oscillator is analyzed again and the updated characteristic equation thus obtained is given in Eq. 3.12.

$$s^2 + \frac{R_2'\left(C_2' R_{eq}\left(R_3' - R_1'\right) + R_1' R_3' C_1'\right)}{R_1' R_3' R_{eq} C_1' C_2'} s + \frac{R_2'\left(R_3' - R_1'\right) + R_1' R_{eq}}{R_1' R_2' R_3' R_{eq} C_1' C_2'} = 0 \tag{3.12}$$

The f_0 and CO are now modified as

$$f_0 = \frac{1}{2\pi}\sqrt{\frac{R_2'\left(R_3' - R_1'\right) + R_1' R_{eq}}{R_1' R_2' R_3' R_{eq} C_1' C_2'}} \tag{3.13}$$

$$\mathrm{CO}: \left(C_2' R_{eq} + R_1' C_1'\right) R_3' \leq C_2' R_{eq} R_1' \tag{3.14}$$

where, $R_1' = R_1//R_{Y1}//R_{-Z-}$, $R_2' = R_2 + R_{X+}$, $R_3' = R_3 + R_{X-}$, $R_{eq} = R_{Y4}//R_{Z+}$, $C_1' = C_1 + C_{Y1} + C_{-Z-}$ and $C_2' = C_2 + C_{Y4} + C_{Z+}$.

For equal value of capacitors ($C_1 = C_2$) and $C_1 \gg C_{Y1} + C_{-Z-}$ and $C_2 \gg C_{Y4} + C_{Z+}$, Eq. 3.14 can be written as follows.

$$\mathrm{CO}: \left(R_{eq} + R_1'\right) R_3' \leq R_{eq} R_1' \tag{3.15}$$

Moreover, if external resistor R_1 used in the design of the proposed circuit is chosen of smaller value such that $R_1' = R_1//R_{Y1}//R_{-Z-} \approx R_1 \Rightarrow R_{eq} + R_1' \approx R_{eq}$, then Eq. 3.15 can be written as follows.

$$\text{CO: } R'_3 \leq R'_1 \tag{3.16}$$

A noteworthy observation from Eqs. 3.13 and 3.16 is that the FDCCII parasitic do not affect f_0 and CO adversely. Moreover, to compensate these parasitic, passive components are selected as follows: $R_1 \ll (R_{Y1}//R_{-Z-})$, $R_2 \gg R_{X+}$, $R_3 \gg R_{X-}$, $C_1 \gg C_{Y1} + C_{-Z-}$ and $C_2 \gg C_{Y4} + C_{Z+}$.

3.3 Simulation Results

The simulation results of circuit of Fig. 3.1 are carried out using HSPICE with 0.18 μm CMOS parameters. For the realization of the proposed quadrature oscillator, FDCCII's CMOS structure is taken from (Chaturvedi et al. 2018). The supply voltages of ±0.9 V and bias currents $I_B = 30$ μA and $I_{SB} = 5$ μA are used in simulations. The resistor and capacitor values chosen for simulation are: $R_1 = 1.1$ kΩ, $R_2 = R_3 = 1$ kΩ and $C_1 = C_2 = 1$ nF. Simulated output voltages and their frequency spectrums are depicted in Fig. 3.2. Similarly, simulated waveforms of output currents and their frequency spectrums are shown in Fig. 3.3. Value of f_0 as observed from simulations is 148 kHz (6.9%). The error in the oscillation frequency is because of parasitic values. If parasitic values are included in the passive components, then theoretical oscillation frequency is found to be same as simulated oscillation frequency. For voltage outputs V_1 and V_2, THD values are 1.27% and 1.12%, respectively, and THDs for the currents, I_1 and I_2 are found to be 1.9% and 1.52%, respectively. Thus, THD for each output is under 2%. Therefore, the proposed circuit exhibits low THD for each output. Additionally, response of the proposed circuit at higher frequency is checked for the following passive component values: $R_1 = 1.1$ kΩ, $R_2 = R_3 = 1$ kΩ and $C_1 = C_2 = 100$ pF. Time-domain waveforms for V_1 and V_2 and their frequency spectrums are shown in Fig. 3.4. The simulated f_0 is 1.4 MHz in Fig. 3.4. The power consumption of the proposed VQO is 0.6 mW.

Moreover, variations of f_0 against R_2 for different values of capacitors are depicted in Fig. 3.5. It is evident that f_0 varies from 148 to 66.7 kHz for $C_1 = C_2 =$ 1nF and from 1.4 to 0.65 MHz for $C_1 = C_2 = 100$ pF, when R_2 is varied from 1 to 5 kΩ at 0.5 kΩ step size.

Furthermore, the effects of capacitor variations and threshold voltage variations in MOS transistors are examined via Monte Carlo (MC) simulations. For both the cases, 10% Gaussian deviation is chosen for simulations. Figures 3.6 and 3.7 show the simulated waveforms for V_1 and V_2 for the variations in capacitor and threshold voltage, respectively. It is to be observed from Fig. 3.7 that the performance of proposed circuit is not adversely affected by the threshold voltage variation.

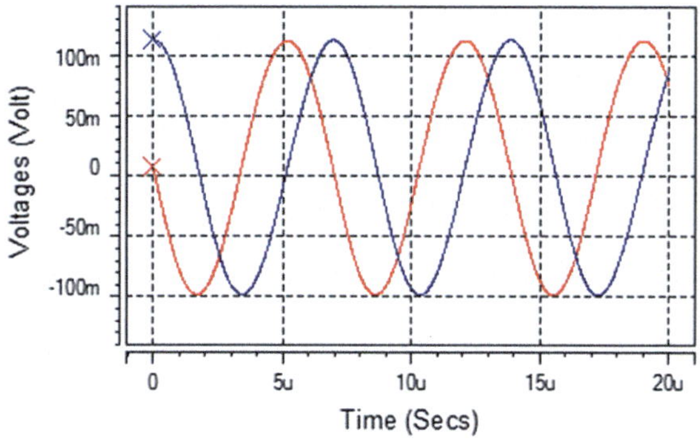

(a)

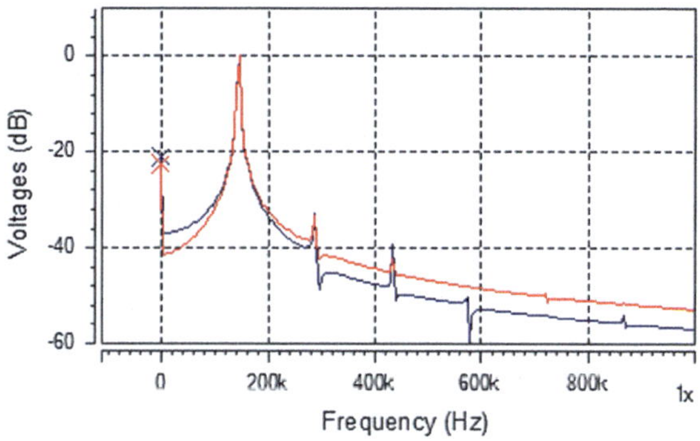

(b)

Fig. 3.2 **a** Simulated waveforms of V_1 and V_2 at 148 kHz, **b** simulated frequency spectrums of V_1 and V_2

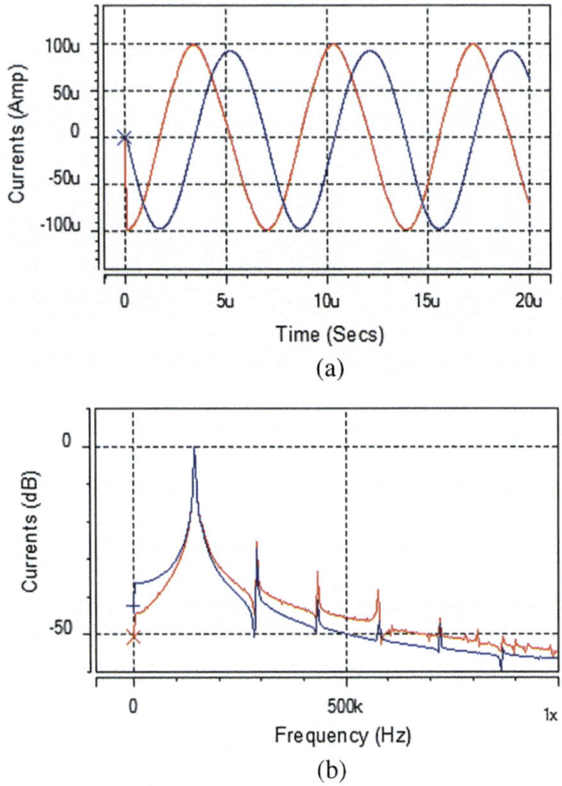

Fig. 3.3 **a** Simulated waveforms of I_1 and I_2 at 148 kHz, **b** frequency spectrums of I_1 and I_2

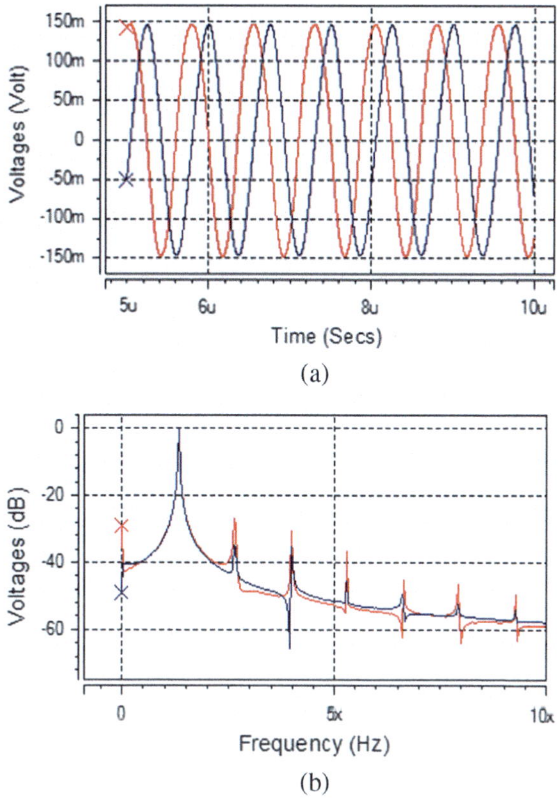

Fig. 3.4 **a** Simulated waveforms of V_1 and V_2 at 1.4 MHz, **b** frequency spectrums of V_1 and V_2

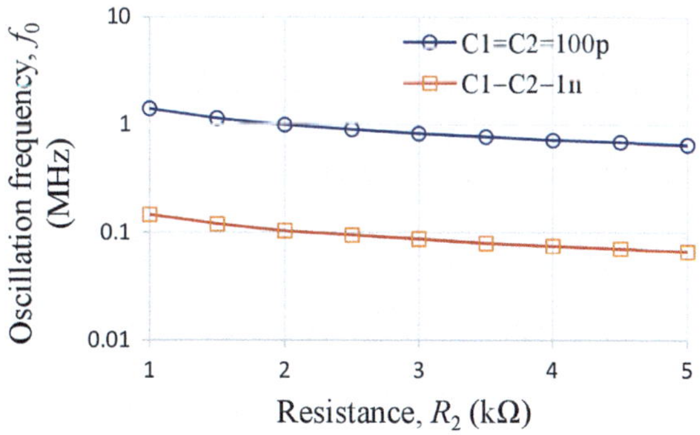

Fig. 3.5 Oscillation frequency variation against resistance, R_2

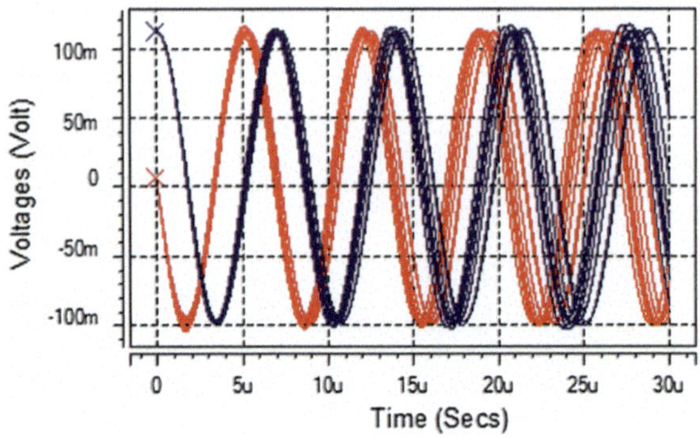

Fig. 3.6 MC simulations showing the simulated waveforms of V_1 and V_2 for capacitance variation

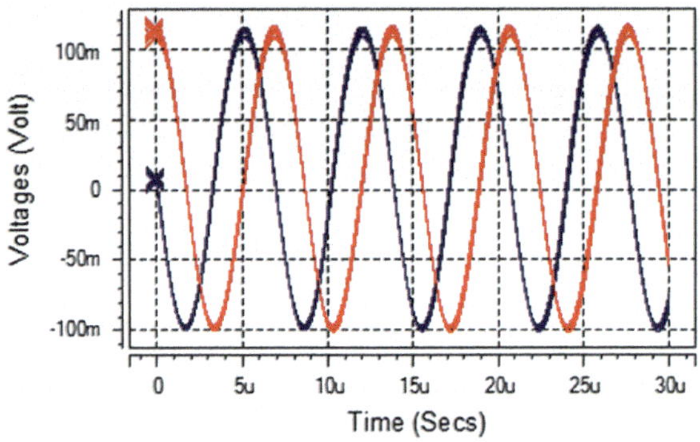

Fig. 3.7 MC simulations showing the simulated waveforms of V_1 and V_2 for threshold voltage variation

3.4 Experimental Verification of the Proposed Quadrature Oscillator

Next, the experimental verifications are done to verify the practical applicability of the proposed VQO. Using the commercial ICs (AD844), a possible practical realization of the proposed VQO is depicted in Fig. 3.8. Supply voltages ±5 V are applied to obtain the experimental results. The passive components used are $R_1 = 11$ kΩ, $R_2 = R_3 = 10$ kΩ and $C_1 = C_2 = 10$ nF. The experimentally obtained waveforms of voltage outputs V_1 and V_2 are shown in Fig. 3.9. Value of the measured frequency corresponding to the experimental results is found to be 1.53 kHz (3.77% error). Quadrature relationship between V_1 and V_2 is evident from the Lissajous pattern as shown in Fig. 3.10. Additionally, experimentally observed waveforms of output voltages, V_1 and V_2, when capacitor value is changed from 10 to 1 nF are shown in Fig. 3.11. Updated value of measured frequency corresponding to the change in capacitor value is 15.56 kHz (2.19% error). Furthermore, the passive components are changed to $R_1 = 1.34$ kΩ, $R_2 = R_3 = 1$ kΩ and $C_1 = C_2 = 1$ nF to obtain the results at higher frequency; Fig. 3.12 shows the corresponding observed waveforms of output voltages, V_1 and V_2. The experimentally measured frequency is 134.9 kHz (15.2% error). The error in the measured frequency is because of the parasitic impedances of IC AD844. Moreover, for getting the minimum error in the measured oscillation frequency, the value of R_2 and R_3 should be considerably high as compared to the parasitic resistance present at inverting terminal of AD844.

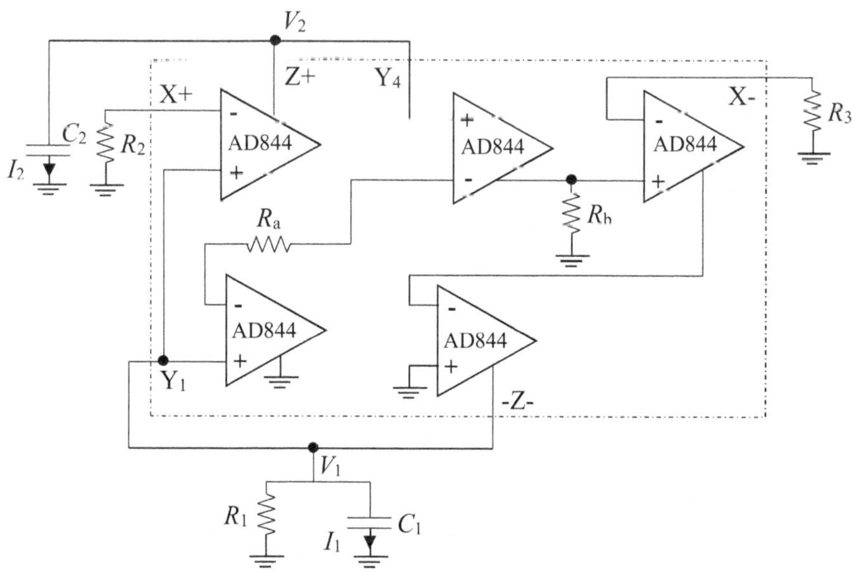

Fig. 3.8 Possible practical realization of the proposed quadrature oscillator using AD844

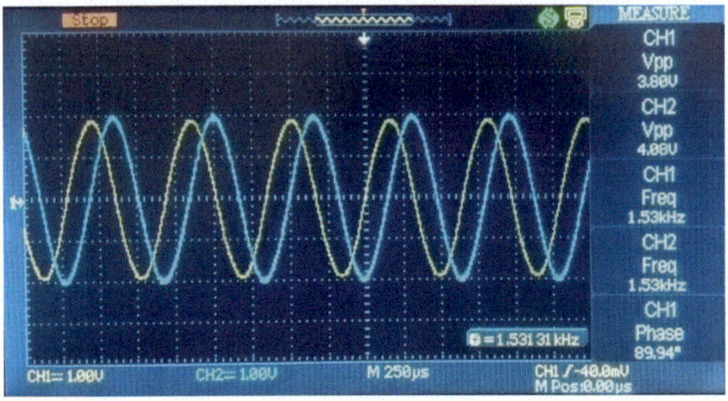

Fig. 3.9 Experimentally observed waveforms of V_1 and $V_2, f = 1.53$ kHz

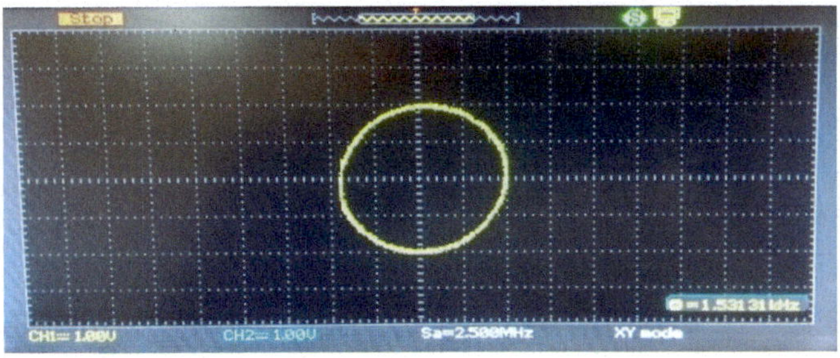

Fig. 3.10 Lissajous pattern obtained for V_1 and V_2

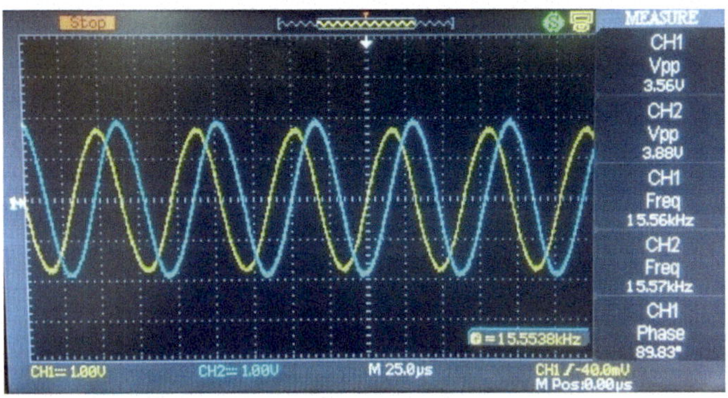

Fig. 3.11 Experimentally observed waveforms of V_1 and $V_2, f = 15.56$ kHz

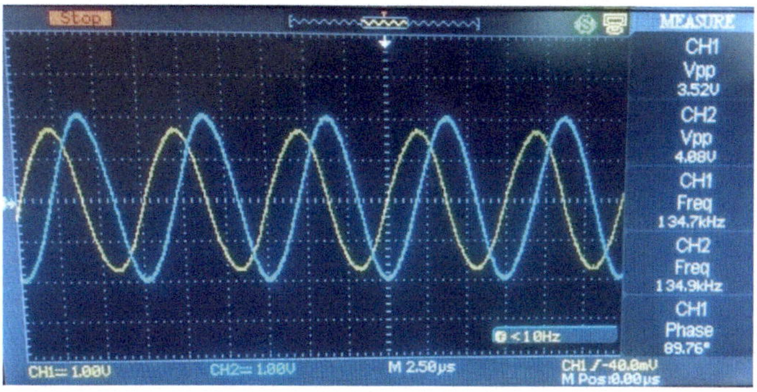

Fig. 3.12 Experimentally observed waveforms of V_1 and $V_2, f = 134.9$ kHz

3.5 Comparison and Discussion

The main features of the proposed circuit compared with some other relevant circuits of quadrature oscillators are given in Table 3.1. Noteworthy advantages with reference to the proposed VQO as observed from Table 3.1 are as follows: (i) single active element-based realization, (ii) use of all grounded passive components, (iii) concurrent delivery of two quadrature output voltages and two quadrature output currents, (iv) orthogonal controllability of oscillation frequency and CO, (v) low power consumption and (vi) low THD.

3.6 Conclusion

A VQO circuit comprising of an FDCCII, three grounded resistors and two grounded capacitors is proposed in this chapter. The key features of the proposed quadrature oscillator are as follows: active and passive sensitivity figures less than unity, orthogonal controllability of oscillation frequency and CO, low operating supply voltages, low THD of each output and low power dissipation. The performance of the proposed VQO is also investigated under non-ideal and parasitic conditions of the used active device. HSPICE simulation results are shown for the verification of the proposed VQO. Moreover, practical realization of the proposed quadrature oscillator using commercially available ICs has been shown.

Table 3.1 Comparison of the proposed quadrature oscillator with other relevant quadrature oscillators

References	Used active element	Active element(s) count	Passive component(s) $(R + C)$ count	All grounded passive components	Number of voltage outputs	Number of current outputs	Orthogonal control of f_0 and CO	THD	Power consumption (mW)	Operating supply voltages (V)	Experimental results shown
Abaci and Yuce (2017)	DVCC	1	2 + 2	No	2	0	No	3.48	1.62	±0.75	No
Maheshwari and Chaturvedi (2012)	DXCCII	2	2 + 2	No	2	0	No	1.2	–	±2.5	No
Sotner et al. (2015)	OTA and CA	2 + 1	0 + 2	Yes	4	0	Yes	2.2	–	±1	No
Chaturvedi et al. (2019a)	DXCCTA	2	0 + 3	Yes	2	0	No	3.3	–	±1.25	No
Yucel and Yuce (2014)	CCII	2	3 + 2	No	2	0	Yes	–	–	±0.75	No
Maheshwari (2007)	DVCC	3	3 + 2	Yes	2	0	No	–	–	±2.5	No
Minaei and Yuce (2010)	DVCC	3	2 + 2	No	2	0	No	–	0.47	±1.5	No

(continued)

Table 3.1 (continued)

References	Used active element	Active element(s) count	Passive component(s) $(R-C)$ count	All grounded passive components	Number of voltage outputs	Number of current outputs	Orthogonal control of f_0 and CO	THD	Power consumption (mW)	Operating supply voltages (V)	Experimental results shown
Maheshwari and Chaturvedi (2011)	DVCC	3	3 + 2	Yes	0	4	No	2	7.5	±2.5	No
Minaei and Ibrahim (2005)	DVCC	2	4 + 2	No	0	2	–	–	–	±2.5	No
Keskin and Biolek (2006)	CDTA	2	4 + 2	No	0	2	Yes	1	–	±2.5	No
Kumar and Chaturvedi (2016)	DXCCII	2	2 + 2	No	0	3	No	1.4	–	±1.25	No
Maheshwari (2003)	CCCII	3	0 + 2	Yes	0	3	Yes	<1	–	±2.5	No
Kumar and Chaturvedi (2018a)	DXCCTA	2	1 + 2	No	0	2	No	–	–	±1.25	No
Prommee and Khateb (2014)	CC-CDCCC	1	0 + 2	Yes	0	2	No	2.33	–	±1.25	No

(continued)

Table 3.1 (continued)

References	Used active element	Active element(s) count	Passive component(s) ($R + C$) count	All grounded passive components	Number of voltage outputs	Number of current outputs	Orthogonal control of f_0 and CO	THD	Power consumption (mW)	Operating supply voltages (V)	Experimental results shown
Kumar and Chaturvedi (2017)	CIDITA	1	1 + 2	No	2	2	Yes	2.8	–	±1.25	No
Mohan et al. (2016)	FDCCII	1	2 + 2	No	2	4	No	–	–	±1	No
Maheshwari (2014)	DVCC	2	2 + 2	Yes	4	4	No	6	–	±2.5	No
Chaturvedi and Mohan (2015)	DD-DXCCII	1	3 + 2	Yes	3	2	Yes	–	0.24	±1	No
Maheshwari (2008)	DVCC	2	2 + 2	Yes	4	4	No	1.6	–	±2.5	No
Yuce (2017)	CCII	2	2 + 2	No	2	2	No	6.06	1	±0.75	Yes
	DVCC	2	2 + 2	Yes	2	1	No	4.96	1	±0.75	No

(continued)

Table 3.1 (continued)

References	Used active element	Active element(s) count	Passive component(s) ($R + C$) count	All grounded passive components	Number of voltage outputs	Number of current outputs	Orthogonal control of f_0 and CO	THD	Power consumption (mW)	Operating supply voltages (V)	Experimental results shown
Maheshwari and Khan (2007)	CDBA	2	3 + 2	No	2	2	Yes	–	–	–	No
Maheshwari (2004)	CCCII	3	0 + 2	No	2	2	Yes	–	–	–	No
Kumar and Chaturvedi (2018b)	DXCCTA	1	1 + 2	No	3	3	No	1.5	–	±1.25	Yes
Chaturvedi and Kumar (2019)	DXCCTA	1	0 + 2	Yes	2	4	Yes	<3	1.47	±1.25	No
Proposed	**FDCCII**	**1**	**3 + 2**	**Yes**	**2**	**2**	**Yes**	**1.9**	**0.6**	**±0.9**	**Yes**

Abbreviations DVCC—differential voltage current conveyor, DXCCII—dual-X second-generation current conveyor, OTA—operational transconductance amplifier, CA—current amplifier, DXCCTA—dual-X current conveyor transconductance amplifier, CCII—second-generation current conveyor, CDTA—current differencing transconductance amplifier, CCCII—second-generation current-controlled conveyor, CC-CDCCC—current-controlled current differencing current copy conveyor, CIDITA—current inverting differential input transconductance amplifier, FDCCII—fully differential second-generation current conveyor, DD-DXCCII—differential difference-dual-X second-generation current conveyor, CDBA—current differencing buffered amplifier, R-resistor, C-capacitor, THD-total harmonic distortion, f_0—oscillation frequency, CO—condition of oscillation, '–'—not given

References

Abaci A, Yuce E (2017) Modified DVCC based quadrature oscillator and lossless grounded inductor simulator using grounded capacitor (s). AEU-Int J Electron Commun 76:86–96

Chaturvedi B, Kumar A (2019) Electronically tunable first-order filters and dual-mode multiphase oscillator. Circuits, Syst, Signal Process 38:2–25

Chaturvedi B, Mohan J (2015) Single active element based mixed-mode quadrature oscillator using grounded components. IU-J Electr Electron Eng 15:1897–1906

Chaturvedi B, Mohan J, Kumar A (2018) A new versatile universal biquad configuration for emerging signal processing applications. J Circuits, Syst Comput 27:28 p

Chaturvedi B, Kumar A, Mohan J (2019a) Low voltage operated current-mode first-order universal filter and sinusoidal oscillator suitable for signal processing applications. AEU-Int J Electron Commun 99:110–118

Chaturvedi B, Mohan J, Kumar A (2019b) Dual-mode quadrature oscillator based on single FDCCII with all grounded passive components. In: Advances in signal processing and communication, Springer, Singapore, pp 317–326

El-Adawy AA, Soliman AM, Elwan HO (2000) A novel fully differential current conveyor and applications for analog VLSI. IEEE Trans Circuits Syst II, Analog Dig Signal Process 47:306–313

Keskin AU, Biolek D (2006) Current-mode quadrature oscillator using current differencing transconductance amplifier (CDTA). IEE Proc-Circuits, Devices Syst 153:214–218

Kumar A, Chaturvedi B (2016) A novel MO-DXCCII based CMQO operated at low voltage. Grenze Int J Eng Technol 2(2):9–17

Kumar A, Chaturvedi B (2017) Novel CMOS current inverting differential input transconductance amplifier and its application. J Circuits, Syst Comput 26. article ID: 1750010

Kumar A, Chaturvedi B (2018a) Realization of novel cascadable current-mode all-pass sections. Iran J Electr Electron Eng 14:162–169

Kumar A, Chaturvedi B (2018b) Novel CMOS dual-X current conveyor transconductance amplifier realization with current-mode multifunction filter and quadrature oscillator. Circuits, Syst, Signal Process 37:2250–2277

Kumar A, Chaturvedi B, Mohan J, Maheshwari S (2017) Single active element based orthogonally controllable MOSFET-C quadrature oscillator. In: IEEE international conference on multimedia, signal processing and communication technologies (IMPACT), pp 151–155

Maheshwari S (2003) Electronically tunable quadrature oscillator using translinear conveyors and grounded capacitors. Act Passive Electron Compon 26:193–196

Maheshwari S (2004) New voltage and current-mode APS using current controlled conveyor. Int J Electron 91:735–743

Maheshwari S (2007) High input impedance VM-APSs with grounded passive elements. IET Circuits Devices Syst 1:72–78

Maheshwari S (2008) High output impedance current-mode all-pass sections with two grounded passive components. IET Circuits Devices Syst 2:234–242

Maheshwari S (2014) Sinusoidal generator with π/4-shifted four/eight voltage outputs employing four grounded components and two/six active elements. Active Passive Electron Compon 2014:7

Maheshwari S, Chaturvedi B (2011) High output impedance CMQOs using DVCCs and grounded components. Int J Circuit Theory Appl 39:427–435

Maheshwari S, Chaturvedi B (2012) High-input low-output impedance all-pass filters using one active element. IET Circuits Devices Syst 6:103–110

Maheshwari S, Khan IA (2007) Novel single resistor controlled quadrature oscillator using two CDBAs. J Active Passive Electron Devices 2:137–142

Minaei S, Ibrahim MA (2005) General configuration for realizing current-mode first-order all-pass filter using DVCC. Int J Electron 92:347–356

Minaei S, Yuce E (2010) Novel voltage-mode all-pass filter based on using DVCCs. Circuits, Syst, Signal Process 29:391–402

Mohan J, Chaturvedi B, Maheshwari S (2016) Low voltage mixed-mode multi phase oscillator using single FDCCII. Electronics 20:36–42

Mohan J, Chaturvedi B, Kumar A, Jitender (2020) Active-C realization of multifunction biquadratic filter and third order oscillator. Radio Sci 55:e2019RS006877

Prommee P, Khateb F (2014) High-performance current-controlled CDCCC and its applications. Indian J Pure Appl Phys (IJPAP) 52(10):708–716

Sotner R, Jerabek J, Herencsar N, Vrba K, Dostal T (2015) Features of multi-loop structures with OTAs and adjustable current amplifier for second-order multiphase/quadrature oscillators. AEU-Int J Electron Commun 69:814–822

Yuce E (2017) DO-CCII/DO-DVCC based electronically fine tunable quadrature oscillators. J Circuits, Syst Comput 26. article ID: 1750025

Yucel F, Yuce E (2014) CCII based more tunable voltage-mode all-pass filters and their quadrature oscillator applications. AEU-Int J Electron Commun 68:1–9

Chapter 4
Low Power Design Techniques for Integrated Circuits

Bipin Chandra Mandi

Abstract It is essential to retain power and energy efficiency in low-power integrated circuits (ICs) over a wide load current/voltage range to reduce the consumption from the battery in portable/non-portable devices. The power/energy efficiency highly depends on voltage and frequency scaling when all the parts of the devices are in operation. There are also power and clock gating when all the parts of the devices are not in operation. The dynamic and static voltage scaling are main part for power gating. The power saving can be done by varying the supply voltage to ICs. The pulse width, pulse skip, depth and frequency modulation are common techniques for clock gating/frequency generation. The pulse width modulation (PWM) is generally used for fixed frequency operation. The pulse frequency modulation (PFM) is generally used for variable frequency operation depending on load voltage and current demands. The pulse skip modulation (PSM) is special technique to skip the pulses for frequency operation depending on IC operation mode (sleep mode and standby mode). In this chapter, all the existing techniques available for power reduction are discussed with the suitable diagram and examples.

Keywords Energy efficiency · Power gating · PWM · PFM · PSM · PT · Simulink model

4.1 Introduction

It is very challenging task to optimize the efficiency for a low-power integrated circuits (ICs) for a wide power range and reduce the power consumption from the battery operated in powered devices (Erickson and Maksimović 2001). The power/energy efficiency highly depends on voltage and frequency scaling when all the part of the devices is in operation (Trescases and Wen 2011). There are also power and clock gating when all the part of the devices are not in operation. The

---

B. C. Mandi (✉)
Electronics and Communication Engineering Department, International Institute of Information Technology, Naya Raipur, Raipur, Chattishgarh 493661, India
e-mail: bipin@iiitnr.edu.in

R. Dhiman and R. Chandel (eds.), *Nanoscale VLSI*,
Energy Systems in Electrical Engineering,
https://doi.org/10.1007/978-981-15-7937-0_4

dynamic and static voltage scaling are main part for power gating. The power saving can be done by varying the supply voltage to ICs. The pulse width, pulse skip, depth and frequency modulation are common techniques for clock gating/frequency generation. The fixed frequency-operated pulse-width modulation (PWM) is generally used for switching the power circuit. The pulse-frequency modulation (PFM) is also generally used for variable frequency operation depending on load voltage and current demands. The pulse skip modulation (PSM) is special technique to skip the pulses for frequency operation depending on ICs operation mode (sleep mode and standby mode) (Kapat et al. 2011). In this chapter, all the techniques available till date for power reduction are discussed with the suitable diagrams.

4.2 Low-Power Design Techniques

This section presents an overview of the existing control techniques that are used in integrated circuits (ICs) operating over a wide range of load power demanded. The power gating and the clock gating are normally used in power reduction in IC/power circuits. Based on switching frequency principles, all the frequency operation techniques can be categorized as PWM, PFM, PSM and pulse train PT control (Trescases and Wen 2011; Kapat et al. 2011). The output voltage feedback is indispensable for all the frequency operation techniques, and there exist numerous methods for implementing individual control schemes. A PWM technique considers a fixed switching frequency throughout, and under varying mostly the load current, and slightly varying the input supply voltage, the output voltage is regulated by adjusting the duty ratio. A PFM technique is aimed to regulate the output voltage by adjusting the switching frequency or switching time period. A PSM technique considers charge as well as skip pulses and regulates the output voltage by varying the number of charge and/or skip cycles. The pulse train (PT) control technique uses high and low pulses with different frequencies and their on-state or off-state durations are set using either predefined values or being generated. More details about individual schemes along with their possible implementation configurations are presented in successive sub-sections.

4.2.1 Conventional Techniques

I. Power Gating

Power gating is a technique which is used in IC/power circuit design to minimize the power consumption, by disabling the current propagation to some portion or sub-circuit blocks of the circuit that are in idle and standby mode as shown in Fig. 4.1. The v_{in}/v_{dd} is the supply/input voltage and i_{in} is the supply/input current to the circuit, respectively. The v_o is denoted as the output voltage and i_o is denoted as the load current of the power circuit, respectively. The d (duty cycle)/f (frequency) is to control the IC or power circuit operation. In addition to decrease the standby/idle or

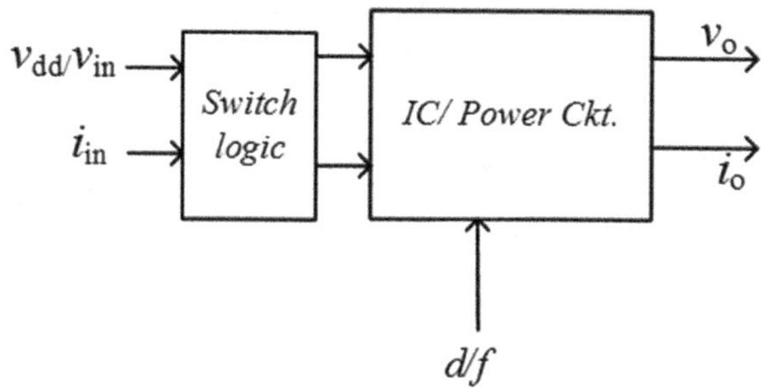

Fig. 4.1 Power gating technique in IC/power circuit as the supply voltage/current is disabled according to the switch logic

leakage power, the power gating has the advantage of enabling the current through the circuit by a switch. The power gating can be classified into two ways:

A. **Dynamic voltage scaling**: In case of dynamic voltage scaling, the supply voltage can be scaled according to different kinds of requirement of the voltage or load current demands.
B. **Static voltage scaling**: In case of static voltage scaling, the supply voltage is fixed but the supply to the corresponding circuits can be disabled according to different kinds of requirement of the voltage or load current demands.

II. Clock Gating

The clock frequency to the circuits can be enabled or disabled according to different kinds of requirement of the clock frequency operation as shown in Fig. 4.2. The clock gating can reduce the power consumption to the power circuit. The clock gating logic can change the clock frequency operation to ICs.

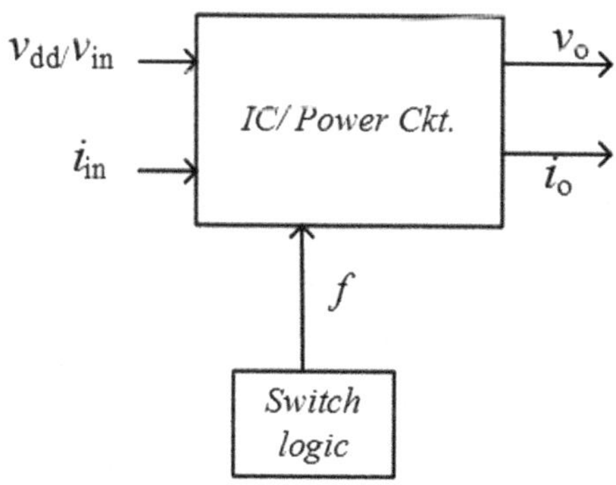

Fig. 4.2 Clock gating technique in IC/power circuit as the clock frequency is disabled according to the frequency operation

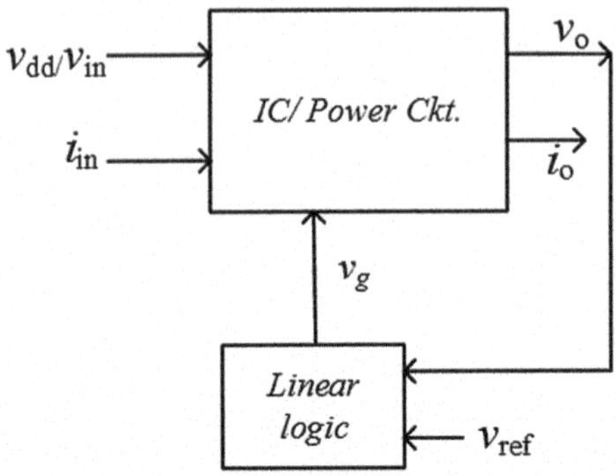

Fig. 4.3 Linear logic operation in IC/power circuit as per the reference voltage

4.2.2 Linear Circuit Operation

For the operation of the linear circuit, where switching mode is not involved, the supply/input voltage is scaled up/down according to different kinds of requirement of the voltage or load power demands as depicted in Fig. 4.3. The v_g is denoted as the gate voltage of the MOSFET in working on linear region. Depending on the reference voltage the gate voltage varies using op-amp with linear logic operation.

4.2.3 Switching Circuit Operation

For the switching mode/circuit operation, the input/supply voltage or current is applied to the power circuit depending on the requirement of the voltage or load current demands in load power side (Mandi 2017).

A. Pulse-Width Modulation (PWM) technique

The PWM control techniques are frequently used in various commercial power management products (Kapat et al. 2016). The schematic diagram of a PWM technique in digital design for the power circuits is depicted in Fig. 4.4. The output/load side voltage as denoted as v_o is to remain closed in comparison with the referred voltage v_{ref}, and the error/subtracted voltage $v_e = (v_{ref} - v_o)$ is fed to an error amplifier which also consists of a voltage controller $G_c(s)$. All signals have been shown in the waveforms of the digital domain. The controller output v_c is then compared with a periodic sawtooth waveform v_r to generate the duty ratio command, and the latter is passed through a latch circuit to generate the gate signal ***d/u*** of the switching frequency F_s for the power MOSFET. The DPWM technique uses a single voltage feedback loop, which makes it simple to implement. It also offers superior load regulation and improved output impedance. The power/energy efficiency can be with proper frequency selection for optimizing conduction loss and switching losses.

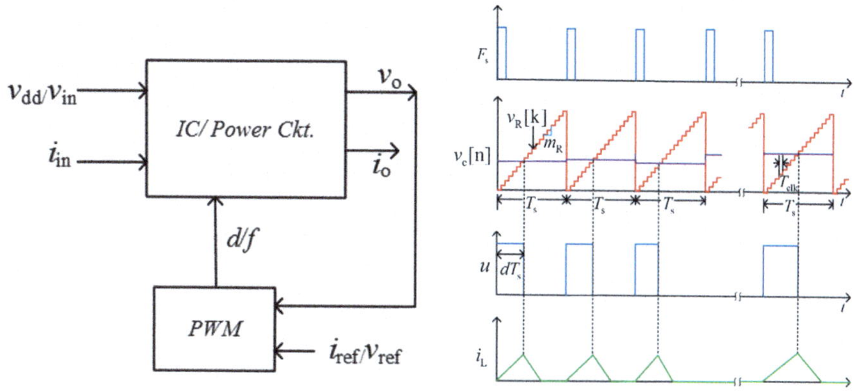

Fig. 4.4 PWM control scheme in IC/power circuit with waveforms as the *d* is the duty ratio of the clock frequency *f*

B. Pulse-Frequency Modulation (PFM)

A PFM technique is displayed in Fig. 4.5 to minimize the power losses due to power switches by varying switching time period (on-time and off-time) with the load power demands. There are many methods existing, but mainly, these three types are common such as the variation of on-time without changing the off-time, the variation of off-time without changing the on-time, or changing both on-time and off-time. Among available PFM methods, the COT-PFM is popular, and its schematic diagram along with the control waveforms is shown in Fig. 4.6. The on-time is kept constant, and the off-time varies with the load current to vary the switching time period/frequency. The 'constant' on-time is triggered when the output voltage v_o goes below the reference voltage v_{ref}, i.e., for $v_o \leq v_{ref}$. A minimum off-time $T_{OFF-min}$ is generally incorporated in order to avoid a complete collapse during a step-up transient in MOT PFM control scheme as shown in Fig. 4.7. This is particularly important for a power circuit, in which $T_{OFF-min}$ helps to partially transfer the inductor energy to the output capacitor to avoid its complete discharge during the slew-up process of the inductor. The nature of switching frequency variation with the load power is different depending on the time period variation of different PFM methods.

For a given input voltage, the energy injected by the source during the on-time using COT-PFM remains constant. Thus, the injected energy would take longer duration to discharge by the output capacitor for decreasing load current, which increases the off-time, thereby decreasing the switching frequency. Power losses may be further minimized by adjusting the on-time through real-time optimization using load current information. Figure 4.6 shows that the effective time period time is varied with the input voltage v_{in}. The ripple of the output voltage and the effective time varies for the constant on-time t_{on}.

The light-load efficiency using the COT-PFM scheme may not be optimized simply by taking a constant on-time throughout, which also results in unacceptably

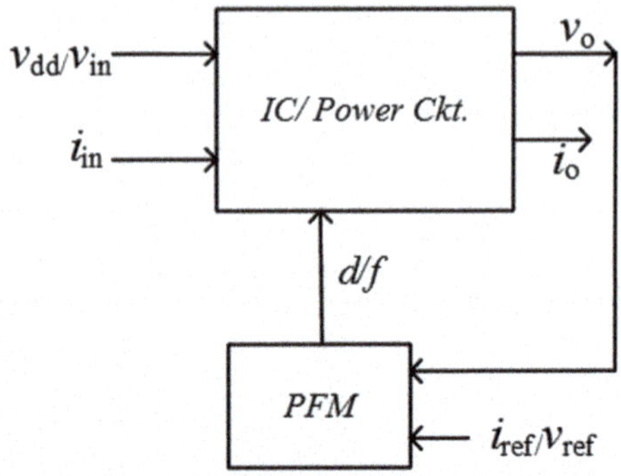

Fig. 4.5 PFM technique in IC/power circuit as the clock frequency *f* varies according to the power load demands

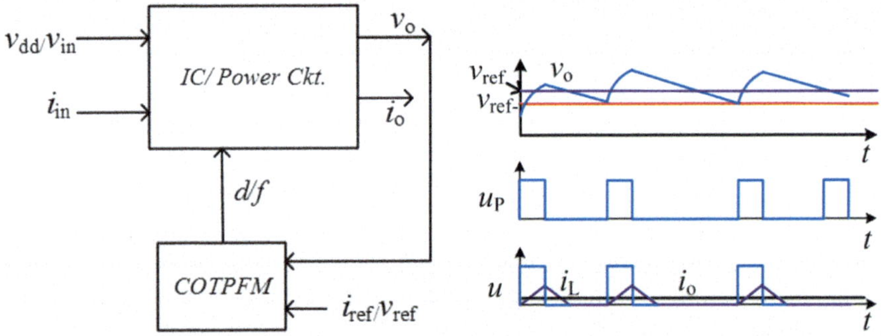

Fig. 4.6 COT-PFM in IC/power circuit with waveforms as the clock frequency *f* varies according to the power load demands

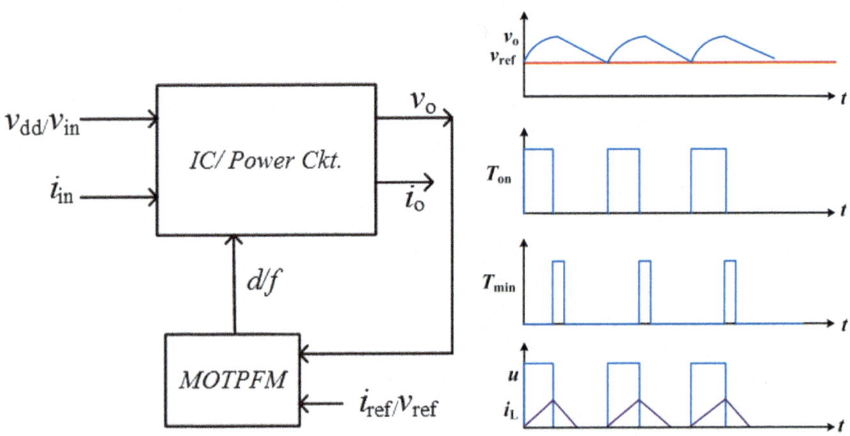

Fig. 4.7 MOT PFM in IC/power circuit with waveforms as the clock frequency *f* varies according to the power load demands

large output voltage ripple as displayed in Fig. 4.6. Thus, an adaptive/adjustment on-time approach is often used, which adjusts the constant on-time parameters based on load current conditions and the ripple requirements. This requires the information of both the supply voltage and the load side current. Alternatively, a voltage hysteresis control can be adapted to keep the output voltage ripple within the specified limit.

The voltage ripple needs to be accordingly adjusted in order to control the effective switching frequency. Nevertheless, majority of the existing PFM techniques apply a ripple-based approach. This makes it difficult to precisely control the switching frequency, thus making it difficult to predict the conducted EMI issues generated at the supply voltage/current (Kapat et al. 2011). A combined PWM/PFM technique is often used in commercial power circuit solutions, which operate over a wide load power. However, both the methods are structurally different and require separate hardware resources along with extra anti-windup arrangements. Further, the selection of the sampling frequency remains a challenge in a digital PFM technique, because of its variable frequency control nature. This makes it difficult for direct digital implementation of a combined PWM/PFM technique.

B.1. Hysteresis PFM scheme

The hysteresis PFM technique in the power circuit operation is displayed in Fig. 4.8. In this control technique, the switching frequency varies from the lower side threshold voltage to upper side threshold voltage. The constant on-time charge pulse is allowed until the upper limit is touched or crossed. The constant on-time pulse is not regenerated until the output voltage touches the lower limit of threshold the output voltage. This scheme is also called the burst PFM control schcme. To minimize the conduction loss and switching losses, the HPFM control scheme is useful for within the output voltage limit.

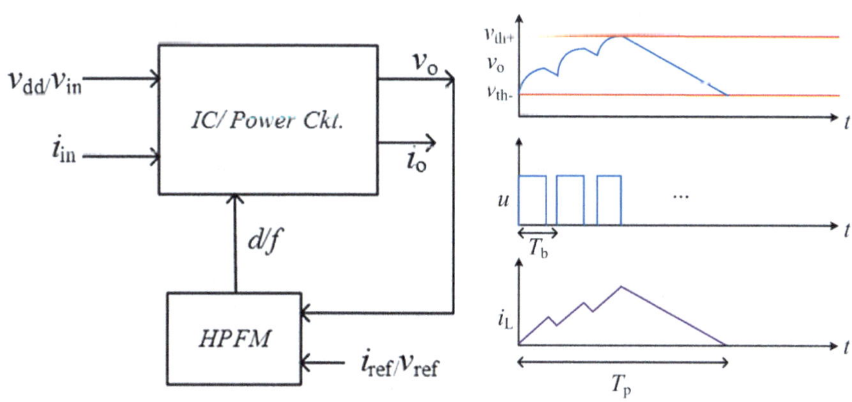

Fig. 4.8 Hysteresis PFM in IC/power circuit with waveforms as the clock frequency f varies according to the power load demands

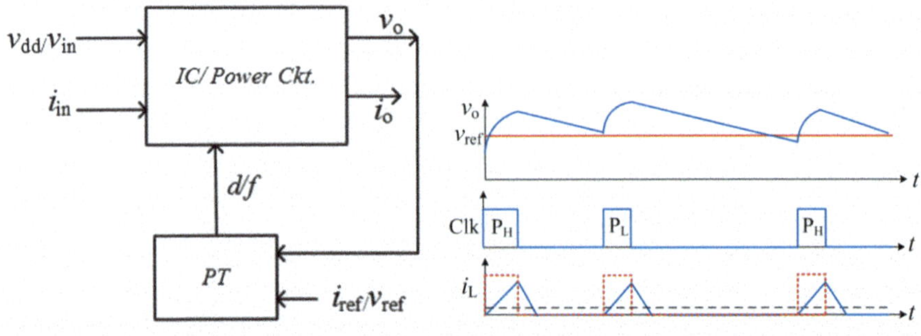

Fig. 4.9 Pulse train in IC/power circuit with waveforms as the two different pulses vary according to the power load demands

B.2. Pulse Train (PT) control technique

Figure 4.9 shows an IC/power circuit controlled by a pulse train (P_T) control technique. There are two different and predefined types of pulses. The first one is high pulse (P_H), which consists of high on-time. The second one is low pulse (P_L), which consists of low on-time. The output voltage (v_o) of the power circuit is to remain closed to the reference voltage (v_{ref}). It the output voltage (v_o) at the rising clock remains higher or equal to the reference voltage (v_{ref}), the high pulse is generated; otherwise, the low pulse is generated as described in Fig. 4.9. The main objective of the pulse train control scheme is to make the power circuit more efficacy under light to nominal load condition. Therefore, the occurrence of the high pulse is more in medium and high load condition. The occurrence of the low pulse is more in light-load condition.

C. Pulse Skipping Modulation (PSM)

The basic principle of the PSM control scheme in the power circuit/IC is stated as shown in Fig. 4.10. In PSM control scheme, some number of charge cycle (duty cycle) is skipped the ripple of inductor current and the ripple of output voltage increases to regulate the output voltage without violating the charge balance rule (Kapat et al. 2011). For example, if two charge pulses are skipped, the effective period is 3T. The existing PSM control scheme skips some charge pulses and generates the corresponding duty cycle to maintain charge balance. There are two well-known PSM approaches—classical or conventional PSM and voltage-controlled PSM. The details of the following PSM control scheme are described as follows.

C.1. Classical PSM

The schematic diagram of a power circuit/ICs governed by a conventional PSM technique is shown in Fig. 4.11. The gate signal d of the MOSFET switch is controlled by a PSM logic which is in synchronism with fixed frequency clock F_{ext} ($1/T_p$) with a fixed duty ratio D. If $v_o \leq v_{ref}$ at the beginning of the nth clock period, $u_{PSM} = 1$ and the MOSFET is controlled by F_{ext} throughout the clock period; otherwise, $u_{PSM} = 0$ and the MOSFET remains disabled for the complete period. The former and latter

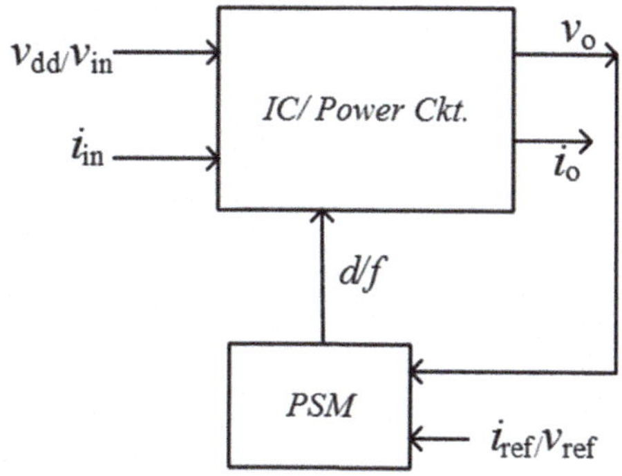

Fig. 4.10 PSM technique in IC/power circuit as the skip cycle varies according to the power load demands

are referred as the charge and skip cycles, respectively. Thus, the switching events can be reduced by skipping/avoiding a few clock pulses to maximize the light-load efficiency. A PSM technique consists of one charge followed by one or more skip cycles. The burst-mode/hysteresis-mode kind of PSM consists of multiple charge and skipped cycles. The total count of charge and skipped cycles can be obtained using capacitor charge balance.

C.2. Voltage-Controlled PSM

The schematic diagram of a voltage-controlled PSM (VCPSM) scheme is shown in Fig. 4.12. This technique retains the same pulse skip mechanism that in Fig. 4.12; however, the mechanism to generate the duty ratio under a charge pulse differs. Unlike using a fixed duty ratio under a classical PSM, the duty ratio in the VCPSM is generated using the feedback voltage loop. Here, the number of skipped cycles can be indirectly controlled by varying the controller gain as well as other feed-forward gain parameters. The *clk* signal is the switching frequency of the PWM control scheme. The u_{PSM} is the switching pulse for the allowing the charge pulse

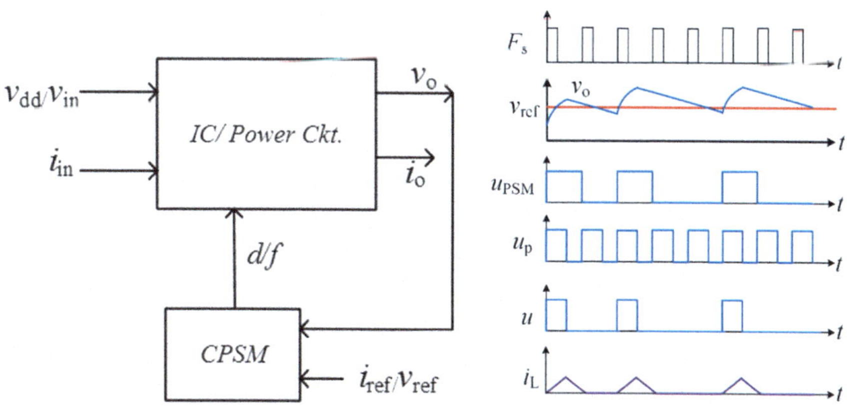

Fig. 4.11 Classical PSM in IC/power circuit with waveforms as the skip cycle varies depending on the load power demands

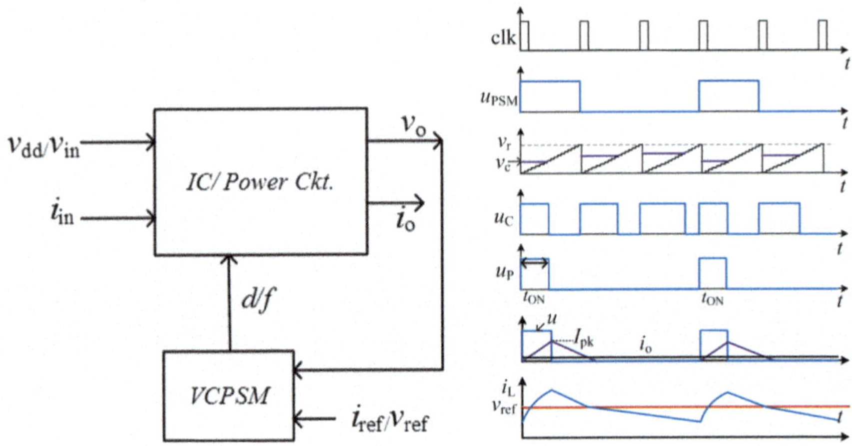

Fig. 4.12 Voltage-controlled PSM in IC/power circuit with waveforms as the skip cycle varies according to the variation of the output voltage variation from the reference voltage

or disallowing the charge pulse as shown in Fig. 4.12. The u_C is the switching pulse width generated from the digital PWM control scheme after comparing the signals shown in Fig. 4.12. The output side voltage (v_o) is to remain close to the reference voltage (v_{ref}). The error voltage $v_e = (v_{ref} - v_o)$ is needed to pass through to an error amplifier which also consists of a voltage controller $G_c(s)$ (Kapat et al. 2016). To make the power circuit/IC more efficacy under light-load power situation, the VPSM technique is highly preferable. The VCPSM needs to be selected proper number of the charge cycles, and/or skipped cycles (u) in the existing PSM schemes cannot be predefined. So that it can overcome the disadvantage of predicted the ripple parameters and ensuring stable periodicity. Thus, it remains a big challenge to customize the sequence as well as the count of charge and skipped pulses with stable periodicity to further optimize the efficiency under light-load condition (Mandi et al. 2018).

C.3. Current-Controlled PSM

The schematic diagram of a current-controlled PSM (CCPSM) scheme is shown in Fig. 4.13. This technique retains the same pulse skip mechanism as VCPSM; however, the mechanism to generate the duty ratio under a charge pulse differs. Unlike using a fixed duty ratio under a classical PSM, the duty ratio in the CCPSM is generated using the feedback current control loop. Here the number of skipped cycles can be indirectly controlled by varying the controller gain as well as other feed-forward gain parameters.

C.4. Fractional Pulse Skipping Modulation

Fractional PSM is a type of the PSM mode. It can be used in the power circuit for changing the *d*/*f* as the different skipped pulses varies according to the power load demands. As it is operating in light-load condition the single fixed charge cycle with

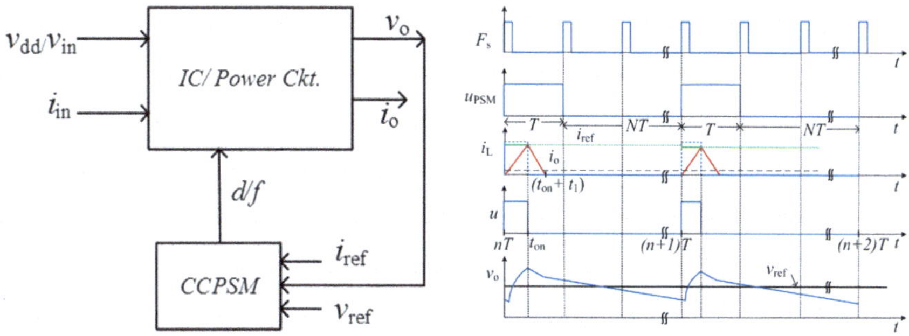

Fig. 4.13 Current-controlled PSM in IC/power circuit with waveforms as the skip cycle varies according to the variation of the output voltage variation from the reference voltage and the current

a time period (T) followed by skip cycles $(n + 1)T$ or $(n + 2)T$. The average skipped cycles are around $(n + 0.5)T$, which is fractional skipped cycles. The FPSM has the flexibility of power spectral spreading to reduce the effect of conducted Electro Magnetic Interference (EMI) by choosing number of skip cycles $(n + 1)T$ and $(n + 2)T$ (Mandi et al. 2016). The FPSM has the periodic stable behavior within the predictable ripple of the output voltage. It is used the existing digital PWM with a little modified features. The main objective is to optimize the light-load/medium load power efficiency in a power circuit with the power spectrum spreading. In the waveforms, two different time periods are there which are $(1 + N_1)$ and $(1 + N_2)$. If the N_1 and N_2 are equal, then the FPSM is equivalent to current-controlled PSM; otherwise, it is current-controlled FPSM. The average of the N_1 and N_2 is giving the fractional value of the fixed period of F_s. The signal u_{PSM} is duty cycle of controlling the charge and skipped cycles (Fig. 4.14).

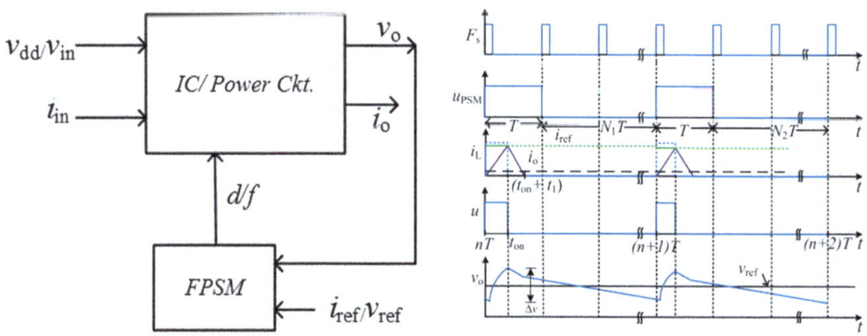

Fig. 4.14 Fractional PSM in IC/power circuit with waveforms as different skipped pulses vary according to the power load demands

4.3 Some Simulation Study with MATLAB/Simulink for a Power Circuit/IC

Several control schemes for a power circuit at light-load condition are described including the working principle, advantage and disadvantage in terms ripple, regulation and efficiency (Mandi 2017). The results for the case study for a power circuit (buck converter) under light load current of 44mA are given in Figs. 4.15, 4.16, 4.17, 4.18, 4.19, 4.20 and 4.21 (Mandi, 2017). Although a PWM scheme in Fig. 4.15

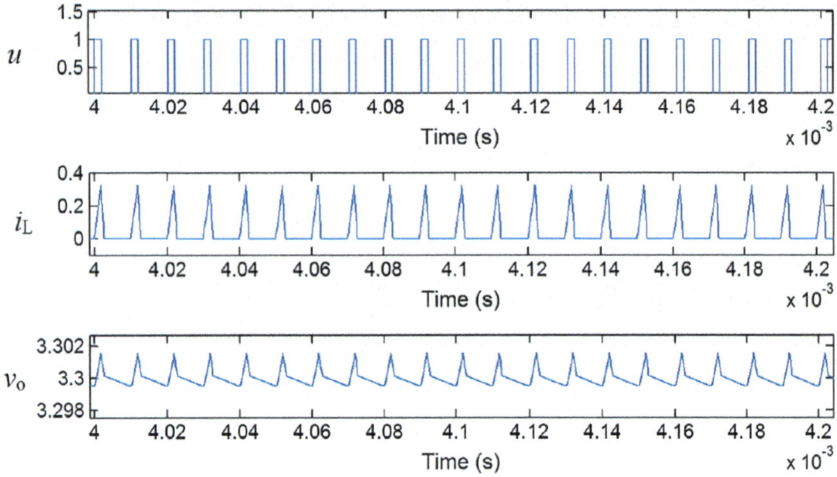

Fig. 4.15 PWM control scheme simulated results in power circuit with waveforms

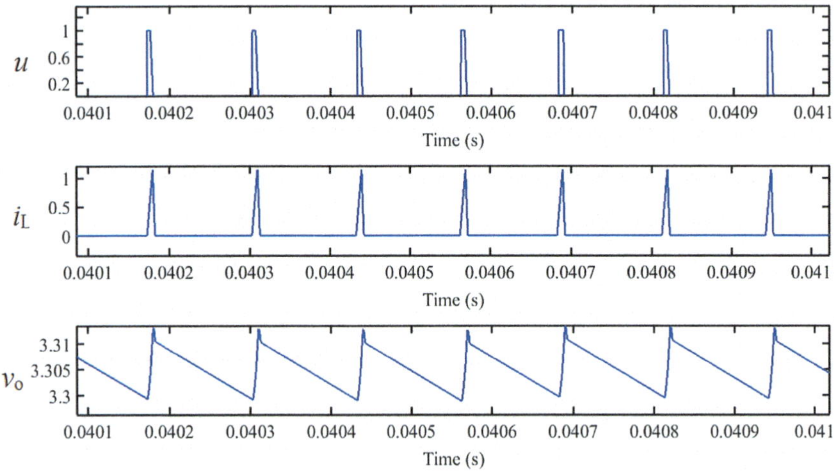

Fig. 4.16 COT-PFM control scheme simulated results in power circuit with waveforms

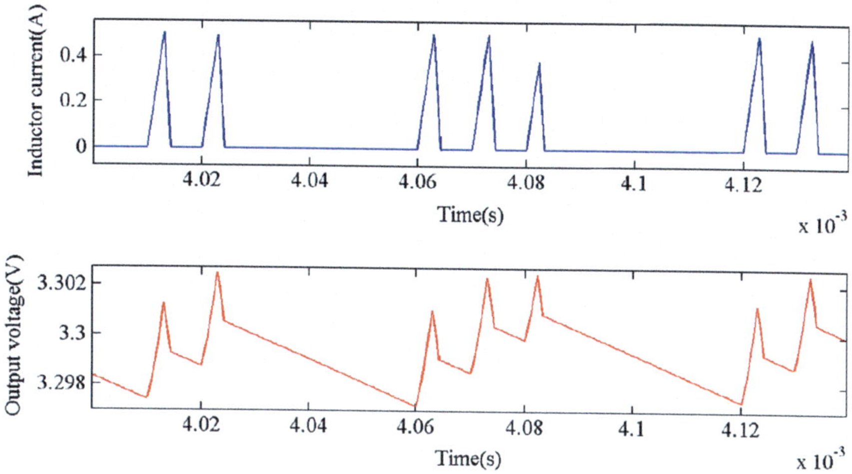

Fig. 4.17 Burst PFM control scheme simulated results in power circuit with waveforms

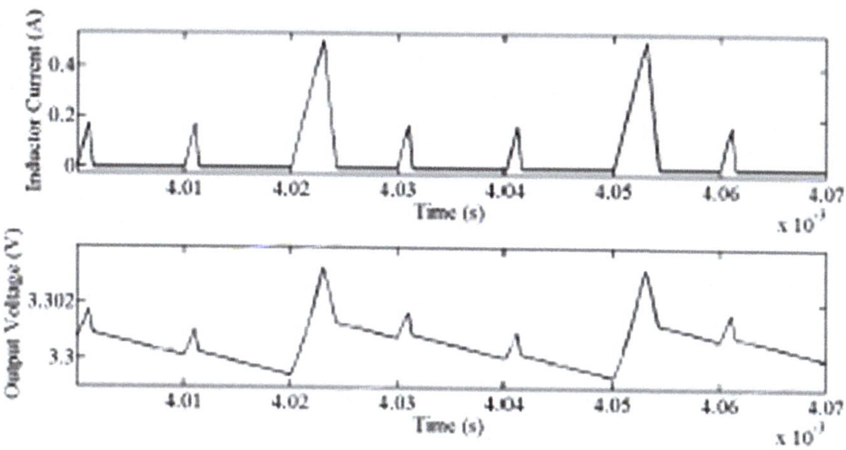

Fig. 4.18 Bi-frequency PFM control scheme in power circuit with waveforms

offers fixed frequency operation, periodic switching events under high-frequency PWM clock result in substantial switching and driving losses. A PFM scheme in power circuit improves the efficiency under light/medium load by reducing transition between the on-time and off-time, i.e., by decreasing the switching period with the variation of the load power; as explained in the following waveform plot results. From Fig. 4.16, for COT-PFM control, the 'ON' time is constant for a particular range of load current as depicted in the simulation results. The off-time of COT-PFM varies with change of the load current information. A burst-mode/hysteresis control in Fig. 4.17 increases the output voltage ripple and results in aperiodic behavior. The pulse train (PT) or the bi-frequency PFM control scheme in Fig. 4.18 uses two

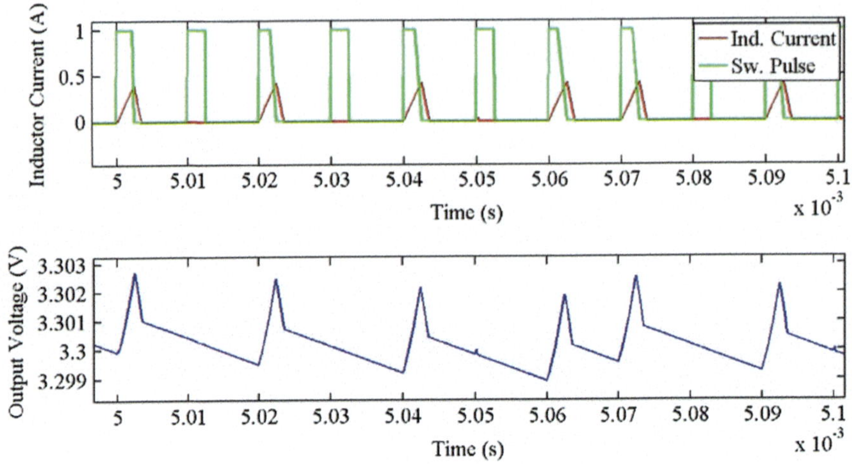

Fig. 4.19 Classical PSM control scheme in power circuit with waveforms

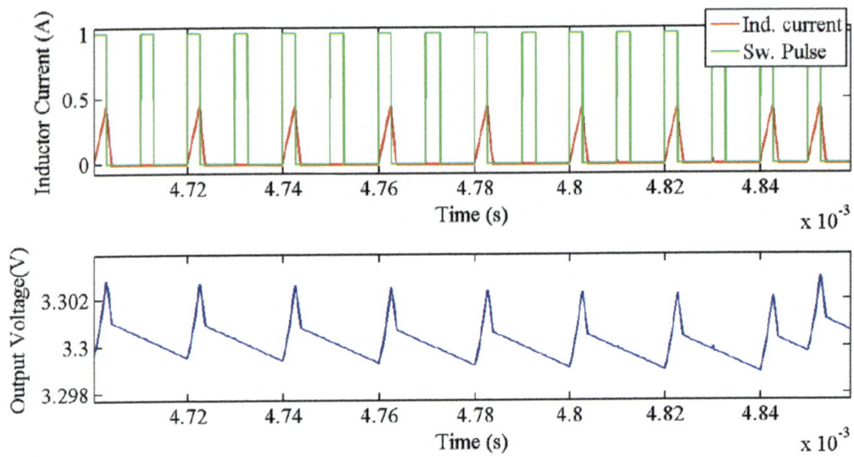

Fig. 4.20 Voltage-controlled PSM control scheme in power circuit with waveforms

different pulses (high and low) with different frequencies. The conventional PSM is with a fixed duty pulse as shown in Fig. 4.19. The charge pulses occur when $v_o <= v_{ref}$, and skipped pulses occur when $v_o > v_{ref}$. The voltage-controlled PSM is with a PWM duty pulse as shown in Fig. 4.20.

The voltage-controlled PSM's charge pulses and skipped pulses similar way of conventional PSM occur when $v_o <= v_{ref}$ and $v_o > v_{ref}$, respectively. The number of skipped cycles depends on input voltage (v_o) and load current [Load resistor (R)] variation. Aperiodic behavior occurs frequently in conventional PSM rather than voltage-controlled PSM as displayed in the results in Fig. 4.20 and in Fig. 4.21. The

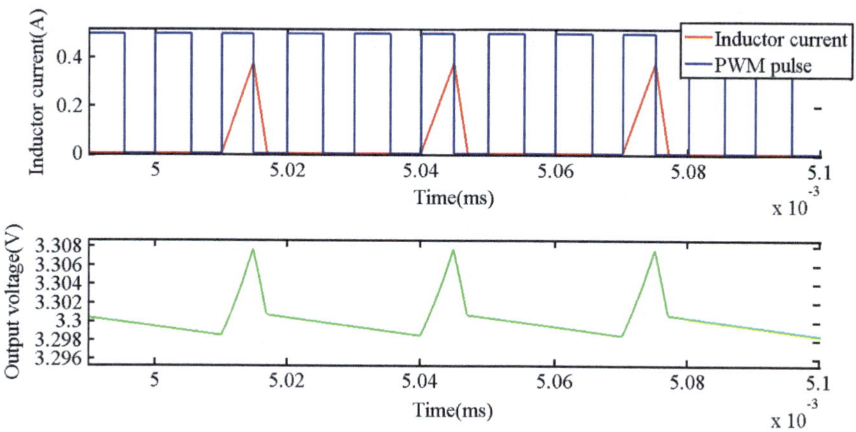

Fig. 4.21 Voltage-controlled PSM control scheme in power circuit with waveforms

number of charge and skipped cycles has no control with the variation of load current and input voltage.

4.4 Summary

The energy efficiency in low-power integrated circuits (ICs) can be reduced using different control techniques over a wide load current/voltage in portable/non-portable devices. The power and clock gating can be applied depending on the device's operation. The power saving can be done by varying the supply voltage to ICs. The pulse-width and pulse-frequency modulation are common techniques for nominal load application depending on design specifications. The hysteresis and pulse train are common techniques for frequency generation. The pulse depth and pulse skipping modulation are used in low-power integrated circuits for idle and standby operation. The classical, voltage/current mode and fractional pulse skipping techniques are more preferred in low-power integrated circuits to achieve high efficiency. The multi-mode and unified control scheme will be more preferable over a wide load power in portable/non-portable devices in the near future.

References

Erickson RW, Maksimović D (2001) Fundamentals of power electronics, 2nd edn. Kluwer, Dordrecht, Netherlands

Kapat S, Banerjee S, Patra A (2011) Achieving monotonic variation of spectral composition in DC-DC converters using pulse skipping modulation. IEEE Trans Circuits Syst I 58(8):1958–1966

Kapat S, Mandi BC, Patra A (2016) Voltage-mode digital pulse skipping control of a dc-DC converter with stable periodic behavior and improved light-load efficiency. IEEE Trans Power Electron 31(4):3372–3379

Mandi BC (2017) Digital modulation techniques in Dc-Dc converters for improved efficiency over a wide load range. Dissertation, IIT Kharagpur, 2017

Mandi BC, Kapat S, Patra A (2016) Fractional pulse skipping in digitally controlled DC-DC converters for improved light-load efficiency and power spectrum. In: Proceedings of the IEEE APEC, pp 2504–2510

Mandi BC, Kapat S, Patra A (2018) Unified digital modulation techniques for DC-DC converters over a wide operating range: implementation, modeling, and design guidelines. IEEE Trans Circuit Syst (TCAS-I), 64(4):1442–1453

Trescases O, Wen Y (2011) A survey of light load efficiency improvement techniques low-power DC-DC converters. In: Proceedings of the IEEE ICPE-ECCE, Korea, 2011, pp 326–333

Part II
Modeling and Simulation for Post-CMOS Devices

Chapter 5
Bilayer Graphene Nanoribbon Tunnel FET for Low-Power Nanoscale IC Design

Vobulapuram Ramesh Kumar, Uppu Madhu Sai Lohith, Shaik Javid Basha, and M. Ramana Reddy

Abstract In the electronics industry, silicon is the primary material of choice to meet the demands. The advancement in the technology led to the involvement of the smaller devices with improved performance. Due to the scaling of silicon MOSFET devices, the complications increases such as tunneling effect, gate oxide leakage, and channel punch through. In order to overcome these issues, new materials with improved characteristics are needed. From the last two decades, researchers are focused to find new nanomaterials which can substitute for renowned silicon in next-generation electronic devices. Graphene is the most promising material that can replace the silicon-based materials because of its outstanding physical and electrical properties. Graphene provides high carrier velocity and high carrier concentration, resulting in large carrier mobility and faster switching capability. Moreover, graphene is a semimetal with a zero bandgap which is the basic requirement for digital integrated circuits. The quantum confinement of graphene sheet in the form of one-dimensional strips known as graphene nanoribbon (GNR). The GNR provides the energy bandgap of several hundred meV that will be helpful for the design of GNR transistor. Considering the ongoing developments in the fabrication of graphene nanoribbon (GNR) with smooth edges, the design of GNR transistor came to exist. The GNR transistors offer high ON/OFF ratios due to small carrier effective mass and direct energy gap. In this chapter, the bilayer graphene nanoribbon tunnel field-effect transistor (BL-GNRTFET) as the low-power device is discussed. Initially, the device performance which includes the study of BL-GNRTFET along with the

V. Ramesh Kumar (✉) · U. Madhu Sai Lohith · S. Javid Basha · M. Ramana Reddy
Department of Electronics and Communication Engineering, Rajeev Gandhi Memorial College of Engineering and Technology, Nandyal, India
e-mail: rameshkumar.nith@gmail.com

U. Madhu Sai Lohith
e-mail: lohith.uppu@gmail.com

S. Javid Basha
e-mail: javidbasha1104@gmail.com

M. Ramana Reddy
e-mail: ramanareddy0106@gmail.com

R. Dhiman and R. Chandel (eds.), *Nanoscale VLSI*,
Energy Systems in Electrical Engineering,
https://doi.org/10.1007/978-981-15-7937-0_5

monolayer graphene nanoribbon tunnel field-effect transistor (ML-GNRTFET) is analyzed. The parameters such as transfer characteristics, drain characteristics and transconductance are explored and compared with the ML-GNRTFET. It has been observed that the performance of the BL-GNRTFET has the better ON current, low sub-threshold swing when compared to the ML-GNRTFET.

Keywords Graphene nanoribbon (GNR) · Low-power devices · Tunnel FETs · Atomic level simulations

5.1 Introduction

Nowadays, electronic devices are replaced with a short time after their launch in the market. This happens because the new invention devices captivated over the launched devices with additional features such as reduced power consumption, high operating speed, low cost, and reduced size. Today's smartphones and laptops are the best examples of this trend. This trend can be continued by integrating a billion numbers of field-effect transistors (FETs) on a single chip.

Traditionally, the metal oxide semiconductor field-effect transistors (MOSFETs) are rapidly used to integrate on the chip because of their advantages like small silicon area, and its fabrication involves fewer processing steps (Pathak 2001; Krausse 2002; Tang and Burkhart 2009). Due to the improvement in technology, it is required to scale the channel length of the MOSFET. Scaling the MOSFET channel length leads to large leakage current, large power density and high complexity (Chaudhury and Sinha 2019).[1] To avoid these effects, the tunnel FETs (TFETs) are utilized because of their small sub-threshold swing (SS) and low OFF current (I_{OFF}) (Seabaugh and Zhang 2010). However, in Si-based TFETs, the low ON current (I_{ON}) is a serious limitation which demands to use novel materials and fabrication process.

Recently, graphene has captivated because of its huge advantages such as planar structure, high conductivity, high mechanical and thermal stability (Dey et al. 2016; Liu et al. 2012). Graphene is a layer of carbon atoms tightly packed into a 2-D honeycomb lattice. Initially, it is zero bandgap material which reduces the transistor performance, I_{ON}, and I_{OFF}, respectively. Thus, the graphene should be patterned into carbon nanotube (CNT) and graphene nanoribbon (GNR) (Han et al. 2007; Zhou et al. 2007). The GNRs are being investigated as the worthy candidate instead of MOSFETs.

GNRFETs are classified as MOS-like GNRFET, Schottky barrier GNRFET (SB-GNRFET) and tunneling GNRFET (T-GNRFET) (Zhao et al. 2009; Chin et al. 2010; Ghoreishi et al. 2017). In MOS-like GNRFET, the ohmic contacts are acquired with the help of heavily doped n^+ to drain and source regions. The sub-threshold swing of these transistors has a theoretical limit of 60 mV/dec as the I_{ON} is a thermionic

[1] https://application.wiley-vch.de/books/sample/352734358X_c01.pdf.

type (Tsividis and McAndrew 2011). In SB-GNRFET, the Schottky contacts such as source and drain are formed by connecting the two metals to an intrinsic channel. In T-GNRFETs, the source and drain are doped with p^+ and n^+ impurities, respectively. When the positive voltage is applied at the gate terminal, the tunneling barrier width reduces at the source-channel junction. This leads to band-to-band tunneling (BTBT) of electrons from the valance band of the source to the conduction band of the channel. Due to different mechanisms of the I_{ON}, the SS of the TFET is not constrained by KT/q (Chin et al. 2010), where T is temperature, q is the electron charge, and K is the Boltzmann constant. This characteristic material science empowers TFET to be compatible for designing low-power applications.

In this work, the GNR-based transistor with TFET for bilayer GNRs is designed to improve the performance of the integrated circuit. The proposed devices are modeled and simulated in the industry standard Synopsis-based quantumwise ATK tool to obtain the transconductance and current and voltage (I-V) characteristics. The obtained simulation results are compared with the monolayer T-GNRFET to show the effectiveness of the work.

The rest of the work is organized as follows: Sect. 5.2 presents the details of MOSFET and the problems due to scaling the channel the length, In Sect. 5.3, the TFET device structure, operation, and its advantages and limitations are described. The details of GNR and its properties are discussed in Sect. 5.4. Section 5.5 presents the proposed T-GNRFETs designs and their results. Finally, summary of the work is described in Sect. 5.6.

5.2 Metal Oxide Semiconductor Field-Effect Transistor

Traditionally, in digital integrated circuits, the MOSFETs are considered as an important building block. Due to the improvement in technology and ease of MOSFET operation, the MOSFET is widely utilized as a switching device. Thus, it is worthy to know the detail information of the MOSFET. In this section, the basic structure and the operation of MOSFET are discussed. Furthermore, the problems due to scaling the MOSFETs are also described.

5.2.1 Device Structure and Operation

The MOSFET is a four-terminal device with gate, drain, source, and bulk terminals. The MOSFET bulk terminal is always connected to the source terminal making the MOSFET a three-terminal device. The MOSFET is classified into two types (a) n-type MOSFET and (b) p-type MOSFET. In n-type MOSFET, the drain and source terminals are heavily doped with n^+ region and substrate is p-type. Whereas, in p-type MOSFET, the drain and source terminals are doped with p^+ region and substrate is n-type. Note that the discussion focuses on n-type MOSFETs.

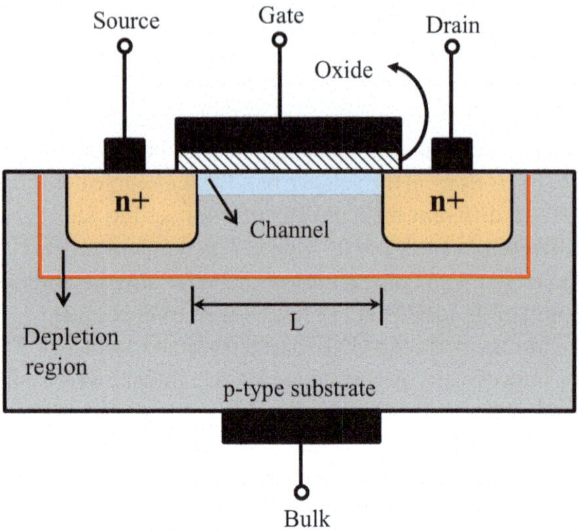

Fig. 5.1 Structure of the MOSFET

Figure 5.1 shows the basic structure of the n-type MOSFET. In this design, the gate terminal is placed on the thin oxide layer between the surface of the drain and source terminals. In this transistor, the n^+ regions are the current conducting terminals. Moreover, the device transistor is symmetrical for the drain and source regions. The current flows due to the negatively charged electrons. The holes under the oxide region pushed downward into the substrate and electrons in the substrate attracted toward the gate when the positive voltage is applied at the gate terminal. The depletion region is filled by the bound negative charges that are affiliated with the acceptor atoms. The electrons reach under the oxide layer form channel. The positive voltage attracts the electrons from the source to drain into the channel. If the positive voltage is applied at the drain terminal, the current flows freely between source and drain terminals.

5.2.2 Problems in Scaling the MOSFETs

Due to the improvement in technology, it is required to increase the performance and reliability of the integrated circuit. This is possible by scaling the dimensions of the MOSFET. Scaling the MOSFET dimensions from the submicron to nanometer range leads to various problems that reduce the performance of the total device. The problems occurred due to scaling the MOSFET are discussed below.

5.2.2.1 High Electric Fields

Scaling the MOSFET increases the strength of the electric field across the gate oxide that can reduce the carrier mobilities. This worst case can breakdown the barrier and increase the leakage currents. The high leakage currents can damage the total device.

5.2.2.2 Hot Carrier

The carriers having an effective temperature higher than the lattice temperature are called hot carriers. These carriers cannot transfer their energies to lattice atoms faster because the carriers are not in thermal equilibrium with the lattice. The hot carrier is generated in the inverted channel when the MOSFET is operated in the linear or saturation. The hot carriers degrade the gate and drain currents, reduce the transconductance, and shift the threshold voltage.

5.2.2.3 Power Supply and Threshold Voltage

Scaling the MOSFET tends to involve a proportional reduction in the voltage supply to maintain the active power and electric field within limits. However, it is not possible to scale the threshold voltage. This ensures since passive power establishes a major portion of the total power consumption in the high performance of CMOS products. The significant power consumed is because of the leakage current. Thus, scaling the threshold voltage is done to remove the increase in I_{OFF}. To obtain the high drive current, it is required to reduce the power supply that results in increasing the active power density.

5.2.2.4 Gate Oxide Tunneling

Since the electron thermal voltage is a constant at room temperature, the ratio between the operating voltage and thermal voltage decreases by scaling the MOSFET that increases the leakage currents. Moreover, by reducing the channel length, it is also required to reduce the oxide thickness. In MOSFETs, due to the thin oxide layer, the quantum mechanical tunneling is subjected which gives rise to gate leakage current.

5.2.2.5 Parasitic Resistances and Capacitances

It is known that the MOSFET has parasitic resistances and capacitances. The resistances and capacitances are reduced as the MOSFET dimensions are scaled down. The influence of the parasitic elements on the current increases expressively. Hence, the parasitic elements will reduce the performance by scaling the MOSFET.

5.2.2.6 Randomness of Dopant Distribution

In small devices, the randomness of the dopant distribution effect is more on MOSFET characteristics. This is because the precise position of the individual dopant atoms cannot be managed. Hence, if the device dimensions are reduced then it is difficult to place the dopant atoms at exact positions.

5.2.2.7 Dissipation of Heat

MOSFETs release their energy in the form of heat in resistive areas. The hot spot is created over the circuit leads to overheating which results in malfunction of the device.

5.2.2.8 Source and Drain Tunneling

In the MOSFETs, as the channel length between the source terminal and drain terminals is reduced, the electrons tunnel from the barrier without applying a voltage at the gate terminal. Thus, scaling device dimensions should be carried out with proper limits.

5.3 Tunnel Field-Effect Transistor

The TFET has a similar structure as MOSFET but differs in the switching techniques, and switching can be done at low voltage than MOSFETs. This device is more useful in low-power electronics as it has low SS, low OFF-state current. Unlike conventional MOSFETs, the short-channel effects are reduced in TFETs because the current is controlled by a tunneling phenomenon. It works on the principle of the band-to-band tunneling which makes it operate at the low SS. TFET is a gated p-i-n diode that works based on reverse bias condition. The source and drain regions are doped heavily with the regions p^+ and n^+.

5.3.1 Structure of TFET and Its Operation

Figure 5.2 shows the TFET structure. The depletion region forms at the junction of the intrinsic region and n^+ doped drain region. The reverse bias condition helps in increasing the depletion region width and helps to produce the swept charge carriers. The produced charge carriers tunnel from intrinsic region to source through the band-to-band tunneling phenomenon. Similar to the MOSFET, The TFET also divided into two types based on the doping profiles. One is n-type in which the source is doped

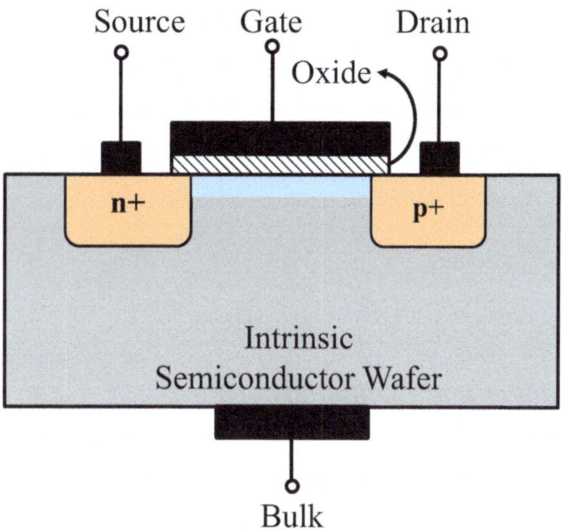

Fig. 5.2 Structure of TFET

with p^+region and the drain is doped with n^+ region. The other is p-type in which the source is doped with n^+ region and the drain is doped with p^+ region.

5.3.2 *Band-to-Band Tunneling*

The band-to-band tunneling represents the current conduction in TFET. The tunneling phenomenon is similar to the working of the tunnel diode. The Fermi level exists in the conduction band of the n-type drain and valence band of the p-type source. Electrons are present in the drain of n-type and holes are present in the source of p-type. Under zero bias condition, the conduction band and valence band of n side and p side come together because of the heavily doped p^+ and n^+ regions as shown in Fig. 5.3a. When a reverse bias is applied, the height of the potential barrier is decreased and the electric current increases shown in Fig. 5.3b. This results in the flow of electrons from the conduction band to the valence band of n-type drain and p-type source, respectively. Thus, the current is increased and the maximum current flows.

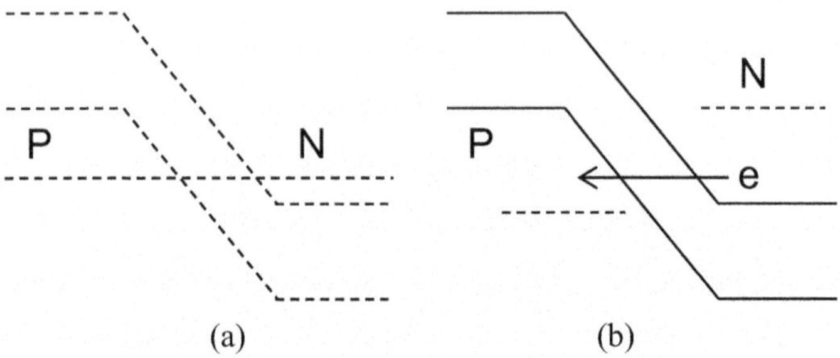

Fig. 5.3 **a** Zero bias condition and **b** reverse bias condition

5.3.3 Advantages and Limitations of TFET

The features of the TFET make it an efficient transistor of the coming future. MOSFET uses the drift and diffusion method for the carrier transport, whereas TFET uses the band-to-band tunneling for the flow of current. This explains that the MOSFET dependency on temperature is higher compared to the TFET. Thus, the sub-threshold potential of the TFET is less than 60 mv/decade. The tunneling width in the TFET is controlled by the gate voltage.

(a) In silicon TFET, the ON current is less as the band-to-band tunneling is not much effective which is required to overcome.
(b) The excessive scaling of the device leads to a very high OFF current that may lead to performance degradation.
(c) The current flows in both directions due to ambipolar nature. It shows p-type behavior with excess holes and n-type with excess electrons. TFETs show the dominant phenomenon of ambipolar nature, in which symmetric structures are maintained and the level of doping of drain and source is done same with single material of gate dielectric.

5.4 Graphene Nanoribbon

In 1996, Mitsutaka Fujita and his team afforded a theoretical method of GNRs to investigate the edge and nanoscale dimension effect in graphene (Fujita et al. 1996; Nakada et al. 1996). Due to the improvement in technology, GNRs have been used in the area of on-chip interconnects, through silicon vias and FETs (Ouyang et al. 2006; Lemme et al. 2007; Echtermeyer et al. 2008; Arsalam et al. 2015). The ballistic transport of the GNR makes its compatible not only for TSVs and interconnects but also for FETs. Utilizing the single GNR sheet, the monolithic system can be developed for both interconnects and transistors. For nanoscale devices, the Si-based FETs are affected by the scaling limitations. It has been investigated that the GNRs will perform better with reduced widths over the traditional MOSFETs. The high-quality GNR has large current densities higher than 10^8 A/cm^2, large carrier mobility

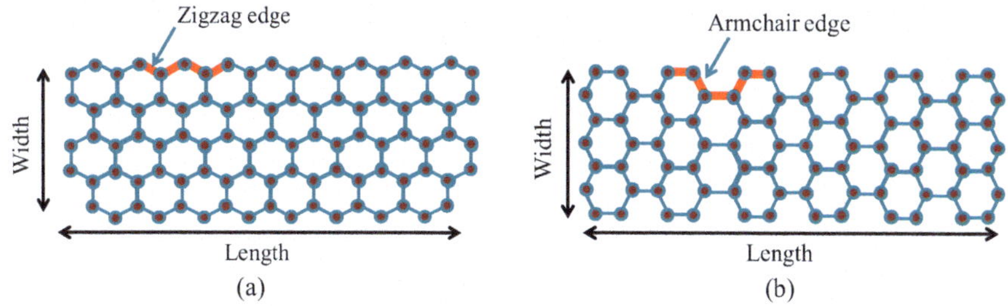

Fig. 5.4 GNR structure **a** zigzag and **b** armchair

3×10^3 cm^2 v^{-1} s^{-1}, and mean free path (MFP) ranging from 1 to 5 μm (Li et al. 2009).

5.4.1 *Structure of Graphene Nanoribbon*

A GNR is a single layer of the graphene sheet that is very thin and narrow which results in a 1-D structure (Kan et al. 2011). Based on the outcome of the GNR width, the GNR is classified into chiral GNR and non-chiral GNR. The chiral GNRs are further categorized as zigzag and armchair GNRs. Figure 5.4a, b shows the zigzag GNRs and armchair GNRs, respectively. Depending on the stacking, the GNRs are divided as single-layer GNR (SLGNR) and multi-layer (MLGNR). The SLGNR acts either as a semiconductor or conductor, whereas the MLGNR acts only as a conductor. Hence, semiconductor the SLGNR is selected for the transistor implementation.

5.4.2 *Semiconducting and Conducting GNRs*

The GNRs act either as a conductor or semiconductor based on the GNR edge patterning. The zigzag GNRs are always conductive, whereas the armchair GNRs act either conductor or semiconductor based on the dimer lines (n) of the GNR. Here, the discussion on the behavior of armchair GNRs is presented.

The armchair GNR acts as semiconductor when $n = 3p$ or $n = 3p + 1$ and conductor when $n = 3p + 2$. To recognize the behavior of the GNRs, it is required to know the GNR bandgap. The GNR acts as a conductor when its bandgap is zero, whereas it acts as semiconductor when the band is higher than zero. The GNR bandgap (E_g) is calculated as

$$E_g = 2|\alpha|\Delta E \tag{5.1}$$

$$\Delta E = \frac{\hbar v_f \pi}{W} \tag{5.2}$$

$$W = (n+1)\frac{\sqrt{3}}{2}a \tag{5.3}$$

where E_g is GNR bandgap and W is the GNR width, respectively, and the remaining parameters are given in Table 5.1.

The calculated values of E_g for $n = 3p$ and $n = 3p + 1$ are given in Table 5.2. For $n = 6$, 7, and 8, the energy band diagram of the GNR sheet is shown in Fig. 5.5. Furthermore, the calculated values are also verified from the quatumwise ATK simulator and shown in Fig. 5.6. From the figure, it is observed that the bandgap is small for $3p + 2$. For the remaining values of n, i.e., $3p$ and $3p + 1$, the bandgap value is higher than zero.

Table 5.1 Parameter description

Parameter	Description	Values
e	Electron charge	1.602×10^{-9} C
$\hbar$	Planck's constant	6.5×10^{-16} eV s
a	C–C bond distance	0.142 nm
v_f	Fermi velocity	10^6 m/s
n	Number of dimer lines	$3p$, $\alpha = 0.27$, $3p + 1$, $\alpha = 0.4$, $3p + 2$, $\alpha = 0.066$ where p is an integer

Table 5.2 Calculated values of dimer lines, GNR width, bandgap, and threshold voltage

Integer multiple of	Dimer lines (n)	GNR width (W) (nm)	ΔE	Bandgap (E_g)	Threshold voltage (V_{th}) (V)
$3p$	3	0.492	4.20	2.27	0.75
	6	0.86	2.40	1.30	0.43
	9	1.23	1.68	0.91	0.30
	12	1.60	1.29	0.70	0.23
	15	1.97	1.05	0.57	0.18
$3p + 1$	4	0.61	3.36	2.69	0.89
	7	0.98	2.10	1.68	0.56
	10	1.35	1.53	1.22	0.40
	13	1.72	1.20	0.96	0.32
	16	2.09	0.98	0.79	0.26

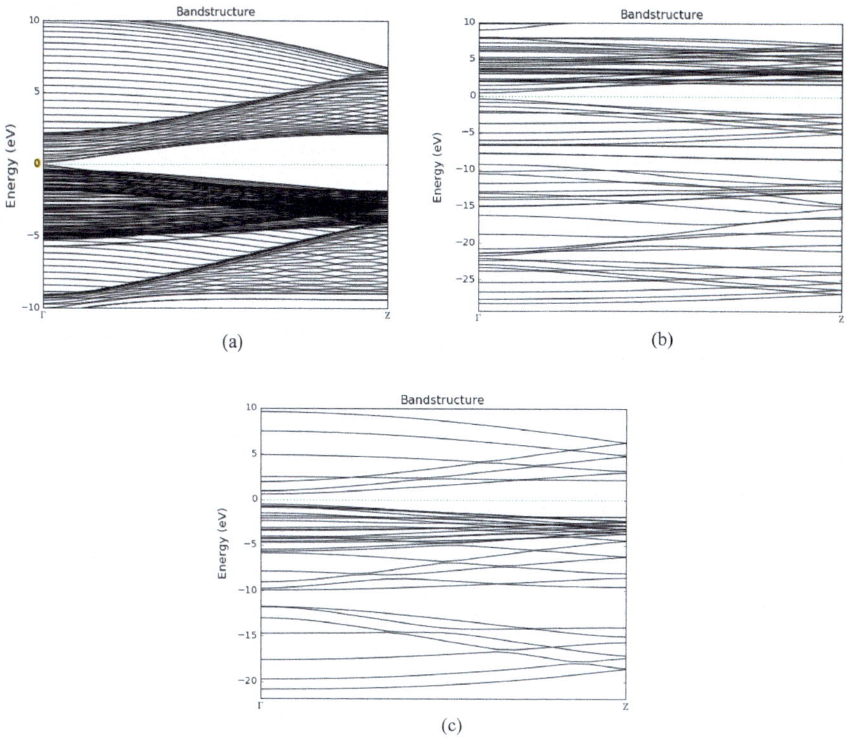

Fig. 5.5 Energy band diagram of GNR when **a** $n = 8$ $(3p + 2)$, **b** $n = 7$ $(3p + 1)$ and **c** $n = 6$ $(3p)$

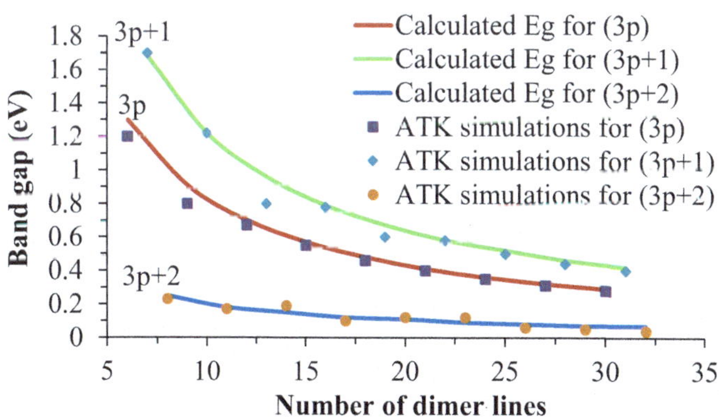

Fig. 5.6 Bandgap of GNRs versus dimer lines

5.4.3 *Properties and Characteristics of GNRs*

Due to atomic organizations of carbon atoms, the GNR has unique electrical, thermal, and mechanical properties. The sp^2 bonding of GNRs is responsible for large conductivity and mechanical strength. The properties of GNRs are described below.

5.4.3.1 Thermal Conductivity and Expansion

Due to the strong in-plane sigma bonds among the carbon atoms, the GNR shows superior conductivity below 20 K and also furnishes the extraordinary strength and stiffness against axial strains. Additionally, the higher interplane and zero in-plane thermal expansion of GNR lead to large flexibility. The GNRs are also suitable for current prospects in nanoscale molecular electronics, reinforcing additive fibers in functional composite materials, sensing, and actuating devices, etc. Hence, it is estimated that the GNR can increase the thermo-mechanical and thermal properties of the composite materials.

5.4.3.2 Aspect Ratio

The GNRs have a high aspect ratio which generalizes that it requires lower load over the other additive materials such as carbon black, chopped carbon fiber, or stainless steel fiber to obtain the same electrical conductivity. The unique electrical conductivity can be acquired because of the high aspect ratio of the GNRs compared to traditional additive materials.

5.4.3.3 Field Emission

The tunneling of the electrons from the conductor tip to vacuum leads to the field emission phenomenon under the application of the strong electric field. Due to the high aspect ratio and lower width of the GNR, the field emission can be acquired. The field emitters are compatible with the application in flat-panel displays. The properties of field emission for MLGNR occurred because of the electrons and light emissions. The light and luminescence emission occurred by the electron field emission and visible part of the spectrum, respectively, for zero potential.

5.4.3.4 Absorbent

Graphene nanoribbon is considered as perspective absorbing materials because of the high flexibility, lightweight, large mechanical strength, and large surface area.

Thus, GNR is a promising candidate for use in air, gas, and water filtration. The various research activities are carried out to use the GNRs instead of charcoal for high purity applications.

5.4.3.5 Strength and Elasticity

In graphite, due to the SP^2 hybridization, every carbon atom is connected via strong sigma bonds to three adjacent atoms. Hence, the GNRs have a strong elastic modulus which is large than the steel that makes high resistance.

5.5 Tunnel Field-Effect Transistors Using GNRs

In this section, the device structure of tunnel field-effect transistor using GNRs for monolayer and bilayer methods is discussed. The quantumwise ATK simulation approaches of these devices are also presented. Furthermore, the simulation results such as transfer characteristics and transconductance of the proposed device are described.

5.5.1 Device Structure

A schematic cross section view of both monolayer T-GNRFET and bilayer T-GNRFET is illustrated in Fig. 5.7a, b, respectively. These devices are simulated in the same conditions. A 20 nm length of armchair GNR with dimer line $n = 10$ is used as the channel material in the proposed designs. The GNR layer is arranged between two layers of the gate oxide, i.e., SiO_2 with the insulating constant of 3.9 and thickness (t_{ox}) of 1 nm. The 2 nm Al_2O_3 material is considered as top and bottom gate oxides. The length of the source and drain is considered as 10 nm. The armchair GNR under the source electrode and drain electrode is doped with the p-type and n-type impurities, whereas the GNR under the gate terminal is considered as intrinsic type. The p-type and n-type regions are doped with a molar fraction of 5×10^{-3}.

5.5.2 Computational Details

The density function theory (DFT) is utilized with local density approximation for the proposed designs to obtain current and voltage curves and transconductance. Figure 5.8 shows the monolayer and bilayer T-GNRFETs designed in the quantumwiseATK. The proposed devices are optimized using an extended Huckel basis set. All the electrical properties of the proposed devices involve 3-D atomic organizations

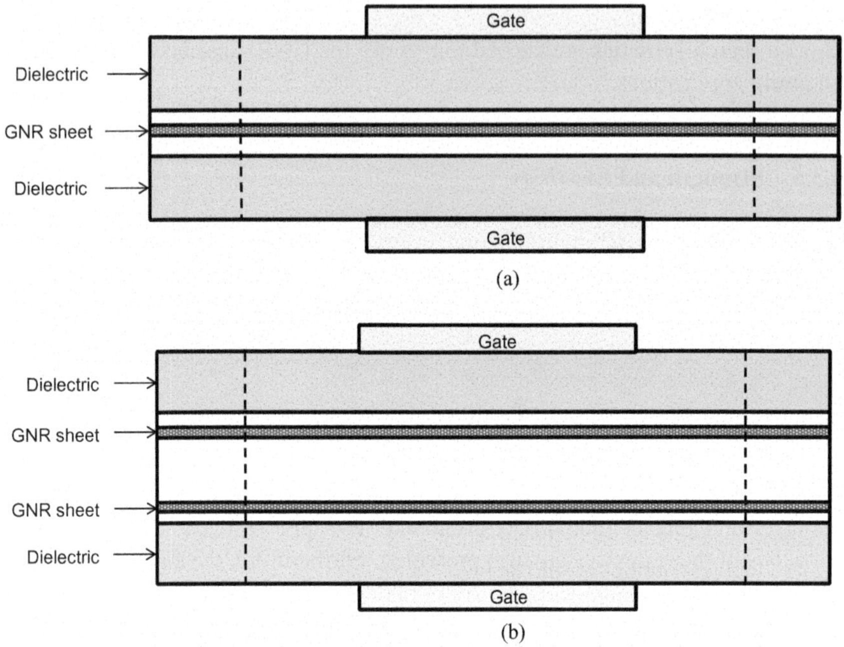

Fig. 5.7 Schematic structure of **a** monolayer GNRFET and **b** bilayer GNRFET

coupled with two semi-infinite electrodes. The properties of the proposed devices are studied by the fully self-consistent DFT integrated with non-equilibrium Green's function equations. The sampling point is considered as $k_a = 1, k_b = 1, k_c = 100$. The temperature and density mesh cutoff are taken as 300 K and 10Ha, respectively. The Hoffman.Hydrogen and Cerda.Carbon (graphite) models are utilized for hydrogen (H) and carbon (C) atoms, respectively. For left and right electrodes, the Dirichelt specifications are used, whereas the Neumann boundary conditions are assumed for the top, bottom, front, and back faces.

The electronic structures of the electrodes are evaluated by using DFT to acquire self-consistent Kohn–Sham potentials and Hamiltonian matrices. This helps to calculate the current in the channel from the transmission spectrum by Landauer–Buttiker equation stated below.

$$I(v) = 2q\left(\int (f_l(e, v) - f_r(e, v))T(e, v)dE\right)/h \tag{5.4}$$

where $f_l(e, v)$ is the Fermi functions for the left electrode, $f_r(e, v)$ is the Fermi functions for the right electrode, q is the electron charge, h is the Planck's constant, and $T(e, v)$ is bias voltage (v) and energy (e) dependent transmission spectrum.

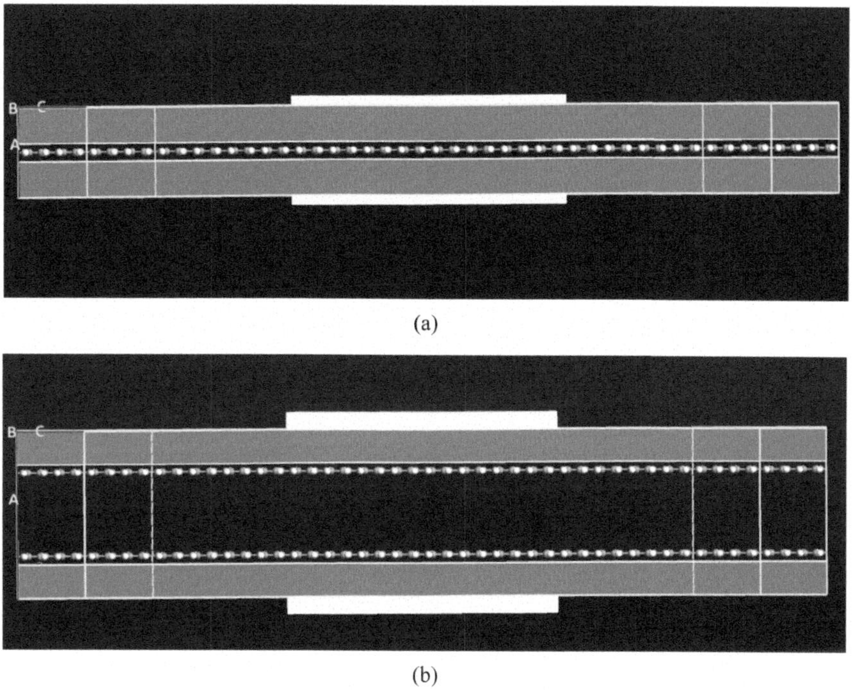

Fig. 5.8 Design of **a** monolayer and **b** bilayer T-GNRFET in ATK

5.5.3 *Transfer Characteristics*

The transfer characteristics of the proposed bilayer T-GNRFET are obtained by simulating in the quantumwise ATK. For comparative study, the monolayer T-GNRFET is also simulated. The obtained I_{DS} versus V_{GS} curves of both monolayer and bilayer GNRFET are shown in Fig. 5.9. Moreover, the I_{DS} versus V_{DS} curves of bilayer T-GNRFET for different V_{GS} are also obtained and shown in Fig. 5.10. From the analysis, it is observed that the bilayer T-GNRFET produces large current compared to the monolayer T-GNRFETs.

5.5.4 *Transconductance*

The transconductance of the proposed devices is also calculated to know the gain. The transconductance can be calculated as

$$g_m = \frac{\partial I_{ds}}{\partial V_{gs}} | V_{ds} \tag{5.5}$$

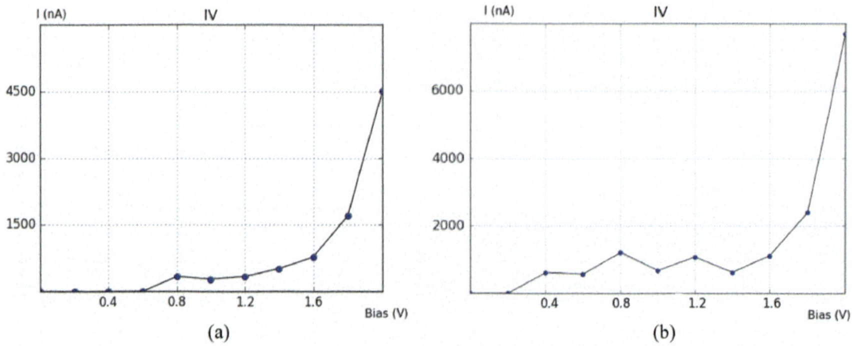

Fig. 5.9 I_{DS} versus V_{GS} curve **a** monolayer T-GNRFET and **b** bilayer T-GNRFET

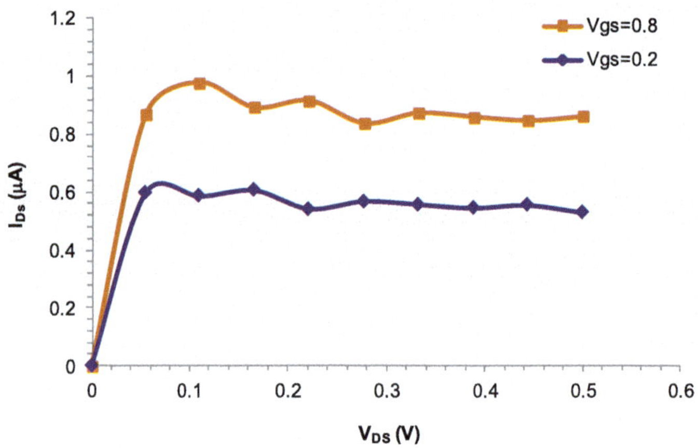

Fig. 5.10 I_{DS} versus V_{DS} curve of the bilayer T-GNRFET

Transconductance is a parameter reporting the device capability to control the barrier height as the gate voltage is employed in the saturation region. The transconductance of monolayer and bilayer T-GNRFETs is also obtained and shown in Fig. 5.11. From the analysis, it is investigated that the transconductance for the bilayer T-GNRFETs is more desirable for future digital systems.

5.6 Summary

The GNRFET-based transistors have captivated the researcher because of its unique trend in device modeling. The GNRFETs reduce the problems obtained by the scaling the channel length of the Si transistors and improve the performance of the digital system. Furthermore, the performance of the digital system can be improved by

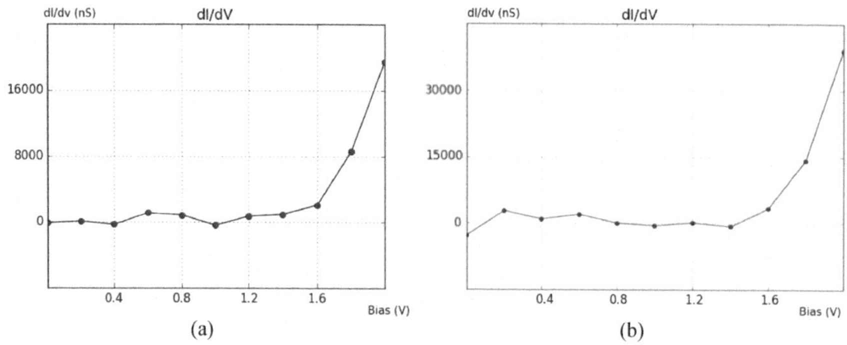

Fig. 5.11 Transconductance of **a** monolayer T-GNRFET and **b** bilayer T-GNRFET

introducing the T-GNRFETs. Thus, in this work, the bilayer T-GNRFET is designed and simulated in quantumwise ATK simulator to obtain the transfer characteristics and transconductance. Additionally, the obtained results are compared with the monolayer T-GNRFET. From the simulation results, it is observed that the bilayer T-GNRFET has large current and transconductance, respectively, over the monolayer T-GNRFET. Hence, utilizing the bilayer T-GNRFETs is optimistic way for implementing the low-power nanoscale ICs.

References

Arsalam A, Manoj KM, Archana K, Vobulapuram RK, Brajesh KK (2015) Performance analysis of single- and multi-walled carbon nanotube based through silicon vias. In: 2015 IEEE 65th Electronic components and technology conference (ECTC). San Diego, CA, pp 1834-1839

Chaudhury S, Sinha SK (2019) Carbon nanotube and nanowires for future semiconductor devices applications. Nanoelectronics 375–398. https://doi.org/10.1016/b978-0-12-813353-8.00014-2

Chin S, Seah D, Lam K, Samudra GS, Liang G (2010) Device physics and characteristics of graphene nanoribbon tunneling FETs. IEEE Trans Electron Devices 57(11):3144–3152

Dey A, Bajpai OP, Sikder AK, Chattopadhyay S, Shafeeuulla Khan MA (2016) Recent advances in CNT/graphene based thermoelectric polymer nanocomposite: a proficient move towards waste energy harvesting. Renew Sustain Energy Rev 53:653–671

Echtermeyer TJ, Lemme MC, Baus M, Szafranek BN, Geim AK, Kurz H (2008) Nonvolatile switching in graphene field-effect devices. IEEE Electron Device Lett 29(8):952–954

Fujita M, Wakabayashi K, Nakada K, Kusakabe K (1996) Peculiar localized state at zigzag graphite edge. J Phys Soc Jpn 65(7):1920–1923

Ghoreishi SS, Yousefi R, Taghavi N (2017) Performance evaluation and design considerations of electrically activated drain extension tunneling GNRFET: a quantum simulation study. J Electron Mater 46(11):6508–6517

Han MY, Özyilmaz B, Zhang Y, Kim P (2007) Energy band-gap engineering of graphene nanoribbons. Phys Rev Lett 98(20)

Kan E, Li Z, Yang J (2011) Graphene nanoribbons: geometric electronic and magnetic properties. In: Mikhailov S (ed) Physics and applications of graphene-theory, INTECH

Krausse GJ (2002) DE-series fast power MOSFET. Direct energy inc. technical note

Lemme MC, Echtermeyer TJ, Baus M, Kurz H (2007) A graphene field-effect device. IEEE Electron Device Lett 28(4):282–284
Li H, Xu C, Srivastava N, Banerjee K (2009) Carbon nanomaterials for next-generation interconnects and passives: physics, status, and prospects. IEEE Trans Electron Devices 56(9):1799–1820
Liu Y, Xie B, Zhang Z, Zheng Q, Xu Z (2012) Mechanical properties of graphene papers. J Mech Phys Solids 60(4):591–605
Nakada K, Fujita M, Dresselhaus G, Dresselhaus MS (1996) Edge state in graphene ribbons: nanometer size effect and edge shape dependence. Phys Rev B 54(24):17954–17961
Ouyang Y, Yoon Y, Fodor JK, Guo J (2006) Comparison of performance limits for graphene nanoribbon and carbon nanotube transistors. Appl Phys Lett 89(20):203107
Pathak AD (2001) MOSFET/IGBT drivers theory and applications. IXYS Corporation application note, IXAN0010
Seabaugh A, Zhang Q (2010) Low-voltage tunnel transistors for beyond cmos logic. Proc IEEE 98(12):2095–2110
Tang T, Burkhart C (2009) Hybrid MOSFET/driver for ultra-fast switching. IEEE Trans Dielectr Electr Insul 16(4):967–970
Tsividis Y, McAndrew C (2011) Operation and modeling of the MOS transistor. Oxford University Press, Oxford
Zhao P, Chauhan J, Guo J (2009) Computational study of tunneling transistor based on graphene nanoribbon. Nano Lett 9(2):684–688
Zhou SY, Gweon GH, Fedorov AV, First PN, de Heer WA, Lee DH, Lanzara A (2007) Substrate-induced bandgap opening in epitaxial graphene. Nat Mater 6(10):770–775

Chapter 6
A Threshold Voltage Model for SiGe Source/Drain Silicon-Nanotube-Based Junctionless Field-Effect Transistor

Anchal Thakur and Rohit Dhiman

Abstract An analytical threshold voltage model for SiGe source/drain silicon-nanotube junctionless field-effect transistor, based on the evanescent-mode analysis, is introduced. With the solution of three-dimensional Poisson equation in cylindrical coordinates, the surface potential along the channel length is determined with suitable boundary conditions. Using these models, the impact of physical device parameters such as core gate radius, oxide thickness, and nanotube thickness on the threshold voltage behavior and drain-induced barrier lowering has been studied. It is shown through extensive analysis that the proposed analytical models are in excellent agreement with TCAD numerical simulation results.

Keywords DIBL · Junctionless (JL) FET · Silicon nanotube · Surface potential · Threshold voltage roll-off

6.1 Introduction

SUB-20 nm metal–oxide–semiconductor field-effect transistor (MOSFET) scaling demands ultrasteep doping profile at the metallurgical junction interfaces, complex thermal budgets for dopant activation, and reduced susceptibility to the short-channel effects (SCEs) (Saurabh and Kumar 2016; Kumar et al. 2016; Dabhi et al. 2019). Numerous pioneering non conventional MOSFET structures, from silicon-on insulator (SOI) junctionless (JL) FETs to multigate architectures, such as double

This work is an outcome of the R&D work undertaken in the project under the Visvesvaraya Ph.D. Scheme of the Ministry of Electronics & Information Technology, Government of India, being implemented by Digital India Corporation (MEITYPHD-3186).

A. Thakur (✉) · R. Dhiman
Electronics and Communication Engineering Department, National Institute of Technology, Hamirpur, Himachal Pradesh 177005, India
e-mail: anchal.thakur31@gmail.com

R. Dhiman
e-mail: rohitdhiman@nith.ac.in

R. Dhiman and R. Chandel (eds.), *Nanoscale VLSI*,
Energy Systems in Electrical Engineering,
https://doi.org/10.1007/978-981-15-7937-0_6

gate (Migita et al. 2014), trigate (Lee et al. 2009; Rios et al. 2011), and gate-all-around (GAA) FETs have been rigorously explored to extend the scaling benefits for very-large-scale integration (VLSI). The SOI JLFET assumes an ultrathin layer (Sahay and Kumar 2016) to enhance the device performance, whereas a vertical cylindrical nanowire (NW) surrounded by a unique circular gate makes the GAA FETs (Dura 2011). A JL NWFET, although, exhibits to have superior immunity against SCEs, substantive technological issues including the high parasitic source/drain resistance and device variability restrain its efficacy for VLSI applications (Gnani et al. 2011). Moreover, the OFF-state current delivered by the device is quite high which degrades its performance in the sub-20 nm regime. Recently, Si/Ge GAA tunnel FET which may be a suitable alternative for future VLSI is demonstrated (Hanna and Hussain 2015; Hanna et al. 2015). The device utilizes a thin and hollow cylindrical Si-nanotube (NT) channel which remains enclosed by the core (inner) and shell (outer) gate stacks. The presence of heterointerfaces increases the barrier height and tunneling width at the drain–channel interface which gives rise to ultimate immunity against SCEs and, therefore, the device has excellent ON- and OFF-state electrical characteristics (Hanna and Hussain 2015; Fahad and Hussain 2013). Our previous work has successfully established that SiGe source/drain Si-NT JLFET can be used to overcome the limitations of the conventional Si-NT FETs (A. Thakur and R. Dhiman, Performance analysis of SiGe source-drain hetero-structure nanotube junctionless FET. in Proc. IEEE TENCON, India, Oct 2019). However, physically correct and compact model elucidating its threshold voltage performance is yet to be developed.

In this work, therefore, we develop a short-channel threshold voltage model for Si-NT JLFET with SiGe source/drain, based on the surface potential solution of evanescent-mode analysis (*EMA*), in the form of Fourier–Bessel series of three-dimensional (3D) Poisson equation. In particular, *EMA* is more accurate unlike other methods such as polynomial exponential and parabolic approximation and accounts for the exponential decay of potential in short-channel devices (Khaveh and Mohammadi 2016; Chang 2012). The improved modeling further illustrates that the proposed hetero-structure Si-NT JLFET shows better immunity against the threshold voltage roll-off and drain-induced barrier lowering (DIBL), instead of Si-NT FETs as reported in (Kumar et al. 2017). To the best of authors' knowledge, it is for the first time that analytical solutions of the surface potential and threshold voltage for SiGe source/drain Si-NT JLFET in sub-20 nm regime have been reported using *EMA*. Finally, the proposed models are validated with the numerical simulation results of Synopsys Sentaurus TCAD.

6.2 Device Structure

The 3D schematic of simulated hetero-structure Si-NT JLFET is shown in Fig. 6.1a, whereas the 2D cross-sectional view is shown in Fig. 6.1b. The outer gate similar to GAA and inner gate surrounded by the oxide layer control the charge carrier transport in Si-NT channel. The core gate is introduced inside a tubular cylindrical nanotube.

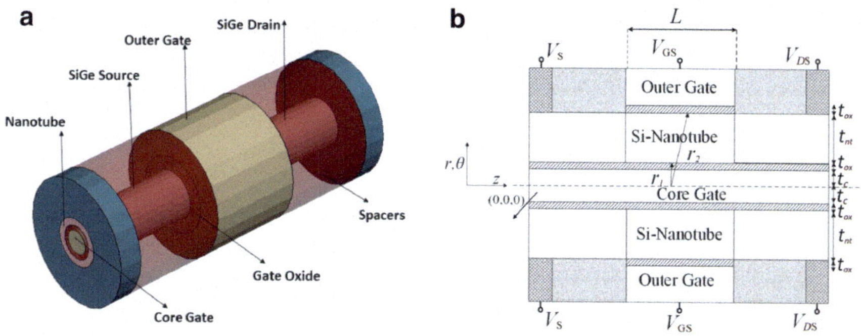

Fig. 6.1 **a** 3D simulated view of the hetero-structure Si-NT JLFET. **b** 2D schematic diagram representation

The source, channel, and drain regions are uniformly doped with doping density $N_d = 1 \times 10^{19}\ \text{cm}^{-3}$. The core gate radius, gate oxide thickness, and nanotube thickness are noted as t_c, t_{ox}, and t_{nt}, respectively. The inner and outer gates are biased with the same gate-to-source voltage, V_{GS}, and work function, $\phi_m = 4.8$ eV. Since side-wall spacers are indispensable to realize hetero-structure NT JLFET, we have used HfO_2 spacer in our design simulations. The standard value of permittivity considered for silicon and HfO_2 as $\varepsilon_{si} = 11.8 \times 8.854 \times 10^{-14}$ F/cm and $\varepsilon_{ox} = 22.5 \times 8.854 \times 10^{-14}$ F/cm.

6.3 Derivation of the Surface Potential

In cylindrical coordinate system, the 3D Poisson equation can be expressed as

$$\frac{1}{r}\frac{\mathrm{d}}{\mathrm{d}r}\left[r\frac{\mathrm{d}(\varphi)}{\mathrm{d}r}\right] + \frac{1}{r^2}\frac{\mathrm{d}^2(\varphi)}{\mathrm{d}\theta^2} + \frac{\mathrm{d}^2(\varphi)}{\mathrm{d}z^2} = -q\frac{N_\mathrm{d}}{\varepsilon_\mathrm{si}} \tag{6.1}$$

where φ is the surface potential which varies along the radial (r) and lateral (z) directions. Since potential variation in the angular direction (θ) is assumed to be zero, we rewrite (6.1) as

$$\frac{1}{r}\frac{\mathrm{d}}{\mathrm{d}r}\left[r\frac{\mathrm{d}}{\mathrm{d}r}(\varphi(r,z))\right] + \frac{\mathrm{d}^2}{\mathrm{d}z^2}(\varphi(r,z)) = -q\frac{N_\mathrm{d}}{\varepsilon_\mathrm{si}} \tag{6.2}$$

A closed-form solution for (6.2) can be obtained using *EMA* which decouples the surface potential into two parts as

$$\varphi(r,z) = \varphi_{1\mathrm{D}}(r) + \varphi_{2\mathrm{D}}(r,z) \tag{6.3}$$

where $\varphi_{1D}(r)$ is the solution for 1D Poisson equation under depletion approximation along the NT thickness, given as

$$\frac{d^2\varphi_{1D}(r)}{dr^2} + \frac{1}{r}\frac{d\varphi_{1D}(r)}{dr} = -q\frac{N_d}{\varepsilon_{si}} \tag{6.4}$$

Furthermore, $\varphi_{2D}(r, z)$ describes 2D variation of the surface potential at the oxide–silicon interface with zero charges and satisfies 2D Laplace equation, which is given as

$$\frac{d^2\varphi_{2D}(r, z)}{dr^2} + \frac{1}{r}\frac{d^2\varphi_{2D}(r, z)}{dr^2} + \frac{d^2\varphi_{2D}(r, z)}{dz^2} = 0 \tag{6.5}$$

Let $\varphi_1(r, z)$ and $\varphi_2(r, z)$ represent the surface potentials for the inner and outer gates, respectively; therefore

$$\varphi_1(r, z) = \varphi(r, z)\big|_{r=t_c+t_{ox}} \tag{6.6}$$

$$\varphi_2(r, z) = \varphi(r, z)\big|_{r=t_c+t_{ox}+t_{nt}} \tag{6.7}$$

The boundary conditions that must be satisfied by (6.4) and (6.5) are as follows:

$$\varphi(r, 0) = V_{bi} \tag{6.8}$$

$$\varphi(r, L) = V_{DS} + V_{bi} \tag{6.9}$$

Here, V_{DS} and V_{bi} represent the drain-to-source bias voltage and built-in potential, respectively, and L is the channel length.

The electric flux density at the silicon–insulator interface is continuous, expressed as

$$C_{ox1}(\varphi_1(r, z) - (V_{GS} - V_{FB})) = \varepsilon_{si}\frac{d}{dr}\varphi(r, z)|r = t_c + t_{ox} \tag{6.10}$$

$$C_{ox2}(\varphi_2(r, z) - (V_{GS} - V_{FB})) = -\varepsilon_{si}\frac{d}{dr}\varphi(r, z)|r = t_c + t_{ox} + t_{nt} \tag{6.11}$$

where C_{ox1} and C_{ox2} are the inner and outer gate capacitances per unit area, respectively, which are given by

$$C_{ox1} = \frac{\varepsilon_{ox}}{t_1}, t_1 = (t_c + t_{ox})\ln\left(1 + \frac{t_{ox}}{t_c + t_{ox}}\right) \tag{6.12}$$

$$C_{ox2} = \frac{\varepsilon_{ox}}{t_2}, t_2 = (t_c + t_{ox} + t_{nt})\ln\left(1 + \frac{t_{ox}}{t_c + t_{ox} + t_{nt}}\right) \tag{6.13}$$

where t_1 and t_2 are the effective oxide thickness of the inner and outer gates, respectively. Substituting (6.8)–(6.14) in (6.6) and (6.7), explicit solutions for the surface potential φ_{1D} (r) and φ_{2D} (r, z) can be obtained as

$$\varphi_{1D}(r) = (V_{GS} - V_{FB}) - \frac{qN_d r^2}{4\varepsilon_{si}} + \frac{qN_d t_{nt}^2}{16\varepsilon_{si}} + \frac{qN_d t_{nt}}{4C_{ox}} \quad (6.14)$$

$$\varphi_{2D}(r, z) = \sum_{n=0}^{\infty} \left[A_n \exp(z\lambda_n) + B_n \exp(-z\lambda_n)\right] J_0(\lambda_0 r) \quad (6.15)$$

In Eq. (6.16), J_n is Bessel function of nth order and A_n, B_n are Fourier–Bessel series coefficients. Since the higher-order Bessel coefficients decay rapidly and the lowest order predicts 2D potential profile quite accurately, we limit (6.16) to the lowest-order mode, $n = 0$. Consequently, the analytical surface potentials for the inner gate ($r = r_1 = t_c + t_{ox}$) and outer gate ($r = r_2 = t_c + t_{ox} + t_{nt}$) are expressed, respectively as

$$\varphi_1(r, z) = \varphi_{1D}(r_1) + \left[A_0 \exp(z\lambda_0) + B_0 \exp(-z\lambda_0)\right] J_0(\lambda_0 r_1) \quad (6.16)$$

$$\varphi_2(r, z) = \varphi_{1D}(r_2) + \left[A_0 \exp(z\lambda_0) + B_0 \exp(-z\lambda_0)\right] J_0(\lambda_0 r_2) \quad (6.17)$$

where

$$A_0 = \frac{(V_{bi} + V_{DS} - \varphi_{1D}(r)) - (V_{bi} - \varphi_{1D}(r)) \exp(-\lambda_0 L)}{2\sinh(\lambda_0 L) J_0(\lambda_0 r)} \quad (6.18)$$

$$B_0 = \frac{(V_{bi} + V_{DS} - \varphi_{2D}(r)) - (V_{bi} - \varphi_{2D}(r)) \exp(-\lambda_0 L)}{2\sinh(\lambda_0 L) J_0(\lambda_0 r)} \quad (6.19)$$

The eigenvalue, λ_0, must satisfy the Poisson equation for continuity at the oxide–silicon interface as

$$\lambda_0 = \left[\frac{-\alpha \pm \sqrt{\alpha^2 - 4\beta\xi}}{2\xi}\right]^{\frac{1}{2}} \quad (6.20)$$

where $\alpha = -\frac{t_{nt}}{2}\left(1 + \frac{t_{nt} C_{ox1(2)}}{2\varepsilon_{si}}\right)$, $\beta = \frac{C_{ox1(2)}}{\varepsilon_{si}}$, $\xi = \frac{t_{nt}^3}{16}$.

The minimum center potential $\varphi_c(z_{min})$ can be obtained by setting $r = r_0 = t_c + t_{ox} + t_{nt}/2$ and is given as

$$\varphi_c(z_{min}) = \varphi_{1D}(r_0) + 2J_0(\lambda_0 r_0)\sqrt{A_0 B_0} \quad (6.21)$$

6.4 Threshold Voltage Model

The threshold voltage, V_T, for SiGe source/drain junctionless transistor is the gate voltage for which $\varphi_c(z_{\min})$ becomes equal to zero (Li et al. 2013), given as

$$V_T = V_{T1} - V_{T2} \tag{6.22}$$

In Eq. (6.23), V_{T1} is the threshold voltage under long-channel approximation, assuming that source and drain have no impact on the channel and V_{T2} is the threshold voltage roll-off due to the effect of source and drain. Considering the long-channel condition and gradual channel approximation, V_{T1} can be obtained based on (6.22) as

$$\begin{aligned} &\varphi_{1D}(r) = 0|_{V_{GS}=V_{T1}} \Rightarrow \\ &V_{T1} = V_{FB} - \frac{qN_d r_0^2}{4\varepsilon_{si}} + \frac{qN_d t_{nt}^2}{16\varepsilon_{si}} + \frac{qN_d t_{nt}}{4C_{ox}} \end{aligned} \tag{6.23}$$

Considering the short-channel condition, V_{T2} is derived as

$$V_{T2} + 2J_0(\lambda_0 r_0) = 0 \tag{6.24}$$

Equation (6.25) can be represented in the form of a second-order polynomial as

$$V_{T2} = \frac{-M_2 + \sqrt{M_2^2 - 4M_1 M_3}}{2M_1} \tag{6.25}$$

where

$$M_1 = \sinh^2(\lambda_0 L) - 2\sinh(\lambda_0 L) \tag{6.26}$$

$$M_2 = 2(2V_{bi} + V_{DS})\sinh(\lambda_0 L) + 2V_{DS} \tag{6.27}$$

$$M_3 = -4\sinh(\lambda_0 L)(V_{bi}(V_{bi} + V_{DS})) + V_{DS}^2 + 2V_{bi}V_{DS} \tag{6.28}$$

Therefore, the analytical closed-form expression for the threshold voltage is expressed as

$$V_T = V_{FB} - \frac{qN_d r_0^2}{4\varepsilon_{si}} + \frac{qN_d t_{nt}^2}{16\varepsilon_{si}} + \frac{qN_d t_{nt}}{4C_{ox}} - \left(\frac{-M_2 + \sqrt{M_2 - 4M_1 M_3}}{2M_1}\right) \tag{6.29}$$

6.5 Results and Discussion

In this section, the proposed analytical models are verified with the numerical simulations which are carried out using Synopsys Sentaurus TCAD. The SiGe source/drain Si-NT JLFET is simulated by including drift-diffusion and bandgap narrowing models to calibrate carrier transport and band bending. To account for mobility degradation, the Lombardi and Philips unified mobility models have been incorporated. In addition, the thermionic model has also been included along with nonlocal BTBT model. Furthermore, the simulation setup was calibrated by replicating the experimental drain current and gate capacitance of the Si-NT MOSFET (Tekleab 2014) due to its topological resemblance to our proposed structure and is shown in Fig. 6.2.

The comparison of inner and outer surface potentials across the channel length L of the proposed device is shown in Fig. 6.3. As can be seen, the surface potential of inner gate is higher than the outer gate, which is mainly attributed to the asymmetric structure of the device. Further, as per (6.13) and (6.14), for the same physical thickness, the outer gate oxide is effectively thicker than the inner gate. Thus, for the same potential at both inner and outer gate electrodes, the potential drop across the inner Si-HfO_2 is smaller compared to the outer gate oxide, resulting in a higher potential at the inner gate. It can also be observed that the analytical model follows the simulation results for surface potential, quite accurately.

The effect of drain and gate voltages on the center potential across the channel length is shown in Fig. 6.4. We observe that the center potential increases as gate voltage V_{GS} increases, resulting in a larger threshold voltage of the device. On the other hand, as drain voltage V_{DS} increases from 0.1 to 0.3 V, the source–drain barrier height reduces which results in lower threshold voltage for the SiGe source/drain Si-NT JLFET. The proposed model and simulation results are in close agreement.

The threshold voltage variation at different channel lengths with core gate radius t_c as a parameter is shown in Fig. 6.5a. As radius of core gate is increased from 5 to 15 nm, the effective oxide thickness of both the inner and outer gates increases. Therefore, channel charge carriers experience weak gate control resulting into high threshold voltage roll-off, since a thicker oxide will resist the vertical electric field from the gate penetrating into the channel (Chiang 2009). Furthermore, the heterostructure JLFET exhibits better immunity to SCEs in comparison with the junctioned Si-NT FET for the same core gate radius ($t_c = 10$ nm). The analytical results are found to be in good proximity with numerical simulation data confirming that threshold voltage decreases with increase in core gate radius.

In Fig. 6.5b, the variation of threshold voltage with respect to channel length, for different gate oxide thickness, is shown. The increase in oxide thickness t_{ox} around the Si-NT channel increases the effective inner and outer oxide thickness. Thus, a thicker gate oxide results in reduced gate control over the channel region which ultimately causes decrease in the threshold voltage. In case of junctioned Si-NT FET, the gate-channel coupling would be much reduced at smaller lengths, whereas in our proposed device, the channel charges are subjected to better gate control as gate

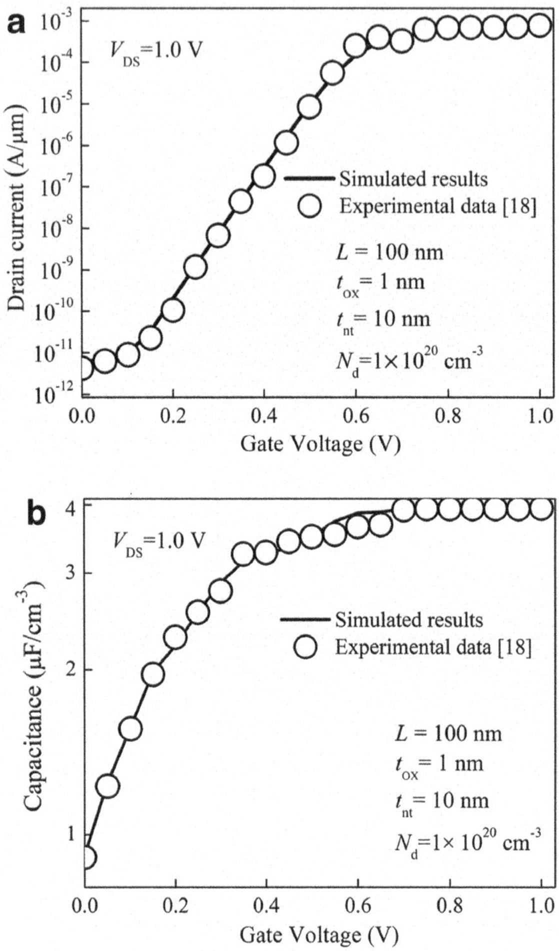

Fig. 6.2 Calibrated versus experimental data of the Si-NT MOSFET **a** drain current, **b** gate capacitance

dimensions are reduced. Consequently, SiGe source/drain Si-NT JLFET experience a smaller threshold voltage roll-off than that of junctioned NT FETs in the sub-20 nm regime.

The threshold voltage for different nanotube thickness t_{nt} is presented in Fig. 6.5c. From the figure, it is clear that threshold voltage decreases as nanotube thickness increases, which is primarily due to the limited space charge between the inner and outer gates and weak electrostatic control of gate over the channel. Thus, device with thinner tube is more immune to SCEs resulting into reduced threshold voltage roll-off at smaller gate lengths.

The variation of DIBL along the channel length with different nanotube thickness is shown in Fig. 6.6. It is noticed that if channel length is reduced below 10 nm,

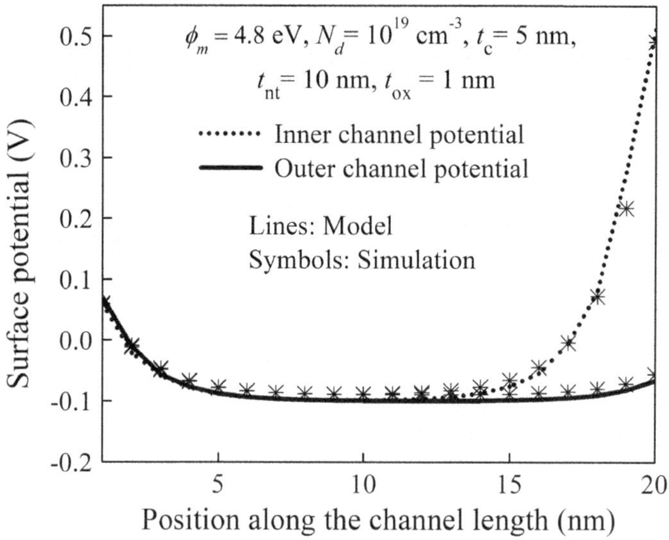

Fig. 6.3 Inner and outer surface potentials versus channel length

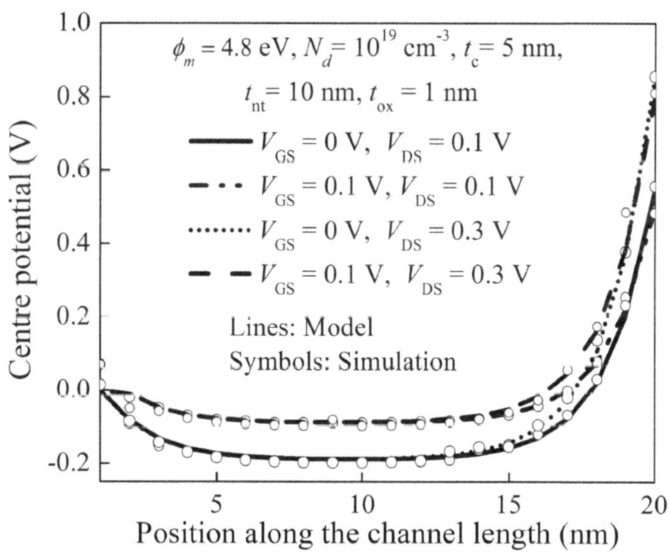

Fig. 6.4 Center potential against the channel length for different values of gate and drain voltages

the DIBL effect becomes significant for the device with thicker tube due to weak gate-channel coupling, which allows the drain to take control over channel carriers. Moreover, the Si-NT JLFET device shows excellent immunity against DIBL, in sharp contrast to the junctioned NT FETs.

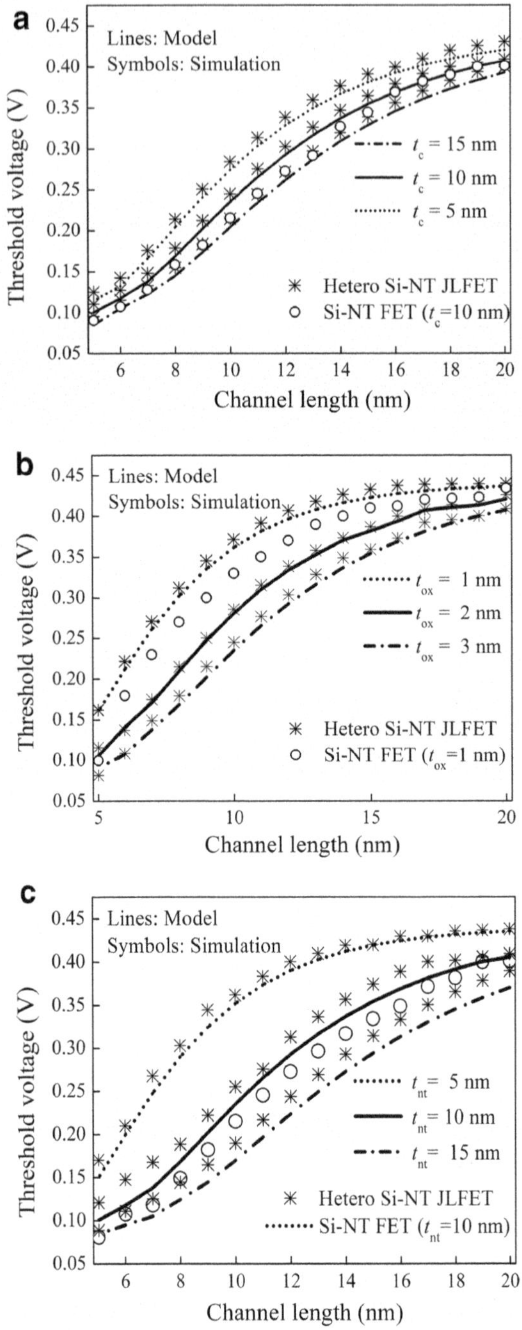

Fig. 6.5 Variation in threshold voltage with channel length for different **a** core gate radius, **b** gate oxide thickness, **c** nanotube thickness

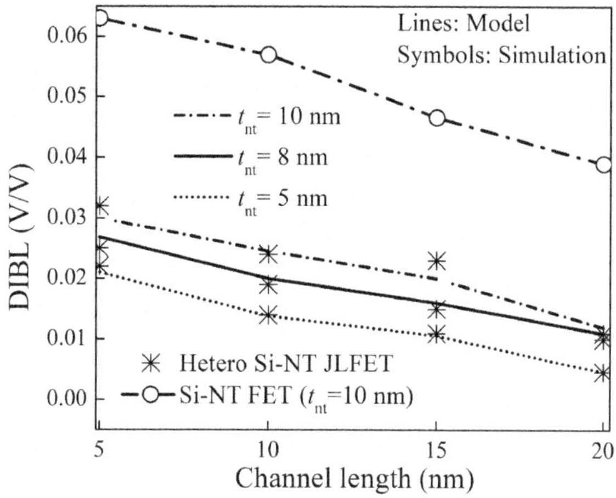

Fig. 6.6 DIBL (V/V) for L = 5–20 nm with nanotube thickness as a parameter

6.6 Conclusion

In this paper, a threshold voltage model for SiGe source/drain Si-NT JLFET has been derived by solving 3D Poisson equation in cylindrical coordinates. It has been found out through extensive modeling and simulation that the proposed models precisely determine the surface potential and threshold voltage for a wide variety of device parameters. In addition, the hetero-structure Si-NT JLFET exhibits improved immunity against SCEs compared to the Si-NT FETs at ultra-scaled dimensions. The models presented in this work provide physical insights into the threshold voltage behavior and may offer basic design guideline for the nanoscale hetero-structure Si-NT JLFETs.

References

Chang T-K (2012) A quasi-two-dimensional threshold voltage model for short-channel junctionless double-gate MOSFETs. IEEE Trans Electron Devices 59(9):2284–2289

Chiang T-K (2009) A new two-dimensional analytical subthreshold behavior model for short-channel tri-material gate-stack SOI MOSFETs. Microelectron Rel 49:113–119

Dabhi CK, Roy AS, Chauhan YS (2019) Compact modeling of temperature-dependent gate-induced drain leakage including low-field effects. IEEE Trans Electron Devices 66(7):2892–2897

Dura J et al (2011) Analytical model of drain current in nanowire MOSFETs including quantum confinement band structure effects and quasi-ballistic transport: device to circuit performances analysis. In: Proceedings of the international conference on simulation of semiconductor processes and devices, pp 43–46

Fahad HM, Hussain MM (2013) High-performance silicon nanotube tunneling FET for ultralow-power logic applications. IEEE Trans Electron Devices 60(3):1034–1039

Gnani E, Gnudi A, Reggiani S, Baccarani G (2011) Theory of the junctionless nanowire FET. IEEE Trans Electron Devices 58(9):2903–2910

Hanna AN, Hussain MM (2015) Si/Ge hetero-structure nanotube tunnel field effect transistor. J Appl Phys 117(1):014310-1–014310-7

Hanna AN, Fahad HM, Hussain MM (2015) InAs/Si hetero-junction nanotube tunnel transistors. Sci Rep 9

Khaveh HRT, Mohammadi S (2016) Potential and drain current modeling of gate-all-around tunnel FETs considering the junctions depletion regions and the channel mobile charge carriers. IEEE Trans Electron Devices 63(12):5021–5029

Kumar MJ, Vishnoi R, Pandey P (2016) Tunnel field-effect transistors (TFET): modelling and simulation. Wiley, West Sussex, UK

Kumar A, Bhushan S, Tiwari PK (2017) A threshold voltage model of silicon-nanotube-based ultrathin double gate-all-around (DGAA) MOSFETs incorporating quantum confinement effects. IEEE Trans Nanotechnol 16(5):868–875

Lee C-W, Afzalian A, Akhavan ND, Yan R, Ferain I, Colinge J-P (2009) Junctionless multigate field-effect transistor. Appl Phys Lett 94(5):053511–053512

Li C, Zhuang Y, Di S, Han R (2013) Subthreshold behavior models for nanoscale short-channel junctionless cylindrical surrounding-gate MOSFETs. IEEE Trans Electron Devices 60(11):3655–3662

Migita S, Morita Y, Matsukawa T, Masahara M, Ota H (2014) Experimental demonstration of ultrashort-channel (3 nm) junctionless FETs utilizing atomically sharp V-grooves on SOI. IEEE Trans Nanotechnol 13(2):208–215

Rios R et al (2011) Comparison of junctionless and conventional trigate transistors with L_g down to 26 nm. IEEE Electron Device Lett 32(9):1170–1172

Sahay S, Kumar MJ (2016) Realizing efficient volume depletion in SOI junctionless FETs. IEEE J Electron Devices Soc 4(3):110–115

Saurabh S, Kumar MJ (2016) Fundamentals of tunnel field-effect transistors. CRC Press, Boca Raton, FL, USA

Tekleab D (2014) Device performance of silicon nanotube field effect transistor. IEEE Electron Device Lett 35(5):506–508

Thakur A, Dhiman R (2019) Performance analysis of SiGe source-drain hetero-structure nanotube junctionless FET. In: Proceedings of the TENCON, India, Oct 2019

Chapter 7
III-V Nanoscale Quantum-Well Field-Effect Transistors for Future High-Performance and Low-Power Logic Applications

J. Ajayan and D. Nirmal

Abstract The III-V compound semiconductor-based quantum-well field-effect transistor (QWFET) is one of the most promising solid-state transistor technologies for future high-speed, low-power logic integrated circuit applications due to their high speed and low-voltage operation. This excellent speed and low voltage operation mainly comes from the unique properties of III-V compound semiconductors such as high electron and hole mobility, high electron velocity saturation and high sheet carrier concentration, etc. High-performance III-V compound semiconductor-based n-channel QWFETs are widely available. But, for implementing high-speed low-power CMOS logic integrated circuits, there is a critical issue of identifying high-performance III-V compound semiconductor -based p-channel quantum-well transistors. In order to fully utilize the potential of high mobility III-V compound semiconductor channel materials, instead of developing large diameter III-V wafers, it is better to couple III-V transistors with traditional silicon wafers. This chapter deals with the architecture and electrical performance of III-V nanoscale quantum-well field-effect transistors for future high-speed and low-power logic integrated circuit applications.

Keywords Compound semiconductor · InAs · InGaAs · InSb · Quantum well

---

J. Ajayan
Department of Electronics and Communication Engineering, SNS College of Technology, Coimbatore, India
e-mail: email2ajayan@gmail.com

D. Nirmal (✉)
Department of Electronics and Communication Engineering,
Karunya Institute of Technology and Science, Coimbatore, India
e-mail: dnirmalphd@gmail.com

R. Dhiman and R. Chandel (eds.), *Nanoscale VLSI*,
Energy Systems in Electrical Engineering,
https://doi.org/10.1007/978-981-15-7937-0_7

7.1 Introduction

According to Moore's law, for every two years the transistor density in an integrated circuit will be doubled and this law has been the driving force for the semiconductor industry for over five decades. In order to keep Moore's law alive, the transistor scaling needs to be continued for performance improvement of integrated circuits (Suman 2007; Ajayan et al. 2018a, 2019a, b; Del Alamo 2011). Silicon CMOS transistor scaling has been the driving force behind the success of semiconductor industry, and the two major challenges faced by today's semiconductor industry are

1. To propel the CMOS technology beyond its chip density and functionality by integrating a new low-power, high-speed and high-density memory technology on to the CMOS process.
2. To extend computing and information processing beyond that can be obtained by traditional silicon CMOS technology with the help of an innovative combination of new semiconductor devices and interconnects.

Traditional silicon CMOS transistor scaling is approaching to the end of the roadmap because of its physical limitations like large electron effective mass, short channel effects and large tunneling current. Therefore, it is high time to develop an alternating semiconductor device to replace traditional silicon CMOS transistors for future logic integrated circuit applications (Gilbert et al. 2008; Iwai 2009; Del Alamo et al. 2016; Ajayan and Nirmal 2015, 2016a, b; Ajayan et al. 2018b).

Downscaling or miniaturization of the transistors is the most important and effective method for achieving high speed and low power in CMOS digital logic integrated circuits. Reducing the supply voltage is the most effective method to minimize the power consumption. But for reducing the supply voltage, the threshold voltage of the transistors should be reduced. However, reduction of threshold voltage leads to the increase of subthreshold leakage current. Therefore, it is not easy to reduce the threshold voltage of the transistor and supply voltage. In 2025, the transistor gate length may reach 5 nm. The time is urged to develop nanoscale transistors which are energy efficient, can be operated at lower supply voltages and should have very low OFF-state leakage power. An innovative combination of transistor architecture and materials is required to meet these challenges. In recent years, III-V compound semiconductor-based quantum-well field transistors (QWFETs) have been emerged as an attractive logic transistor technology for next-generation high-speed area-efficient and low-power logic integrated circuit applications due to its high switching speed and low operating voltage. III-V compound semiconductors such as GaAs, InAs, InSb and InGaAs are considered as most promising semiconductor materials for future high-speed low-power digital logic integrated circuit applications due to their unique characteristics such as high electron mobility, high sheet charge density, high electron saturation velocity. (Gerben and Matthias 2010; Yang et al. 2011; Ajayan et al. 2017a, b, 2018a, b; Ajayan and Nirmal 2017a, b). III-V QWFETs can provide a low energy delay product compared to the state-of-the-art silicon CMOS transistors for a supply voltage of less than 0.7 V (Ajayan et al. 2019;

Gilbert et al. 2008; YounHo et al. 2009; Kumar et al. 2017). The effective carrier velocity (V_{eff}) of III-V QWFETs can be calculated as (Gilbert et al. 2008).

$$V_{\text{eff}} = \frac{g_{\text{mi}}}{W_{\text{g}} C_{\text{gi}}} \tag{7.1}$$

g_{mi} intrinsic transconductance
W_{g} width of the gate
C_{gi} intrinsic gate capacitance/unit area

The drive current of the QWFETs is directly proportional to V_{eff}, W_{g} and channel charge density (N_{S})

$$N_{\text{S}} = \frac{C_{\text{gi}}(V_{\text{g}} - V_{\text{t}})}{q} \tag{7.2}$$

V_{t} threshold voltage of the transistor
V_{g} gate voltage

The material properties of some of the most important semiconductors which can be used as channel materials in transistors are given in Table 7.1.

The QWFETs can be either depletion mode (D-Mode) devices or enhancement mode (E-Mode) devices. E-Mode QWFETs exhibit a positive threshold voltage, whereas D-Mode QWFETs exhibit a negative threshold voltage. Even though D-Mode QWFETs are superior in performance in terms of high on current and transconductance, E-Mode QWFETs are desirable for developing high-performance digital logic integrated circuits due to their single power supply operation and outstanding reliability. The schematic of D-Mode and E-Mode QWFETs is shown in Fig. 7.1. The distance between gate and quantum well is large in D-Mode QWFETs.

Table 7.1 Properties of semiconductors at room temperature (300 K)

Parameters	Si	GaAs	$In_{0.53}Ga_{0.47}As$	InAs	InSb
Mobility of electrons (cm^2/Vs) ($n_s = 10^{12}/cm^2$)	600	4600	7800	20,000	30,000
Electron velocity saturation (10^7 cm/S)	1	1.2	0.8	3.5	5
Energy band gap (eV)	1.12	1.42	0.72	0.36	0.18
Ballistic mean free path (nm)	28	80	106	194	226

(Suman 2007; Ajayan et al. 2017a, b, 2018a, b, c, d, 2019a, b, c; Del Alamo 2011; Gilbert et al. 2008; Iwai 2009; Del Alamo et al. 2016; Ajayan and Nirmal 2015, 2016a, b, 2017a, b; Gerben and Matthias 2010; Yang et al. 2011; Gilbert et al. 2008; YounHo et al. 2009; Kumar et al. 2017; Xue et al. 2012; Radosavljevic et al. 2008; Bolognesi et al. 1999; Suman et al. 2007; Lee et al. 2014; Lin et al. 2014, 2015a, b, 2019; Kim et al. 2015; Tae-Woo et al. 2015; Ashley et al. 2007; Hwang et al. 2011; John et al. 1994; Jianqiang et al. 2016; Taewoo and Dae-Hyun 2015; Jaydeep and Kaushik 2008; Kharche et al. 2011)

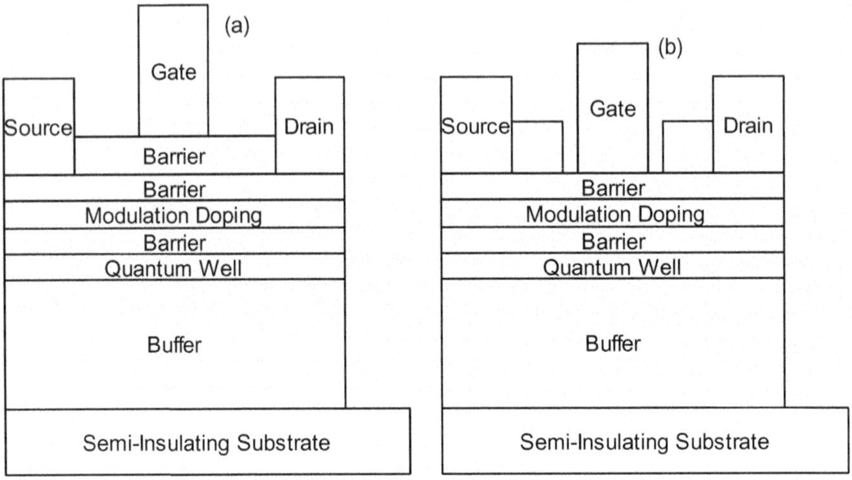

Fig. 7.1 Schematic of **a** depletion mode QWFET **b** enhancement mode QWFET (Suman 2007)

The QWFETs can be built on wafers like GaAs, InP, GaN, Si or SiC, etc. Buffer layers are employed to reduce the effects of lattice mismatches between the quantum well and the wafer. It consists of a wide band gap semiconductor barrier layer to minimize the effect of narrow bandgap channel materials on the breakdown voltages and leakage current. For low-power digital logic integrated circuit applications, transistors with low threshold voltages are highly preferable and the threshold voltage of the transistors is directly proportional to gate length. Therefore, reducing the transistor gate length is essential for achieving the low threshold voltage. Al_2O_3, HfO_2 and ZrO_2 are considered as the suitable high-k dielectric materials for developing III-V QWFETs, and a high-quality metal gate/high-k interface is needed to achieve high performance. Atomic layer deposition (ALD) process can be used to obtain a good quality metal gate/high-k interface (Xue et al. 2012; Radosavljevic et al. 2008; Bolognesi et al. 1999; Suman et al. 2007; Lee et al. 2014; Lin et al. 2014). In digital logic integrated circuit applications, the transistors functions as switches. Therefore, high switching speed, low switching energy, low ON-state resistance and high OFF-state resistance are the key parameters used to determine the suitability of transistors for digital logic integrated circuit applications. Modern digital logic integrated circuits are based on a combination of FETs with complementary electrical characteristics.

High on current and high transconductance is required for achieving high switching speed, and to reduce power consumption, OFF-state leakage current of the transistors should be minimum. A high sheet charge concentration in the quantum well is essential for obtaining high on current. The OFF-state leakage current depends on subthreshold swing which measures the sharpness of the reduction of the drain current below threshold. A thin channel is desirable for achieving a lower subthreshold swing (Kim et al. 2015; Lin et al. 2015a, b, 2019; Tae-Woo et al. 2015; Ashley et al. 2007; Hwang et al. 2011; John et al. 1994; Jianqiang et al. 2016; Taewoo and Dae-Hyun 2015). Subthreshold swing and drain induced barrier lowering (DIBL) are the two key parameters which are used for analyzing the short channel effects of QWFETs. It is found that both subthreshold swing and DIBL increase with decrease in gate length of the transistor that indicates the fact that as the transistor size becomes smaller and smaller short channel effects become severe (Taewoo and Dae-Hyun 2015; Jaydeep and Kaushik 2008; Kharche et al. 2011).

In short channel devices, the apparent channel mobility can be calculated using Matthiessen's law (Yang et al. 2011; Jianqiang et al. 2016) which is given below.

$$\mu_{\text{app}} = \frac{\mu_{\text{eff}}\mu_{\text{B}}}{\mu_{\text{eff}} + \mu_{\text{B}}} \tag{7.3}$$

μ_{app} apparent mobility
μ_{B} ballistic mobility
μ_{eff} effective carrier mobility
μ_{app} can also be expressed as (Yang et al. 2011)

$$\mu_{\text{app}} = \frac{L_{\text{eff}}}{(R_{\text{ON}} - R_{\text{EXT}})qN_{\text{S}}} \tag{7.4}$$

The on resistance of QWFET can be calculated as (Yang et al. 2011).

$$R_{\text{ON}} = R_{\text{EXT}} + \frac{L_{\text{eff}}}{\mu_{\text{app}}qN_{\text{S}}} \tag{7.5}$$

R_{ON} on resistance
R_{EXT} external parasitic resistance
L_{eff} length of the flat portion of the gate directly above the channel
N_{S} Carrier density in the quantum well

7.2 InSb QWFETs

The heterostructure of n-channel and p-channel InSb QWFETs is shown in Fig. 7.2. The heart of InSb QWFET is an InSb quantum well. InSb material has a band gap energy of 0.18 eV and a room temperature electron mobility of over 30,000 cm^2/Vs at a sheet charge density of $10^{12}/cm^2$. Some of the important properties of InSb are given in Table 7.2. Se, S and Te can be used as donor impurities in InSb or AlInSb materials. Be, Cd and Cr can be used as acceptor impurities in InSb or AlInSb materials. InSb QWFET can be grown on a silicon or GaAs wafer. AlInSb layer can be used as buffer and interfacial layers which are used to reduce the effects of lattice mismatches between quantum well and wafer materials. AlInSb spacer layer isolates the dopants

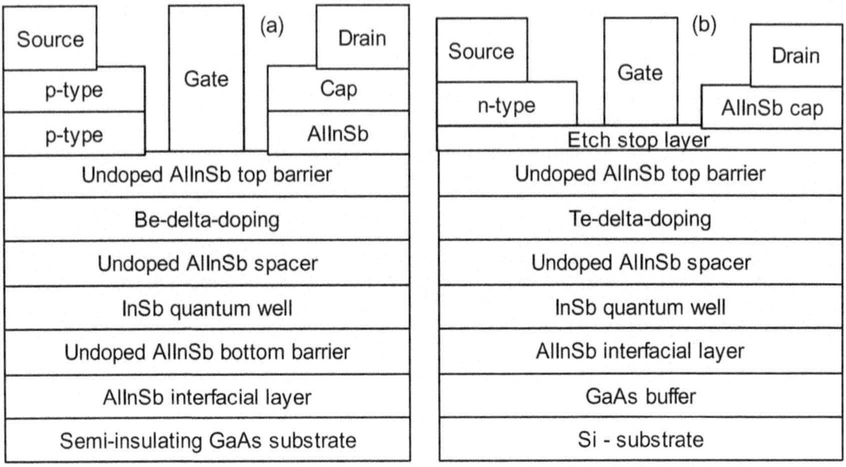

Fig. 7.2 Schematic of **a** InSb p-channel QWFET (Radosavljevic et al. 2008) **b** InSb n-channel QWFET

Table 7.2 Important properties of InSb material at room temperature

Crystal structure	Zinc blende
Electron effective mass	$0.014\ m_0$
Hole effective mass	$0.43\ m_0$
Electron affinity	4.59 eV
Lattice constant	6.479 Å
Energy gap	0.18 eV
Energy separation between Γ and L valleys	0.51 eV
Energy separation between Γ and X valleys	0.83 eV
Intrinsic carrier concentration	$2 \times 10^{16}/\text{cm}^3$
Intrinsic resistivity	4 mΩ cm
Effective conduction band density of states	$4.2 \times 10^{16}/\text{cm}^3$
Effective valence band density of states	$7.3 \times 10^{18}/\text{cm}^3$

from the quantum well which results in the enhancement of electron mobility in the quantum well. Ti/Au metal stack can be used for making the source, drain and gate contacts. In 2008, M. Radosavljevic et al. from Intel Corporation, Technology and Manufacturing Group, USA, reported a high-performance p-channel InSb QWFET that features a compressively strained InSb quantum well.

The characteristics of p-channel InSb QWFET is shown in Fig. 7.3. The drain current of InSb QWFET can be computed as (Jaydeep and Kaushik 2008).

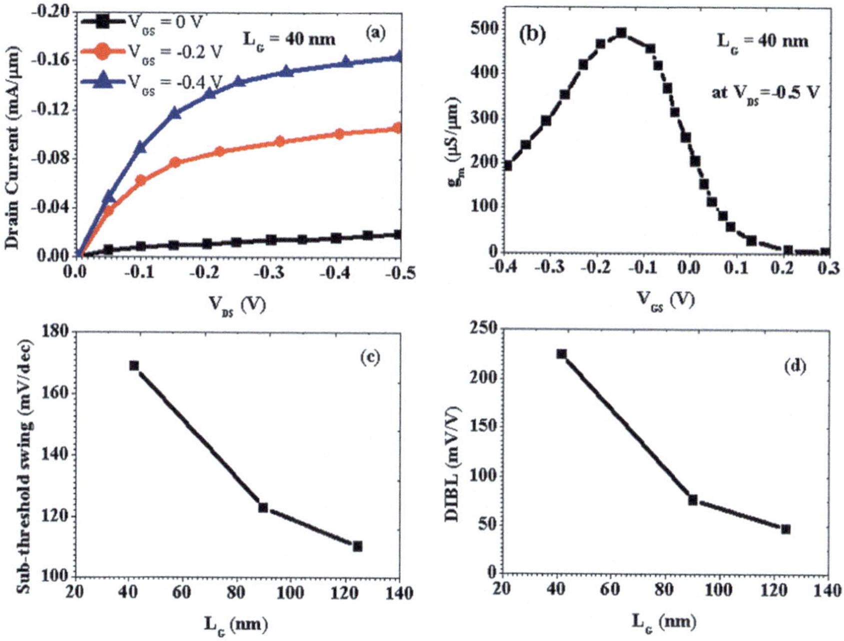

Fig. 7.3 $L_G = 40$ nm p-channel InSb QWFET **a** output characteristics **b** transconductance characteristics **c** influence of gate length scaling on subthreshold swing **d** influence of gate length scaling on DIBL (Radosavljevic et al. 2008)

$$I_{ds} = \frac{g_{ch} V_{ds}(1 + \lambda V_{ds})}{\left[1 + \left(\frac{V_{ds}}{V_{sat}}\right)^m\right]^{1/m}} \tag{7.6}$$

I_{ds} drain current
g_{ch} channel conductance including the resistances of source and drain
V_{ds} voltage between drain and source
V_{sat} saturation voltage at drain
λ fitting parameter related to the finite output conductance in the saturation region
m parameter that determine the shape of the output characteristics in the knee region

$$g_{ch} = \frac{g_{chi}}{1 + g_{chi}(R_S + R_D)} \tag{7.7}$$

$$g_{chi} = \frac{q n_{stot} W \mu}{L} \tag{7.8}$$

q electron charge
g_{chi} intrinsic channel conductance
R_S source resistance
R_D drain resistance
n_{stot} total surface electron sheet density
W channel width
L length of the channel
μ mobility at low field

$$n_{stot} = \frac{n_s}{\left[1 + \left(\frac{n_s}{n_{max}}\right)^{\gamma}\right]^{1/\gamma}} \tag{7.9}$$

$$n_s = 2n_0 \ln\left[1 + \frac{1}{2}\exp\left(\frac{V_{GS} - V_T}{\eta V_{th}}\right)\right] \tag{7.10}$$

γ fitting parameter for the transition to the saturation region
n_s sheet charge density in the channel
n_{max} maximum sheet carrier density
n_0 sheet charge density at threshold
V_T threshold voltage
η body effect parameter
V_{th} thermal voltage

The drain current of InSb QWFETs isdirectly proportional to sheet charge density in the quantum well, width of the gate, gate voltage and inversely proportional to source and drain parasitic resistances and gate length. Therefore, downscaling is essential to increase the drain current of QWFETs.

Meyer's capacitance model can be used for computing gate to source capacitance (C_{GS}) and gate to drain capacitance (C_{GD}) of the InSb QWFETs. These parasitic capacitances play a very important role in determining the switching speed of the QWFETs. For using InSb QWFET as a logic transistor, its parasitic resistances (R_S and R_D) and parasitic capacitances (C_{GS} and C_{GD}) should be minimum (Jaydeep and Kaushik 2008).

$$C_{GS} = \frac{2}{3}C_{ch}\left[1 - \left(\frac{V_{GS} - V_T - V_{DSe}}{2(V_{GS} - V_T) - V_{DSe}}\right)^2\right] \tag{7.11}$$

$$C_{GD} = \frac{2}{3}C_{ch}\left[1 - \left(\frac{V_{GS} - V_T}{2(V_{GS} - V_T) - V_{DSe}}\right)^2\right] \tag{7.12}$$

$$V_{DSe} = V_{DS}, \quad \text{for } V_{DS} < V_{GS} - V_T$$

$$V_{DSe} = V_{GS} - V_T, \quad \text{for } V_{DS} > V_{GS} - V_T$$

$$C_{ch} = WLq\frac{dn_s}{dV_{GS}} \approx \frac{C'_{ch}}{\left[1 + (n_s/n_{max})^{\gamma}\right]^{1/\gamma}} \tag{7.13}$$

$$C'_{ch} = C_i\left[1 + 2\exp\left(-\frac{(V_{GS} - V_T)}{\eta V_{th}}\right)\right]^{-1} \tag{7.14}$$

$$C_i = \frac{WL\varepsilon_i}{d_i} \tag{7.15}$$

C_i insulator capacitance
ε_i permittivity of barrier layer
d_i thickness of barrier layer

InSb QWFETs are considered as the most suitable devices for future digital logic integrated circuit applications (Radosavljevic et al. 2008; Ashley et al. 2007; Jaydeep and Kaushik 2008).

7.3 InGaAs QWFETs

High drain current, high transconductance, large I_{ON}/I_{OFF} ratio, high cut off and maximum oscillation frequencies, low noise, low leakage current, low subthreshold swing and low DIBL are the major requirements of a transistor that can be used for digital logic integrated circuit applications. InGaAs channel-based QWFETs are considered as one of the most desirable transistor technologies for future high-performance digital logic integrated circuit applications due to their excellent scalability, good immunity to short channel effects, outstanding drain current and transconductance, low operating voltage, high-speed and low-noise characteristics. Figure 7.4 shows the ITRS pointing out the need for III-V compound semiconductor-based devices for future high-speed low-power applications. The epitaxial layer structure of InGaAs QWFET is shown in Fig. 7.5. The realization of a reliable T-gate is the critical challenge in the development of InGaAs QWFETs. InGaAs QWFET can also be realized using high-k dielectric materials under the gate to reduce the leakage current. The typical gate structures used in InGaAs QWFETs are T-gate, Γ-gate, Y-gate and rectangular gate. Among the above-mentioned gate structures, T-gate is suitable for high-performance applications due to reduced parasitic effects. The on-state performance of InGaAs QWFETs depends on indium concentration in the channel. Devices with higher indium content provide high drain current and transconductance. Length and width of the gate, indium concentration in the InGaAs channel, distance between source and drain, thickness of the channel layer (d_c), thickness of the InAlAs barrier (t_{ins}) and side recess spacing (L_{side}) are the key parameters that significantly influence the on-state and OFF-state performance of InGaAs QWFETs.

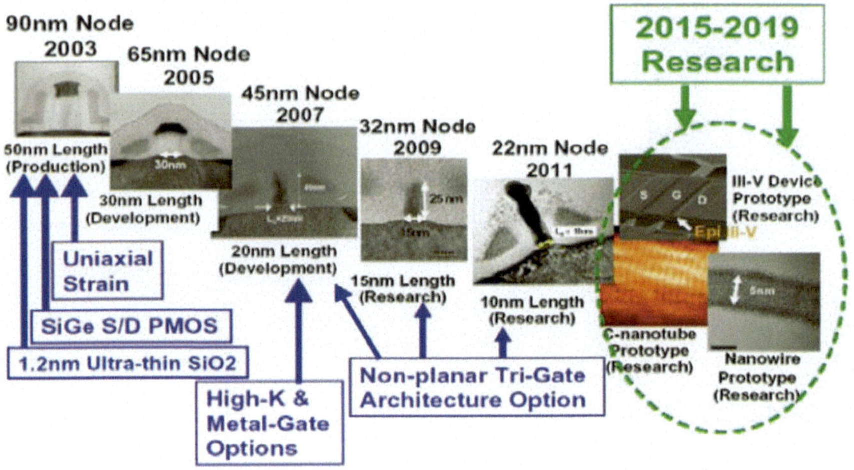

Fig. 7.4 International technology roadmap for semiconductors (ITRS) showing the research interest for the period 2015–2019

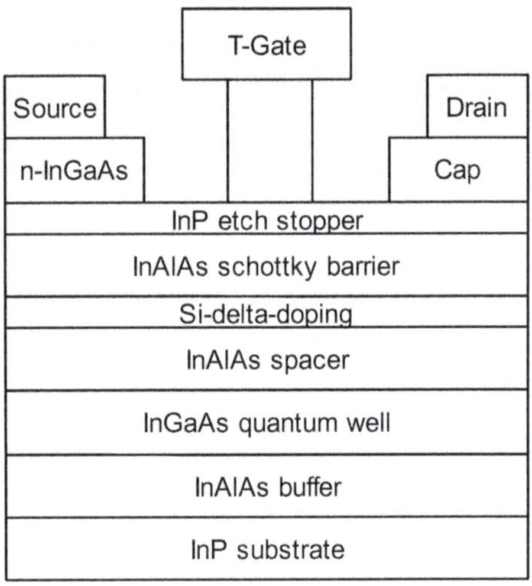

Fig. 7.5 Heterostructure of InGaAs channel-based QWFET

GaAs and InPare are the two widely used substrates which can be used for growing InGaAs QWFETs. The properties of GaAs and InPsemiconductors are given in Table 7.3.

InGaAs and InAs are considered as the suitable channel materials for future logic transistors, and their properties are given in Table 7.4.

Kim et al. (2010) studied the logic characteristics of 40 nm gate length QWFET and found that the channel thickness significantly affects the logic behavior of the

Table 7.3 Important properties of GaAs and InP semiconductors (at 300 K)

Parameter	GaAs	InP
Dielectric constant (static)	12.9	12.5
Dielectric constant (at high frequency)	10.89	9.61
Effective mass of electron (m_e)	0.063 m_0	0.08 m_0
Effective mass of hole (m_h)	0.51 m_0	0.6 m_0
Electron affinity (χ)	4.07 eV	4.38 eV
Lattice Constant	5.653 Å	5.868 Å
Electron mobility (μ_n)	8500 cm^2/Vs	5400 cm^2/Vs
Hole mobility (μ_p)	400 cm^2/Vs	200 cm^2/Vs
Band gap (E_g)	1.424 eV	1.344 eV
Intrinsic carrier concentration	$2.1 \times 10^6/cm^3$	$1.3 \times 10^7/cm^3$
N_C	$4.7 \times 10^{17}/cm^3$	$5.7 \times 10^{17}/cm^3$
N_V	$9 \times 10^{18}/cm^3$	$1.1 \times 10^{19}/cm^3$
Diffusion coefficient of electron (D_n)	200 cm^2/S	130 cm^2/S
Diffusion coefficient of holes (D_p)	10 cm^2/S	5 cm^2/S
Radiative recombination coefficient	7.2×10^{-10} cm^3/S	1.2×10^{-10} cm^3/S
Auger recombination coefficient	10^{-30} cm^6/S	9×10^{-31} cm^6/S
Thermal expansion constant	5.73×10^{-60} C^{-1}	4.60×10^{-60} C^{-1}
Intrinsic resistivity	3.3×10^8 Ω cm	8.6×10^7 Ω cm
Refractive index	3.3	3.1

QWFETs. The effect of channel thickness scaling on the logic performance of QWFET on InP substrate is shown in Fig. 7.6. A higher channel thickness provides high drain current and transconductance but exhibits severe short channel effects. But reducing channel thickness significantly reduces the subthreshold swing and DIBL which are highly desirable for logic transistors.

Li-Dan et al. (2014) derived an expression for calculating the transconductance of a InGaAs QWFET on InP substrate which is given in Eqs. (7.16) and (7.17).

$$g_{m\_int} = \frac{\mu_n W \varepsilon_n \varepsilon_0}{L_g.d_{GC}}.V_{DS} + \frac{\mu_n W \varepsilon_n \varepsilon_0}{L_g.d_{GC}}(V_{GS} - V_T) \tag{7.16}$$

$$g_{m\_ext} = \frac{g_{m\_int}}{1 + g_{m\_int}.R_S + g_d(R_S + R_D)} + \frac{g_{m\_int}}{1 + g_{m\_int}.R_S} \tag{7.17}$$

where

g_{m_int} internal transconductance
g_{m_ext} external transconductance
W width of the gate
ε_n dielectric constant of the semiconductor between channel and gate
ε_0 dielectric constant of air

Table 7.4 Important properties of $In_{0.7}Ga_{0.3}As$ and InAs semiconductors (at 300 K)

Parameter	$In_{0.7}Ga_{0.3}As$	InAs
Dielectric constant (static)	13.42	15.15
Dielectric constant (at high frequency)	11.32	12.3
Effective mass of electron (m_e)	0.0345 m_0	0.023 m_0
Effective mass of hole (m_h)	0.48 m_0	0.41 m_0
Electron affinity (χ)	4.319 eV	4.9 eV
Lattice constant	5.93 Å	6.05 Å
Electron mobility (μ_n)	20000 cm^2/Vs	40000 cm^2/Vs
Hole mobility (μ_p)	400 cm^2/Vs	500 cm^2/Vs
Band gap (E_g)	0.62 eV	0.36 eV
Intrinsic carrier concentration	$7 \times 10^{11}/cm^3$	$1.0 \times 10^{15}/cm^3$
N_C	$1.5 \times 10^{17}/cm^3$	$8.7 \times 10^{16}/cm^3$
N_V	$7.5 \times 10^{18}/cm^3$	$6.6 \times 10^{18}/cm^3$
Electron diffusion coefficient (D_n)	188.7 cm^2/S	1000 cm^2/S
Hole diffusion coefficient (D_p)	0.58 cm^2/S	13 cm^2/S
Coefficient of radiative recombination	0.96×10^{-10} cm^3/S	1.1×10^{-10} cm^3/S
Coefficient of Auger recombination	7×10^{-29} cm^6/S	2.2×10^{-27} cm^6/S
Thermal expansion constant	4.56×10^{-60} C^{-1}	4.60×10^{-60} C^{-1}
Refractive index	3.398	3.51

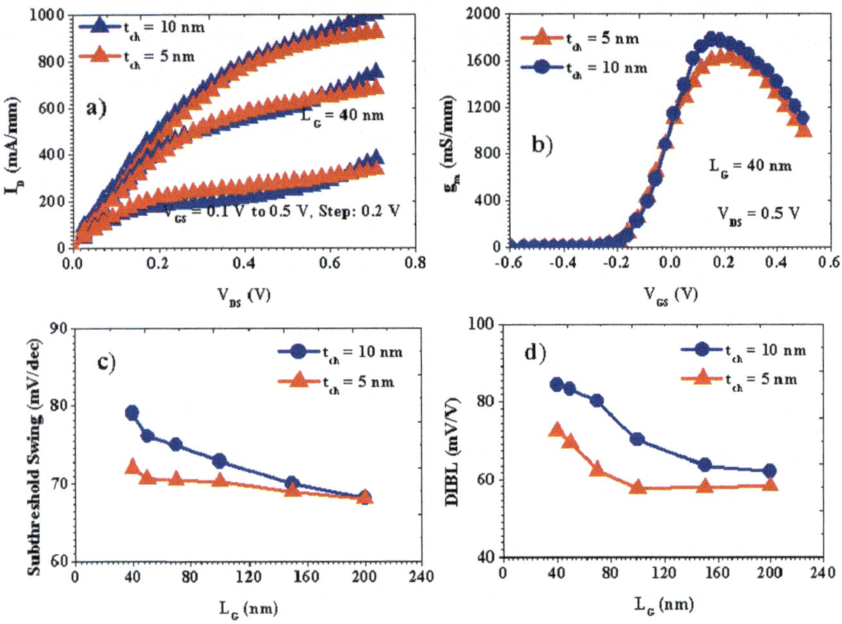

Fig. 7.6 Influence of channel thickness on the performance of QWFET (Kim et al. 2010)

L_g length of the gate
d_{GC} distance between channel and gate
R_S resistance of the source
R_D resistance of the drain

Gate to channel spacing and source/drain parasitic resistances also significantly affect the transconductance of the InGaAs QWFETs. A lower gate to channel spacing is required to improve the transconductance. The speed of QWFET can be measured in terms of cutoff frequency (f_T). A logic transistor with high f_T is desirable for digital logic integrated circuit applications. From Eq. (7.18), it is understood that devices with higher transconductance is required for obtaining high f_T. The cutoff frequency also depends on the gate to source and gate to drain parasitic capacitances. Reducing the parasitic resistances and capacitances associated with source, gate and drain is essential for improving the speed of the transistor.

$$f_T = \frac{g_m}{2\pi(C_{GS} + C_{GD})} \tag{7.18}$$

$$g_m = g_{m\_int} + g_{m\_ext} \tag{7.19}$$

Side recess spacing in the transistor also plays a vital role on the logic characteristics of the InGaAs QWFETs. A reduced side recess spacing is highly suitable for achieving higher transconductance and drain current. However, it also increases the subthreshold current and gate leakage current which results in the increase of power dissipation. Therefore, the side recess spacing must be optimized to improve the logic characteristics of the transistor. The influence of side recess spacing on the logic performance of InGaAs QWFET is shown in Fig. 7.7. The channel aspect ratio of QWFETs can be computed as

$$\alpha = \frac{L_G}{d_{GC} + t_{ch}} \tag{7.20}$$

where

α channel aspect ratio
L_G length of the gate
d_{GC} spacing between gate and channel
t_{ch} channel thickness

For better logic performance, the aspect ratio must be greater than unity.

Resistance of the side recess region (R_{side}) can be calculated as (Suemitsu et al. 1999).

$$R_{side} = \frac{L_{side}}{q.\mu_n.n_s} \tag{7.21}$$

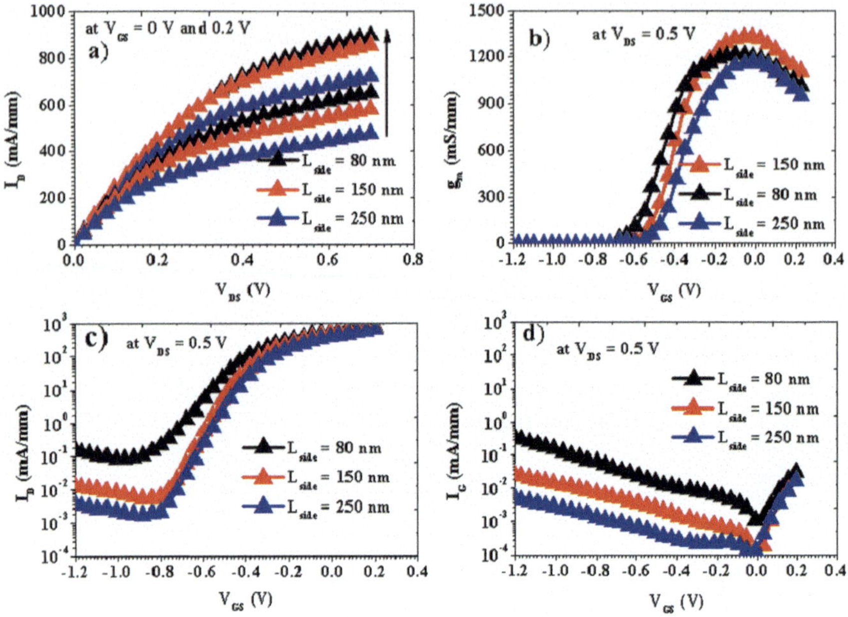

Fig. 7.7 Influence of side recess spacing on the performance of QWFET (Kim et al. 2006)

Side recess region resistance is directly proportional to the side recess spacing. Therefore, a low side recess spacing is preferable for achieving high drain current and transconductance which are the key requirements of a logic transistor. The barrier thickness (t_{ins}) significantly affects the behavior of InGaAs QWFETs. Reducing the barrier thickness results in the increase of subthreshold leakage current and gate leakage current (Fig. 7.8a, b). However, reduction of barrier thickness along with the downscaling of transistor size is found to be effective in improving drain current and transconductance. Therefore, an optimized barrier thickness is required for logic transistors. Decreasing the gate length results in the increase of DIBL and subthreshold swing (SS). The reduction of barrier thickness along with gate length down scaling can effectively minimize the DIBL and SS (Fig. 7.9a, b). Reducing the gate length and barrier thickness also helps to improve the peak transconductance of the QWFETs (Fig. 7.9c). However, decreasing the barrier thickness has the disadvantage of increased threshold voltage (V_{T}) (Fig. 7.9d).

Another requirement of logic transistor is low noise, and the parameter which can be used for measuring the noise performance of a transistor is called minimum noise figure (NF_{min}).

The NF_{min} of a QWFET can be computed as (Takahashi et al. 2012)

$$\text{NF}_{\text{min}} = 10\log\left(1 + 2\pi.K_{\text{f}}.f(C_{\text{GS}} + C_{\text{GD}})\sqrt{(R_{\text{G}} + R_{\text{S}})/g_{\text{m}}^{\text{int}}}\right) \tag{7.22}$$

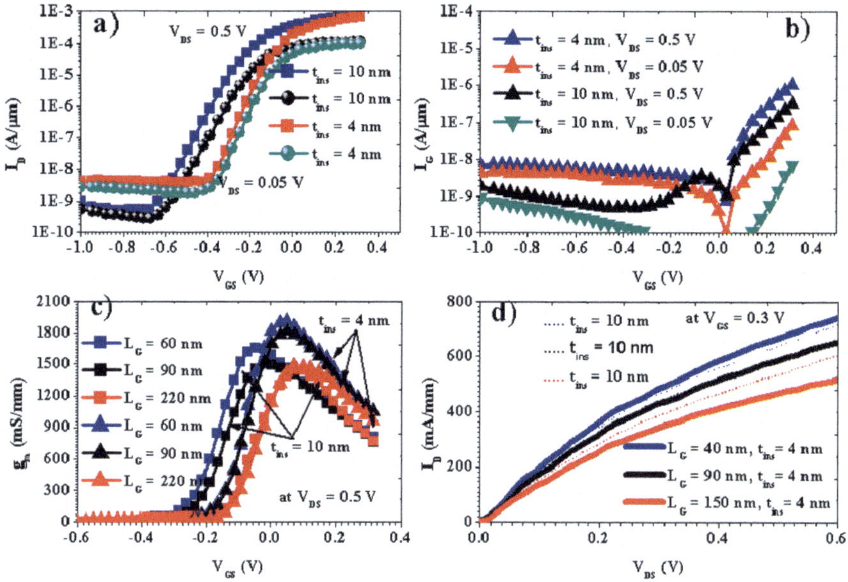

Fig. 7.8 Influence of barrier thickness on the performance of QWFET (Kim and Del Alamo 2007, 2010)

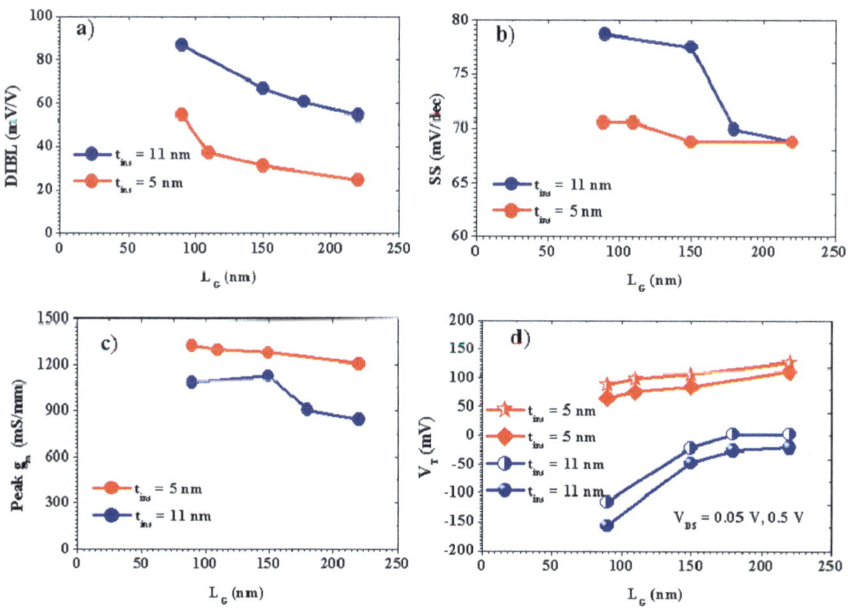

Fig. 7.9 Influence of gate length scaling on the performance of QWFET (Waldron et al. 2007)

K_f fitting factor
f operating frequency
R_G gate resistance

Noise performance of the transistors can be improved by minimizing the device parasitic. By introducing a cavity structure in the gate region, the parasitic capacitances (C_{GS} and C_{GD}) can be effectively minimized. The employment of a cavity structure at the gate has the additional benefit of increased cutoff frequency. Increasing the cutoff frequency leads to the improvement of speed of operation of logic transistor.

Figure 7.10a depicts the influence of gate to channel spacing on the cutoff frequency of InGaAs QWFETs, and from the plot, it is evident that a low gate to channel spacing is required to achieve higher cutoff frequencies. This is because the reduction of gate to channel spacing significantly improves the carrier velocity in the quantum well. When the gate to channel spacing is reduced, the gate has a better control over the channel due to the increased electric field across the gate-channel area. The analytical expression for computing the threshold voltage of a QWFET is given below.

$$V_T = \Phi_B - \frac{\Delta E_C}{q} - \frac{q.N(\delta).d_{GC}}{\varepsilon} \tag{7.23}$$

Φ_B schottky barrier height
ΔE_C conduction band discontinuity between quantum well and the barrier layer
$N(\delta)$ delta doping concentration
d_{GC} gate to channel spacing

E-Mode transistors are highly suitable for digital logic integrated circuit applications, and these transistors exhibit a positive threshold voltage. The factors determining the threshold voltage of a QWFET are schottky barrier height, delta doping

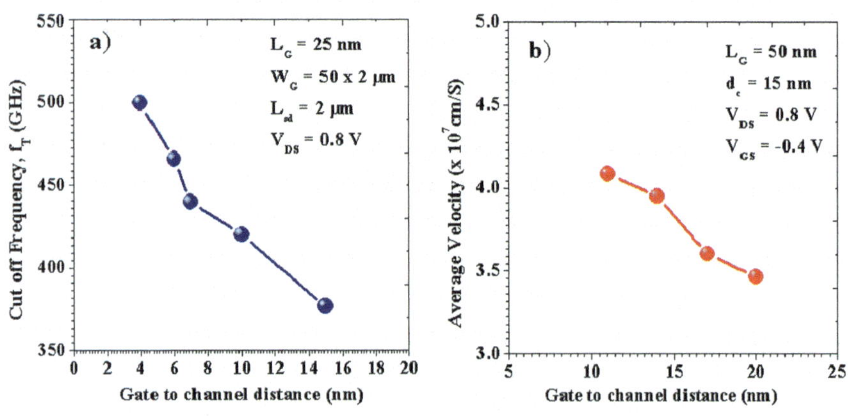

Fig. 7.10 Influence of gate to channel spacing on the performance of QWFET (Endoh et al. 2003)

concentration, gate to channel spacing and conduction band discontinuity. For obtaining positive threshold voltage, schottky barrier height should be maximum, ΔE_C should be minimum, and doping density and gate to channel spacing also should be minimum. However, reducing the doping concentration leads to the degradation of peak transconductance and drain current. Gate metals with higher work function can provide high schottky barrier height. Some of the important gate metals and their work functions are given in Table 7.5.

The resistance between source and drain of the QWFETs can be computed as (Suemitsu et al. 1999)

$$R_{SD} = R_{Sheet}(L_{SG} + L_{GD}) + 2R_{contact} + 2R_{side} + R_{sd,gate} \tag{7.24}$$

where

R_{Sheet} resistance of sheet
$R_{contact}$ resistance at contact
L_{SG} length between source and gate
L_{GD} length between gate and drain
R_{side} side etched region resistance
$R_{sd,gate}$ resistance of the intrinsic channel

For a logic transistor, the ON resistance should be minimum and OFF resistance should be maximum. R_{SD} contributes the major part of ON resistance.

$$f_T = \frac{g_m}{2\pi} \cdot \frac{1}{(C_{GS} + C_{GD})\left(1 + \frac{R_S + R_D}{R_{SD}}\right) + g_m . C_{GD}(R_S + R_D)} \tag{7.25}$$

The relationship between cutoff frequency and device parasitics is given in Eq. (7.25) (Yamashita et al. 2002; Suemitsu et al. 1998; Saranovac et al. 2017). Equation (7.25) reveals the fact that for obtaining high speed, the parasitic resistances and capacitances of the logic transistor should be minimum. Figure 7.11a, b shows the output and transconductance characteristics of E-Mode InGaAs QWFETs. Figure 7.11 also pointed out that smaller transistors are suitable for future digital logic integrated circuit applications due to their high drain current and transconductance.

Table 7.5 Gate metals and their work functions for InGaAs QWFETs

Gate metal	Work function (eV)
Platinum (Pt)	5.65
Gold (Au)	5.2
Nickel (Ni)	5.15
Palladium (Pd)	5.1
Molybdenum (Mo)	4.6
Titanium (Ti)	4.1

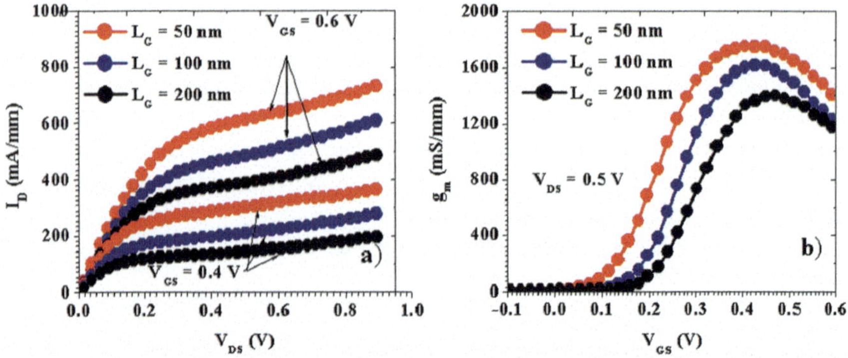

Fig. 7.11 Influence of down scaling on the performance of E-mode QWFET (Kim et al. 2010)

The expected threshold voltages of future E-Mode logic transistors are 0.1 V or below for high-speed operation.

7.4 Buried Platinum Technology and Composite Channels

InGaAs, InAs and InSb are considered as the most suitable channel materials for high-speed low-noise and low-power digital logic integrated circuit applications. The band gap of InAs and InSb is 0.36 and 0.18 eV, respectively. The band gap of InAs and InSb is very low compared with silicon (1.12 eV) and $In_{0.53}Ga_{0.47}As$ (0.72 eV). The poor band gap of InAs and InSb leads to increased leakage currents in QWFETs. Therefore, a composite channel structure can be employed to address this issue. An example of a QWFET using composite channel is shown in Fig. 7.12 which consists of InGaAs upper sub-channel, InAs core channel and a InGaAs lower sub-channel. This composite channel structure combines the advantages of both InGaAs and InAs channel materials. High-K dielectric materials like HfO_2, Al_2O_3 or ZrO_2 can be placed beneath the gate to further reduce the gate leakage. Adding a highly doped vertical source and drain regions can further improve the electrical performance of the QWFETs. The doped source and regions are found to be effective in minimizing source and drain parasitic resistances which result in the increase of drain current, peak transconductance, and it also helps to reduce the noise effects. The vertical source and drain regions also introduce a lateral strain in the channel because of the lattice mismatch between composite channel layers and the vertical source and drain regions. This lateral strain significantly improves the electron mobility in the quantum well. The performance of the composite channel QWFETs can be further improved by adopting a double delta (δ) doping process. The use of δ-doping layers on either side of the composite channel can effectively increase the sheet charge density in the quantum well which results in the increase of drain current and peak transconductance.

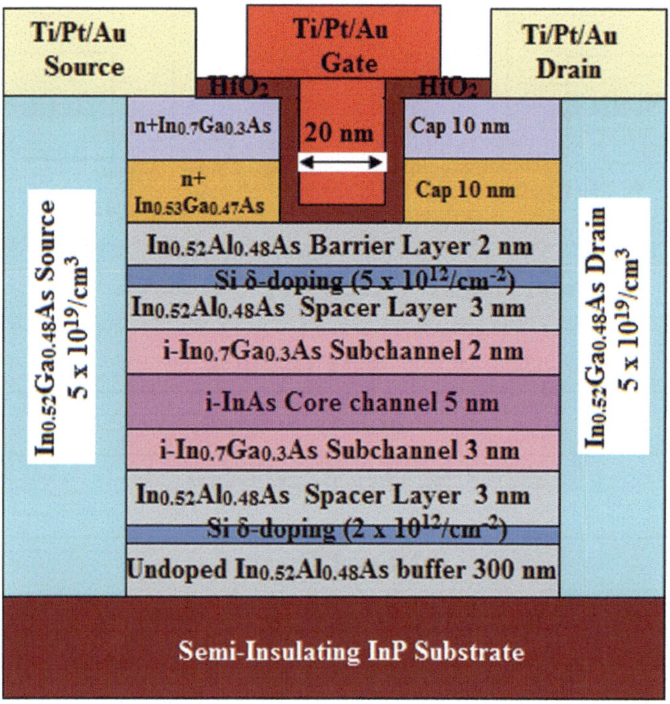

Fig. 7.12 Structure of composite channel QWFET (Ajayan et al. 2017)

III-V QWFETs have been considered as one of the most attractive transistor technology for sub-10 nm CMOS node due to their unique characteristics such as low operating voltage and high switching speed. These outstanding characteristics come from the lower electron effective mass in InAs and InSb quantum wells. InAs QWFETs have the disadvantage of high OFF-state leakage current due to the narrow band gap of InAs channel material. At room temperature, InAs material can offer an electron mobility of over 20,000 cm^2/Vs. The band-to-band tunneling (BTBT) and impact ionization are the two major reasons for this high OFF-state leakage current. In order to reduce the OFF-state leakage current, the following techniques can be used.

1. The lateral spacing between drain and gate can be increased. But this method has the disadvantage of low integration density.
2. Use of an intrinsic vertical spacer layer between the quantum well and the source/drain regions.
3. Raised or regrown source/drain technology can effectively minimize the OFF-state leakage current (see Fig. 7.13).
4. Buried platinum technology (see Fig. 7.14).

The type of doping profile has a strong influence on the BTBT rate and the bipolar gain behavior of the QWFETs.

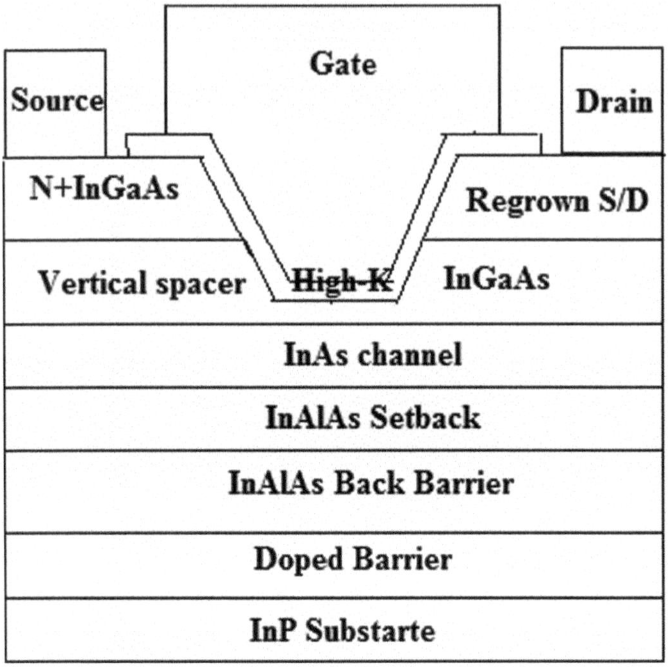

Fig. 7.13 QWFET with regrown source/drain technology

$$\text{Bipolar current gain, } \beta = \frac{I_e}{I_{BTBT}} \tag{7.26}$$

I_e electron current in the channel

δ-doping and uniform doping are the two popular doping types used in QWFETs. Self-aligned fabrication technique can be used for fabricating InAs or InGaAs QWFETs. The popular self-aligned architectures are

1. Recessed gate structure
2. Implanted source and drain
3. Regrown source and drain
4. Metallic source and drain

Among the above-mentioned self-aligned device architecture, recessed gate structure is highly desirable in manufacturing InAs or InGaAs QWFETs due to the ease of fabrication and outstanding scalability. Recessed gate structure can also minimize parasitic which helps to increase the speed of the device. Molecular beam epitaxy (MBE), metal organic chemical vapor deposition (MOCVD) and atomic layer deposition (ALD) are the various techniques which can be used for growing the epitaxial layers of QWFETs. Platinum sinking process or buried platinum metal gate technology provides many advantages like higher Schottky barrier height, reduced gate

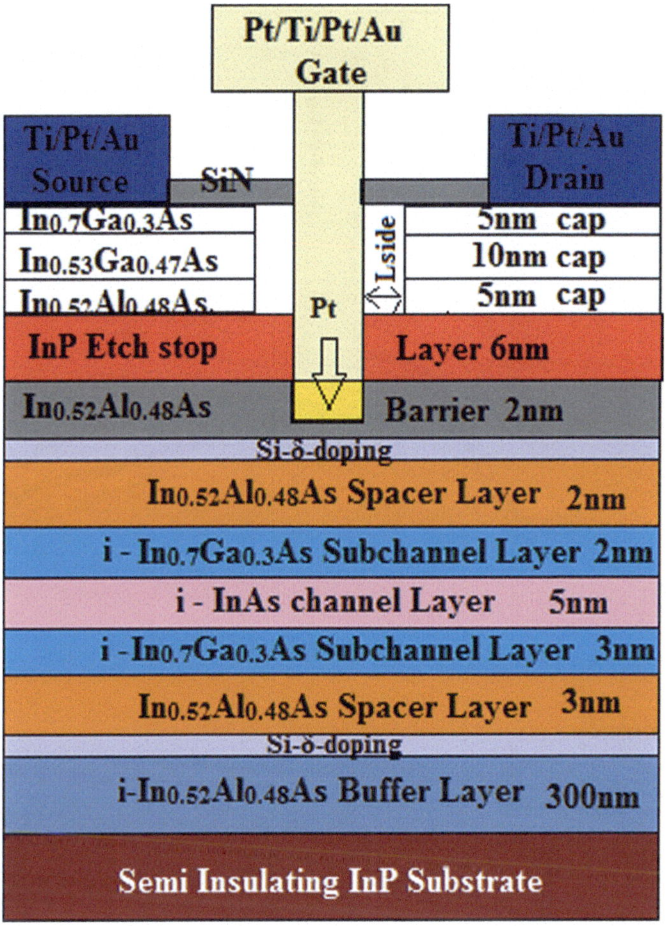

Fig. 7.14 QWFET with buried platinum metal gate technology (Ajayan and Nirmal 2016)

to channel spacing, reduced impact ionization and elimination of kink effect in the output characteristics of QWFETs. Since platinum metal is buried in the InAlAs barrier layer, the effective gate to channel separation is reduced. The impact of various gate metals on the electrical characteristics of QWFETs is shown in Fig. 7.15. Buried platinum metal gate technology provides reduced subthreshold swing, low subthreshold current, low gate leakage current and high peak transconductance which are the key requirements of a logic transistor.

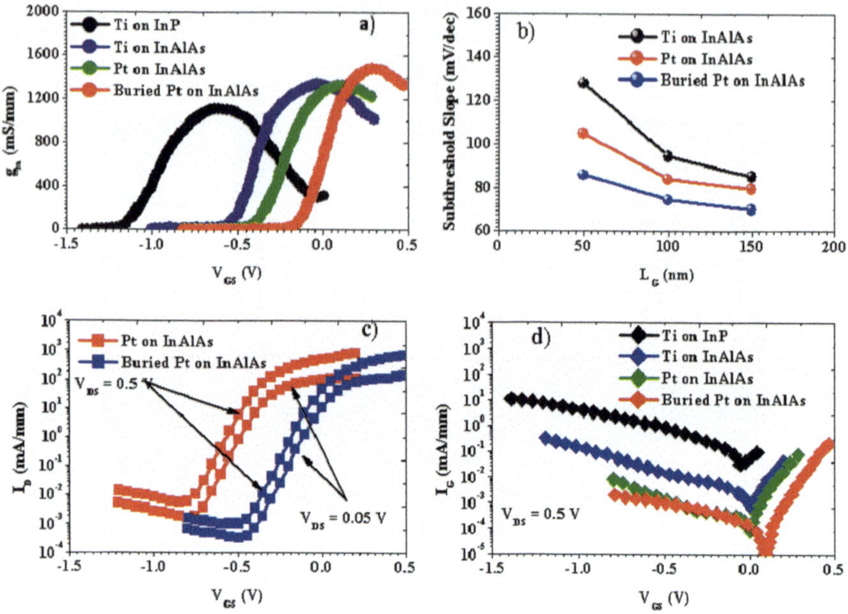

Fig. 7.15 Impact of buried platinum metal gate technology on QWFETs (Kim et al. 2007)

7.5 Summary

This chapter highlights the significance of III-V nanoscale QWFETs for future high-speed low-noise and low-power digital logic integrated circuit applications. As expected, the conventional silicon CMOS scaling is approaching to the end of the roadmap. A simple method of restricting the increasing power consumption that arises from the increase in transistor density in an integrated circuit is to scale down the power supply while maintaining the speed performance. III-V QWFETs are gaining tremendous attention because of their high switching speed and low noise at low operating voltages enabled by outstanding carrier transport properties of new III-V channel materials such as InGaAs, InAs and InSb.

References

Ajayan J, Nirmal D (2015) A review of InP/InAlAs/InGaAs based transistors for high frequency applications. Superlattices Microstruct 86:1–19

Ajayan J, Nirmal D (2016a) 20-nm T-gate composite channel enhancement-mode metamorphic HEMT on GaAs substrates for future THz applications. J Comput Electron 15:1291–1296

Ajayan J, Nirmal D (2016b) 20 nm high performance enhancement mode InP HEMT with heavily doped S/D regions for future THz application. Superlattices Microstruct 100:526–534

Ajayan J, Nirmal D (2017a) 22 nm In0:75Ga0: 25As channel-based HEMTs on InP/GaAs substrates for future THz applications. J Semiconductors 38:27–32

Ajayan J, Nirmal D (2017b) 20-nm enhancement-mode metamorphic GaAs HEMT with highly doped InGaAs source/drain regions for high-frequency applications. Int J Electron 104:504–512

Ajayan J, Subash TD, Kurian D (2017a) 20 nm high performance novel MOSHEMT on InP substrate for future high speed low power applications. Superlattices Microstruct 109:183–193

Ajayan J, Nirmal D, Prajoon P, Pravin JC (2017b) Analysis of nanometer-Scale InGaAs/InAs/InGaAs composite channel mosfets using high-K dielectrics for high speed applications. AEU-Int J Electron Commun 79:151–157

Ajayan J, Ravichandran T, Mohankumar P, Prajoon P, Pravin JC, Nirmal D (2018a) Investigation of DC and RF performance of novel MOSHEMT on silicon substrate for future submillimetre wave applications. Semiconductors 52(16):1991–1997

Ajayan J, Ravichandran T, Mohankumar P, Prajoon P, Pravin JC, Nirmal D (2018b) Investigation of DC-RF and breakdown behaviour in Lg = 20 nm novel asymmetric GaAs MHEMTs for future submillimetre wave applications. AEU-Int J Electron Commun 84:387–393

Ajayan J, Ravichandran T, Prajoon P, Pravin JC, Nirmal D (2018c) Investigation of breakdown performance in Lg = 20 nm novel asymmetric InP HEMTs for future high-speed high-power applications. J Comput Electron 17(1):265–272

Ajayan J, Nirmal D, Ravichandran T, Mohankumar P, Prajoon P, Arivazhagan L, Chandan KS (2018d) InP high electron mobility transistors for submillimetre wave and terahertz frequency applications: a review. Int J Electron Commun 94:199–214

Ajayan J, Nirmal D, Dheena K, Mohankumar P, Arivazhagan L, Augustine Fletcher AS, Subash TD, Saravanan M (2019a) Investigation of impact of gate underlap/overlap on the analog/RF performance of composite channel double gate MOSFETs. J Vac Sci Technol B 37(6):06221

Ajayan J, Nirmal D, Mohankumar P, Arivazhagan L, Saravanan M, Saravanan S (2019b) LG = 20 nm high performance gaas substrate based metamorphic metal oxide semiconductor high electron mobility transistor for next generation high speed low power applications. J Nanoelectron Optoelectron 14(8):1133–1142

Ajayan J, Nirmal D, Mohankumar P, Dheena K, Augustine F, Arivazhagan L, Santhosh Kumar B (2019c) GaAs metamorphic high electron mobility transistors for future deep space-biomedical-millitary and communication system applications: a review. Microelectron J 92:104604

Ashley T, Buckle L, Datta S, Emeny MT, Hayes DG, Hilton KP, Jefferies R, Martin T, Phillips TJ, Wallis DJ, Wilding PJ, Chau R (2007) Heterogeneous InSb quantum well transistors on silicon for ultra-high speed, low power logic applications. Electron Lett 43(14):1–3

Bolognesi CR, Martin WD, David HC (1999) Impact ionization suppression by quantum confinement: effects on the DC and microwave performance of narrow-gap channel InAs/AlSb HFET's. IEEE Trans Electron Devices 46(5):826–832

Del Alamo JA (2011) Nanometre-scale electronics with III–V compound semiconductors. Nature 479:317–323

Del Alamo JA, Antoniadis DA, Lin J, Wenjie L, Alon V, Xin Z (2016) Nanometer-scale III-V MOSFETs. J Electron Devices Soc 4(5):205–214

Endoh A, Yamashita Y, Shinohara K, Hikosaka K, Matsui T, Hiyamizu S, Mimura T (2003) InP based high electron mobility transistors with a very short gate-channel distance. Jpn J Appl Phys 42:2214–2218

Gerben D, Matthias P (2010) Benchmarking of III–V n-MOSFET maturityand feasibility for future CMOS. IEEE Electron Device Lett 31(10):1110–1113

Gilbert D, Mantu KH, Kangho L, Ravi P, Willy R, Marko R, Titash R, Robert C (2008) Carrier transport in high-mobility III–V quantum-well transistors and performance impact for high-speed low-power logic applications. IEEE Electron Device Lett 29(10):1094–1097

Hwang E, Mookerjea S, Hudait MK, Datta S (2011) Investigation of scalability of $In_{0.7}Ga_{0.3}As$ quantum well field effect transistor (QWFET) architecture for logic applications. Solid-State Electron 62:82–89

Iwai H (2009) Roadmap for 22 nm and beyond. Microelectron Eng 86:1520–1528

Jaydeep PK, Kaushik R (2008) Technology circuit co-design for ultra fast InSb quantum well transistors. IEEE Trans Electron Devices 55(10):2537–2545

Jianqiang L, Yufei W, del Alamo JA, Antoniadis DA (2016) Analysis of resistance and mobility in InGaAs quantum-well MOSFETs from ballisticto diffusive regimes. IEEE Trans Electron Devices 63(4):1464–1470

John KZ, Agis AI, Stephen AR, Masselink WT (1994) Transistor performance and electron transport properties of high performance InAs quantum-well FET's. IEEE Electron Device Lett 15(12):489–492

Kharche N, Klimeck G, Kim DH, Del Alamo JA, Luisier M (2011) Multiscale metrology and optimization of ultra-scaled InAs quantum well FETs. IEEE Trans Electron Devices 58(7):1963–1971

Kim D-H, Del Alamo JA (2007) Logic performance of 40 nm InAs HEMTs. In: IEEE international electron devices meeting, IEDM 2007. IEEE, pp 629–632

Kim D-H, Del Alamo JA (2010) Scalability of sub-100 nm InAs HEMTs on InP substrate for future logic applications. IEEE Trans Electron Devices 57:1504–1511

Kim D-H, Del Alamo JA, Lee J-H, Seo K-S (2006) The impact of side-recess spacing on the logic performance of 50 nm InGaAs HEMTs. In: International conference on indium phosphide and related materials conference proceedings. IEEE, pp 177–180

Kim D-H, Del Alamo JA, Lee J-H, Seo K-S (2007) Logic suitability of 50-nm $In_{0.7}Ga_{0.3}As$ HEMTs for beyond-CMOS applications. IEEE Trans Electron Devices 54:2606–2613

Kim T-W, Kim D-H, Del Alamo JA (2010a) Logic characteristics of 40 nm thin-channel InAs HEMTs. In: International conference on indium phosphide & related materials (IPRM). IEEE, pp 1–4

Kim D-H, del Alamo JA, Chen P, Ha W, Urteaga M, Brar B (2010b) 50-nm E-mode $In_{0.7}Ga_{0.3}As$ PHEMTs on 100-mm InP substrate with f max > 1 THz. In: IEEE international electron devices meeting (IEDM). IEEE, pp 30.6.1–30.6.4.34

Kim TW, Koh DH, Shin CS, Park WK, Orzali T, Hobbs C, Maszara WP, Kim DH (2015) Lg = 80-nm trigate quantum-well $In_{0.53}Ga_{0.47}As$ metal–oxide–semiconductor field-effect transistors with Al_2O_3/HfO_2 gate-stack. IEEE Electron Device Lett 36(3):223–225

Kumar MP, Hu CY, Walke AM, Kao KH, Chao TS (2017) Improving the electrical performance of a quantum well FET with a shell doping profile by heterojunction optimization. IEEE Trans Electron Devices 64(9):3563–3568

Lee S, Huang CY, Cohen-Elias D, Thibeault BJ, Mitchell W, Chobpattana V, Stemmer S, Gossard AC, Rodwell MJ (2014) Highly scalable raised source/drain InAs quantum well MOSFETs exhibiting ION = 482 μA/μm at IOFF = 100 nA/μm and VDD = 0.5 V. IEEE Electron Device Lett 35(6):621–623

Li-Dan W, Peng D, Yong-Bo S, Jiao C, Bi-Chan Z, Zhi J (2014) 100-nm T-gate InAlAs/InGaAsInP-based HEMTs with fT = 249 GHz and fmax = 415 GHz. Chin Phys B 23:038501

Lin J, Antoniadis DA, del Alamo JA (2014) Off-state leakage induced by band-to-band tunneling and floating-body bipolar effect in InGaAs quantum-well MOSFETs. IEEE Electron Device Lett 35(12):1203–1205

Lin J, Antoniadis DA, del Alamo JA (2015a) Physics and mitigation of excess off-state current in InGaAs quantum-well MOSFETs. IEEE Trans Electron Devices 62(5):1448–1455

Lin J, Antoniadis DA, del Alamo JA (2015b) Impact of intrinsic channel scaling on InGaAs quantum-well MOSFETs. IEEE Trans Electron Devices 62(11):3470–3476

Lin J, Zhao X, Clavero IM, Antoniadis DA, del Alamo JA (2019) A scaling study of excess OFF-state current in InGaAs quantum-well MOSFETs. IEEE Trans Electron Devices 66(3):1208–1212

Radosavljevic M, Ashley T, Andreev A, Coomber SD, Dewey G, Emeny MT, Fearn M, Hayes DG, Hilton KP, Hudait MK, Jefferies R, Martin T, Pillarisetty R, Rachmady W, Rakshit T, Smith SJ, Uren MJ, Wallis DJ, Wilding PJ, Robert C (2008) High-performance 40 nm gate length InSb P-channel compressively strained quantum well field effect transistors for low-power (VCC = 0.5 V) logic applications. Proceedings of IEDM, pp 727–730

Saranovac T, Hambitzer A, Ruiz DC, Ostinelli O, Bolognesi C (2017) Pt gate sink-in process details impact on InP HEMT DC and RF performance. IEEE Trans Semicond Manuf 30:462–467

Suemitsu T, Enoki T, Yokoyama H, Ishii Y (1998) Improved recessed-gate structure for sub-0.1-μm-gate InP-based high electron mobility transistors. Jpn J Appl Phys 37:1365–1372

Suemitsu T, Yokoyama H, Umeda Y, Enoki T, Ishii Y (1999) High-performance 0.1-μm gate enhancement-mode InAlAs/InGaAs HEMT's using two-step recessed gate technology. IEEE Trans Electron Devices 46:1074–1080

Suman D (2007) III-V field-effect transistors for low powerdigital logic applications. Microelectron Eng 84:2133–2137

Suman D, Dewey G, Fastenau JM, Hudait MK, Loubychev D, Liu WK, Radosavljevic M, Rachmady W, Chau R (2007) Ultrahigh-speed 0.5 V supply voltage $In_{0.7}Ga_{0.3}As$ quantum-well transistors on silicon substrate. IEEE Electron Device Lett 28(8):685–687

Taewoo K, Dae-Hyun K (2015) Scaling and carrier transport behavior of buried-channel $In_{0.7}Ga_{0.3}As$ MOSFETs with Al_2O_3 insulator. Solid State Electron 111:218–222

Tae-Woo K, Hyuk-Min K, SeungHeon S, Chan-Soo S, Won-Kyu P, Eddie C, Manny R, Jae Ik L, Dmitry V, Tommaso O, Dae-Hyun K (2015) Impact of H_2 high-pressure annealing onto InGaAs quantum-well metal–oxide–semiconductor field-effect transistors with Al_2O_3/HfO_2 gate-stack. IEEE Electron Device Lett 36(7):672–674

Takahashi T, Sato M, Nakasha Y, Hirose T, Hara N (2012) Improvement of RF and noise characteristics using a cavity structure in InAlAs/InGaAs HEMTs. IEEE Trans Electron Devices 59:2136–2141

Waldron N, Kim D-H, del Alamo JA (2007) 90 nm self-aligned enhancement-mode InGaAs HEMT for logic applications. In: IEEE international electron devices meeting, IEDM. IEEE, pp 633–636

Xue F, Jiang A, Zhao H, Chen YT, Wang Y, Zhou F, Lee J (2012) Channel thickness dependence of InGaAs quantum-well field-effect transistors with high-κ gate dielectrics. IEEE Electron Device Lett 33(9):1255–1257

Yamashita Y, Endoh A, Shinohara K, Hikosaka K, Matsui T, Hiyamizu S, Mimura T (2002) Pseudomorphic $In_{0.52}/Al_{0.48}/As/In_{0.7}/Ga_{0.3}/As$ HEMTs with an ultrahigh fT of 562 GHz. IEEE Electron Device Lett 23:573–575

Yang L, Mathieu L, Mark SL (2011) Temperature dependence of the transconductancein ballistic III–V QWFETs. IEEE Trans Electron Devices 58(6):1804–1808

YounHo P, Hyun Cheol K, Kyung Ho K, Hyung-Jun K, Suk-Hee H (2009) Spin interaction effect on potentiometric measurements in a quantum well channel. IEEE Trans Magn 45(6):2389–2392

Chapter 8
FinFET: A Beginning of Non-planar Transistor Era

Kajal and Vijay Kumar Sharma

Abstract Aggressive scaling of metal–oxide–semiconductor field-effect transistor (MOSFET) is a barrier in the progress of very large-scale integration (VLSI) technology, and new innovative devices and techniques are always required to boost the electronics industry. Fin-shaped field-effect transistor (FinFET) is the appropriate device to eliminate the limitations of MOSFET devices. FinFET is a three-dimensional (3D) multi-gate transistor with improved channel stability, less short channel effects (SCEs) and excellent isolation compared to the MOS transistor. The best qualities of FinFET that attracts research designers are better SCEs, improved subthreshold slope, less random doping fluctuation and independent gating. Process, voltage and temperature (PVT) variation is one of the scaling problems in MOSFET devices, and due to PVT variations, the circuit shows abnormal power consumption and performance degradation. In this chapter, we concentrate on the influence of PVT variations on different FinFET-based circuits. PVT variations can cause deviation in power consumption, delay and leakage current which finally degrade the performance of FinFET devices.

Keywords FinFET · PVT variations · Ultra-low-power VLSI technology · CMOS

8.1 Introduction

Improvement in VLSI technology is necessary for the betterment of electronic devices. MOSFET device dominated the entire VLSI technology from many years, but now due to further scaling of MOS devices it leads the severe SCEs, subthreshold leakage, more standby power dissipation and reliability variations which drastically affects the circuit performance and reliability of the system (Turi and Delgado-Frias 2017). The main challenge faced by future bulk MOS scaling is process and material technology limitations. Continuous efforts are made by the researchers to expand

Kajal · V. K. Sharma (✉)
School of Electronics and Communication Engineering, Shri Mata Vaishno Devi University, Katra, Jammu and Kashmir 182320, India
e-mail: vijay.buland@gmail.com; 18dec002@smvdu.ac.in

R. Dhiman and R. Chandel (eds.), *Nanoscale VLSI*,
Energy Systems in Electrical Engineering,
https://doi.org/10.1007/978-981-15-7937-0_8

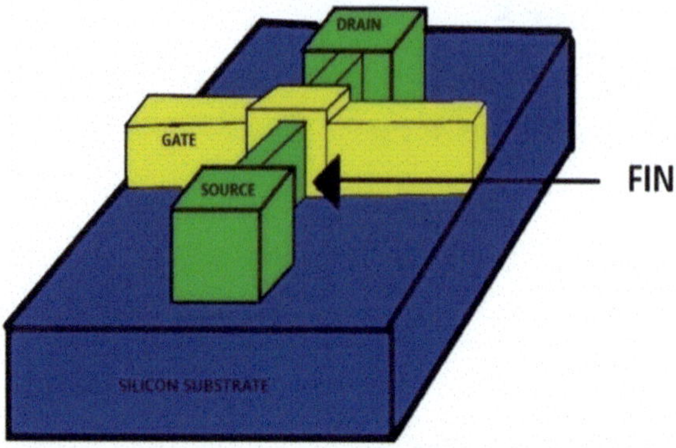

Fig. 8.1 Basic FinFET structure

the silicon scaling results into innovative material and device structures to overcome the limitations of bulk MOSFET. A FinFET is one of these innovations, and FinFET becomes a popular transistor due to its front and back gate structure. The transistor's current and threshold voltage can monitor by biasing these gates properly which helps to manage the problem of standby power dissipation. Figure 8.1 represents the structure of FinFET device (Gupta et al. 2019). Multi-gate transistors are a considerable option for nanoscale VLSI technology. FinFET gains the limelight among all multi-gate MOSFET devices due to its better control to SCEs, lower leakage, excellent isolation and more driving capability for both low-power and fast speed applications.

Most of FinFETs are double-gate devices with vertical fins in the gate. In FinFET, channels are created on both sides of the fin and at the top end. There are no free carriers available because of the finlike structure, so this particular FinFET structure is the main reason for suppressing SCE in FinFET (Zimpeck et al. 2015). Better subthreshold slope, excellent SCE control, independent gating and less random doping fluctuation are the best qualities of FinFET that makes it more superior to MOS technology (Bagheriye et al. 2018). The front and back gate of FinFET provides better control over the channel which in turn reduces the leakage current and SCE, so FinFET is the suitable device to replace the MOS technology in the future VLSI technology (Taghipour and Asli 2017; Mukhopadhyay et al. 2018). Due to the low leakage power of FinFET, it becomes a very popular choice for memories. Memories are used most commonly in digital systems, and a large amount of power is saved in memories with FinFET devices.

8.1.1 Scaling Challenges in MOSFET

Aggressive scaling of MOSFET causes various challenges in VLSI technology, and one of the most prominent drawbacks of MOS scaling is SCE. In a deep submicron region, when the channel length of device is less than 100 nm, SCEs start to degrade the circuit performance and are also known as second-order effects. The key SCEs are hot carrier effect, threshold voltage variations, gate-induced barrier lowering, velocity saturation.

Due to short channel length, subthreshold or weak conduction current occurs between the drain and source in MOS transistor when the gate voltage (V_{GS}) is less than the threshold voltage (V_T). This small leakage current is known as subthreshold leakage current and affects the performance of the transistor. Detail of scaling challenges and its impact on CMOS performance is studied in reference (Jacob et al. 2017). Most portable devices, such as mobile devices, laptops and various communication devices, have long downtime and run in standby mode if not in use. But there is a small leakage current flow through the circuit due to short channel length which causes the standby power dissipation. The researcher had suggested various techniques for overcoming the shortcomings of CMOS transistors (Sharma and Pattanaik 2014). Figure 8.2 represents the main scaling challenges in MOS technology.

Currently, one of the extremely challenging areas of research is to minimize the leakage power consumption, mostly in on-chip devices which are doubling in every two years. It is more challenging to minimize the static leakage power than the dynamic leakage power because, in dynamic power, the leakage power depends on transistors count, their operating status and type without taking into consideration the switching operation. On the other hand, when a transistor is in the OFF state, there are no input applied to a transistor, it has reached a stable state, and a small amount of leakage current flows through the transistor and causes power dissipation

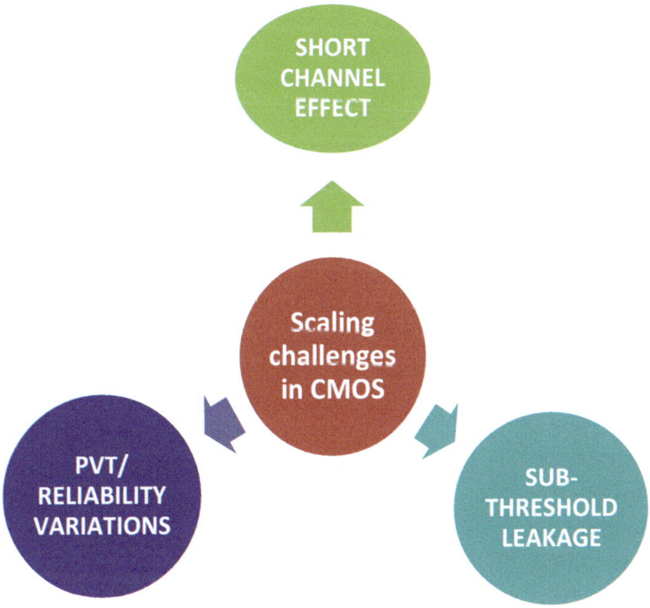

Fig. 8.2 Scaling challenges in MOS technology

(Upasani et al. 2010). There are many advantages of scaling like the compact size of the devices and high speed. Despite this, there are some limitations of scaling in terms of SCEs which cause leakage current hence increasing power dissipation.

Subthreshold leakage current harms the characteristics of the devices and affects the reliability of the devices. PVT variability and reliability effect are major issues in present VLSI technology. One of the most critical and common problems of reliability is negative temperature bias instability (NBTI). NBTI directly challenges the reliability of digital VLSI devices. As a result, the circuit delay exceeds the design specification and there may be timing violations or logic failure (Mahapatra et al. 2013; Khoshavi et al. 2017). Nowadays, electronic devices are facing a problem of PVT variations and it affects the various performance parameters. The electronics industry is moving from MOS transistor to FinFET, but the problem of PVT variations is still present (Sharma and Pattanaik 2014; Yang and Jha 2014).

In this chapter, we are focusing on PVT variations and consider the impact of PVT variability on FinFET devices and various techniques or methodologies adopted by the researcher to mitigate the PVT variability effect.

8.1.2 FinFET Structure and Operation

The further scaling of MOS transistors is not much profitable for both the research community and the VLSI industry. FinFET is an appropriate transistor to take the position of the MOS transistor. Dr. Chenming Hu has been known as the father of the 3D transistor because he has proposed the concept of FinFET in 1999 (Gupta et al. 2019). FinFET is a type of non-planar or 3D multi-gate transistor in which the channel has a thin vertical fin and the gate is fully enclosed around the channel between the drain and the source. It looks like a fin of fish when viewed so its name has been derived from this fact. FinFET channels are created at the topside and two sidewalls of a fin which provide better control on the channel and give better electrostatic control and electrical characteristics (Zimpeck et al. 2015). High channel doping is required for low leakage current in MOS devices, but this degrades the carrier mobility of a transistor. Gate dielectric with high ON current and good channel control is highly on demand for low-power applications. The gate leakage current through thermal oxide becomes escalated as oxide thickness approaching 2 nm (Rosner 2003).

The working operation and fabrication process of FinFET are almost identical to MOS except for some modifications (Walke et al. 2017; Chen et al. 2018). One of the challenges of manufacturing the FinFET is doping of the drain–source junction in the fin region. Uniformly distributed doping is needed along the fin height and width so that angled implantation is required on the side of the fin. In the case of planar device junction formation, there are various standard techniques for analyzing and monitoring the implantation of dopant on the planar surface, but these methods are not suitable for FinFET junction formation due to the 3D structure of a fin (Pham et al.

2006; Lee et al. 2010). Fin height is the most important parameter for the FinFET fabrication process because it determines the minimum FinFET width (W_{min}). A minimum transistor width of two gates FinFET is given below (Gupta et al. 2019):

$$W_{min} = 2H_{fin} + T_{fin} \tag{8.1}$$

Here, H_{fin} is the fin height and T_{fin} silicon body thickness. A fin height has more impact on transistor width than the T_{fin} component as seen from Eq. (8.1). H_{fin} is fixed in a FinFET, so to increase the FinFET width we can create multiple parallel fin structures. The total physical transistor width (W_{total}) of a tied FinFET gate with n parallel fins can be calculated as shown in Eq. (8.2) (Gupta et al. 2019)

$$W_{total} = nW_{min} = n(2H_{fin} + T_{fin}) \tag{8.2}$$

FinFET is designed with multiple parallel fins to achieve larger channel widths (Colinge 2008). The number of fins at FinFET should be increased to increase the current through the transistor (Sinha et al. 2012; Tawfik et al. 2007). A multiple fin structure of FinFET achieved superior performance but increases the device degradation due to hot carrier effect. When FinFET has multiple fins, then coupling effect in the steep and silicon fin decreases the conduction of the inversion channel carrier and degrades the FinFET performance (Yeh et al. 2018). Double-gate FinFET structure is more preferable due to this reason because it improves the electrostatic integrity, reduces the SCE and minimizes leakage current (Yang and Jha 2014).

The three types of FinFET structures are: shorted gate (SG) FinFET, independent gate (IG) FinFET and asymmetric gate work function shorted gate (ASG) FinFET. In SG FinFET, two gates at the top are shorted together and provide a large drive current. ASG FinFET is the same as SG FinFET in case of a layout area, but ASG FinFET having a different work function for both the gates. ASG FinFET provides a lower leakage current but degrades around 26% of ON state current (I_{ON}). If both FinFET gates are controllable independently, then FinFET is called IG FinFET. IG FinFET has less leakage current than SG FinFET, but it increases the layout area and causes severe degradation in ON state current (Bhattacharya et al. 2015; Yang and Jha 2013).

8.2 PVT Variations

PVT variations are one of the scaling challenges faced by the FinFET technology, show abnormal power consumption due to PVT variations and accelerate the degradation of the circuits (Zimpeck et al. 2015). Variations are classified into two categories: process variation and environmental variation. Additional environmental variability involves variation in temperature and supply voltage across the circuit. The key source of variation in supply voltage variation and temperature variation is voltage (IR) drop in power grid and switching activity deviation across the chip, respectively.

The principal cause of process variation (PV) is variations in the physical parameters of devices that take place during a manufacturing process.

8.2.1 Process Variations

PV is introduced during the fabrication process due to unavoidable errors. As VLSI technology moves toward the deep submicron regime, integrated circuits (ICs) become more sensitive to PV. Process variation is divided into two parts: non-systematic and systematic. A variation in the electrical characteristics of two transistors with the same length and width is recognized as systematic variations and can be adjusted by detailed layout analysis during the manufacturing process. On the other hand, non-systematic variation is a non-predictable part of process variation, and these variations are an unexplainable component of the fabrication process. Non-systematic variations are further classified into two parts: intra-die variation and inter-die variation. Inter-die variation is due to the lack of manufacturing control and induced by technical constraints. Deviation in some parameter value over nominally equivalent manufactured dies refers to inter-die variation. Inter-die variations may occur on the different wafers, or same wafer or different lots (Ezz-Eldin et al. 2015). Figure 8.3 shows the classification of variations. Among all these variations, voltage,

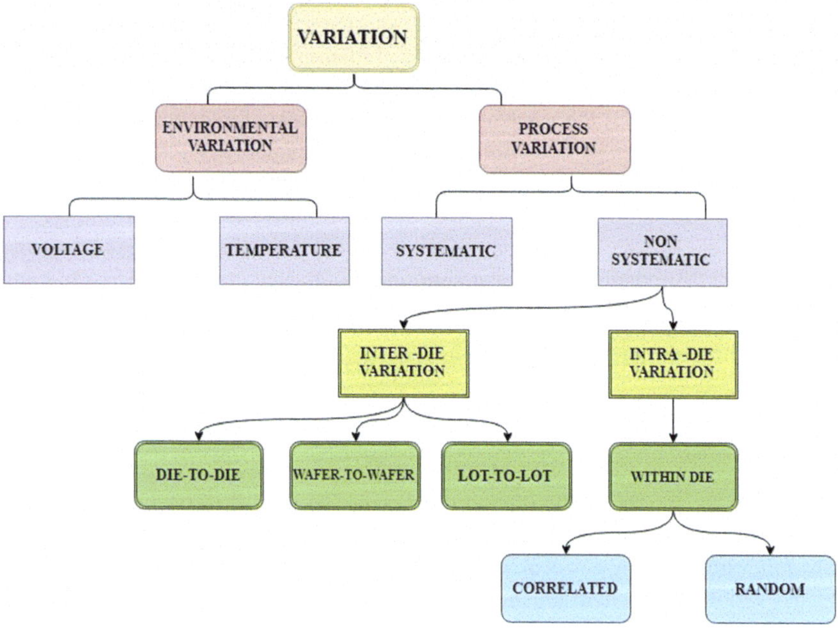

Fig. 8.3 Classification of variation

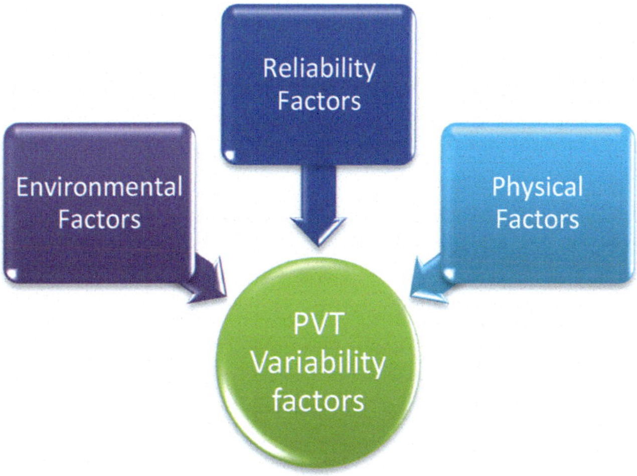

Fig. 8.4 Main factor of PVT variation

temperature and systematic variations can be evaluated and improved by researchers. On the other hand, non-systematic variations are difficult to identify and become unpredictable parts of variations. Intra-die variation affects the various devices on the same die differentially and is further categorized into correlated and random variation. Correlated variation depends on the location of the devices. This closely spaced device has more similar variations than those located far apart. Etching, layout and lithographic information can be required to design, estimate and reimburse for correlated variation. Random variation is considered statistically independent of all other variation components.

Random variation results from edge roughness of gate line and fluctuation of random dopant. Figure 8.4 shows the main factor of PVT variation. During the manufacturing process lithography phase, process variations are mostly induced and the variability in PVT can be divided into three factors:

- Environmental factors: Power supply and temperature fluctuations are the main causes of environmental variations and mainly appeared during the circuit operations.
- Reliability factors: Mainly caused by a transistor aging and the high electrical field in modern circuits.
- Physical factors: Variations in geometric and electrical parameters which induce a lag in transistor performance also trigger process variation.

8.2.2 Supply Voltage Variation

Voltage drops or noisy power sources are the main source of supply voltage variations. Supply voltage variations have a great impact on leakage power, dynamic power and logic gate timing (Yang and Jha 2013). One of the most important parameters of the circuit is supply voltage because it affects the system performance. The gate delay depends upon the saturation current, and saturation current depends upon the supply voltage. FinFET technologies use high-*k*/metal gate stack to boost gate control over channel region, the main source of statistical variations is metal gate granularity, and this contributes to grain orientations that have different work functions. These imperfections can influence the various parameters of FinFET, and the entire block of cells compromises due to variations in transistor structure. Therefore, circuits also suffer from some electrical deviations (Zimpeck et al. 2018; Ban et al. 2014).

8.2.3 Temperature Variations

The temperature of the blocks in IC depends on the power consumption of a block itself and on lateral heat transfer; it also depends on adjacent blocks. A temperature variation comes under the environmental variation factor and mainly causes due to deviations in switching activities of the device. Fluctuations in temperature are dictated by the leakage current and timing characteristics. Due to the unpredictable dopant fluctuation and the sub-wavelength lithography, nanodevices are more susceptible to variability effect. PV directly affects the threshold voltage of FinFET varying various aspects of transistor cells (Almeida et al. 2018). PVT variations are inherent, and essential steps must be taken during the early design step. Figure 8.5 presents the geometric parameters of FinFET which include drain, source and gate (Lee and Jha 2014). To reduce leakage current (I_{OFF}) and improve ON current (I_{ON}), fin engineering is the most essential part during the fabrication process (Yang and Jha 2014). A previous study shows that fin width and gate length have a major impact on I_{ON} and I_{OFF}, but the greatest variance in both currents is due to work function fluctuations (WFFs) that creates a significant deviation in total power that must be considered in the design of VLSI.

PV is inherent in the fabricating processes of semiconductors and impacts on circuit performance and reliability. It is becoming more difficult to determine the circuit performance with the constant change in the circuit elements (logic gates and interconnections). Interconnect variation and gate variation appear to be considering in random variations. Uncertainties in metal line dimensions lead to interconnect variations. A variation in gate process causes change in MOS parameters which create the gates manufactured different from the ones designed. Gate width (W_{GATE}), gate oxide thickness (T_{OX}), gate length (L_{GATE}) and threshold voltage (V_T) are mostly affected parameters by the process variation during the fabrication process (Zimpeck et al. 2015). PV impact translates into variation in device and interconnects electrical

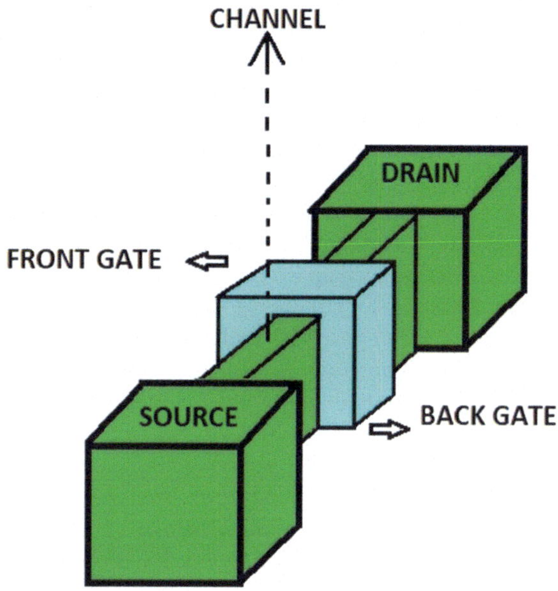

Fig. 8.5 Geometric parameters of FinFET

parameters such as delay, throughput and leakage power variation. FinFET is one of the newest transistors in VLSI technology, and many works are going on.

8.3 Literature Review

Continuous scaling of the MOSFET leads to an increase in aging effect, leakage current and soft error that compels the VLSI technology to move toward the multi-gate devices. FinFET is the best multi-gate device because of its outstanding isolation and high driving capacity for both low-power and high-speed applications. PVT variation is a challenging problem in FinFET and degrades the circuit performance, so researchers have adopted various techniques and methodologies to alleviate this effect. In this part, we are discussing the impact of PVT variations on FinFET devices and various techniques/methodologies adopted by researchers to improve the performance of FinFET devices.

8.3.1 Impact of PVT Variations on FinFET-Based Memories

Memories are always a big part of VLSI digital technology, and applying the FinFET technology to memories introduces an evolution in digital technology due to a huge amount of power-saving. Static random access memory (SRAM) is one of the commonly used memories in VLSI technology, and it always demanded faster design and lower power consumption. Data stability is the biggest problem in the SRAM cell, and this problem becomes more severe with scaling of MOSFET in the sub-nanometer regime. Intra-die and inter-die variations are the main cause of instability in SRAM, so multi-gate devices like FinFET become a better choice for SRAM cell (Kushwah et al. 2016). The impact of FinFET technology on SRAM cells, back gate biasing strategies and performance of SRAM cell under temperature, voltage and parameter variations can be seen in Turi and Delgado-Frias (2017). Leakage power can be reduced up to 65X in a six-transistor FinFET SRAM cell (Tawfik and Kursun 2008).

Researchers proposed IG FinFET SRAM cells that used back gate biasing and PMOS access transistors to achieve high stability performance (Bagheriye et al. 2018). Designers find that FinFET is more appropriate than MOS in deep submicron region especially after 22 nm because of its excellent SCEs, improved sub-threshold slope, independent gating, and less random doping fluctuations. Researchers proposed an architecture-level approach to improve the array robustness, but this type of approach results in area overhead and makes complex circuit design. Researchers also suggested design of FinFET SRAMs based on asymmetric structures like asymmetric drain and source, having different work functions and oxide thickness for the FinFET front and back gate. Those structures are highly sensitive to fluctuations in process parameters. There are various techniques to improve SRAM performance in reference (Bagheriye et al. 2018) and given as:

- Front and back gates are operating independently as they offer flexibility in design as a substitute for threshold voltage control for improving the cell stability of FinFET SRAM.
- A write static noise margin (WSNM) and read static noise margin (RSNM) are enhanced by decreasing and increasing the threshold voltage of an access transistor, respectively, using the independent gate.
- To dynamically increase the RSNM without increasing the area overhead, built-in feedback technique is used in which the back gate of the access transistor is connected to corresponding nodes.
- In some approaches, the p-channel metal–oxide–semiconductor (PMOS) is used in place of the n-channel metal–oxide–semiconductor (NMOS) to improve circuit stability and reduce the risk of leakage current.
- To reduce an access time, strain effect is incorporated with PMOS access transistor in the SRAM cell.
- Schmitt's trigger-based feedback system was used to increase the RSNM, WSNM and tolerance to PV in the subthreshold region, but these cells suffer from low read current and area overhead in this procedure.

The author proposed two cells in reference (Bagheriye et al. 2018); first cell consumes low power and enhances the read and write margins. The second cell provides high write and read margins with high read current. Aging effect due to BTI influences, PVT variations and single event upset is the main issue in nanometer IC design. Read noise margin is the most sensitive SNM and is deeply affected by

PVT variation and aging effect (Almeida et al. 2018). Read operation is performed by a P-type gate and write operation performed with the help of transmission gate to achieve high switching activity in 7T FinFET SRAM. This type of configuration of SRAM provides up to 60.8% of supply voltage reduction (Sneha et al. 2017).

Standard MOS scaling technology driven by higher operating speed, integration density and lower power dissipation has faced many barriers. Now, it is facing a severe variability problem. The researcher introduces a technique for the design of SRAM cell that is aware of variability. The proposed cell's architecture is identical to that of the regular 6T SRAM cell apart from that the access pass gates are replaced with transmission gates. The impact of variation on most of SRAM cell's design metrics degrades circuit efficiency. Comparative analysis based on Monte Carlo simulation shows that the proposed design is capable of greatly mitigating the impact of variation (Islam and Hasan 2012). The detailed comparative investigation of CMOS and FinFET-based 10T SRAM cells is given and can be studied in reference (Pal et al. 2014).

On the other hand, content addressable memory (CAM) used for lookup table based application which enables high speed parallel search operations. Researchers evaluate the design space for FinFET CAMs for symmetric and asymmetric gate work functions (Bhattacharya et al. 2015). Researchers proposed diverse designs and conducted their transient and DC analysis for various mismatch conditions using the computer-aided design (CAD) tool with 22-nm FinFET devices. CAMs are often used in signal processing and in wireless sensor network applications requiring very low power consumption and extremely high speed so that this can be accomplished by developing a CAM using 22-nm and 14-nm PMOS access transistors based on the PTM-MG transistor model. From the available literature, we conclude that IG FinFET displays increased speed and lower power consumption (Arulvani and Mohamed Ismail 2018).

8.3.2 *FinFET Standard Cells Under PVT Variability*

Researchers evaluate the impact of PVT variations on power off predictive standards and timing at 20-nm FinFET technology node. The main factors in PVT variations are environmental factors, physical factors and reliability factors. Variations in electrical and geometrical parameters also provoke a delay in a transistor's performances. Fin height, gate length, fin thickness and metal gate work function fluctuations are the main causes of process variability in FinFET devices. Researchers use more than 10,000 Monte Carlo simulations with work function parameters for PVT variability investigation. Work function fluctuations have a huge impact on OFF state leakage current, and it causes a significant deviation on the static power of standard cells (Zimpeck et al. 2015).

8.3.3 Flip-Flop Performance in FinFET Technology

The impact of aging effect and PVT variations on different flip-flops in FinFET and CMOS technologies and their comparative performance analysis can be seen in reference (Taghipour and Asli 2017). Hot carrier injection (HCI) and bias temperature instability (BTI) mainly affect the heftiness of high-performance FinFET. Researchers acknowledged that temperature and VDD variations are the main causes of power–delay product (PDP) degradation and propagation delay for different FinFET structures.

An average increase of performance is obtained from the following equation (Taghipour and Asli 2017)

$$\text{Average increase}(\%) = \frac{\text{aged} - \text{Fresh}}{\text{Fresh}} \times 100 \tag{8.3}$$

The long-term model can be utilized to estimate the V_T variations, and then the updated V_T is applied to the transistor model file to evaluate the BTI and HCI aging mechanism for reliability analysis. Continuous scaling of a transistor increases the consequence of process variations and aging in circuits. The effect of PV and NBTI aging over the years on the WNMs and the consequent statistical occurrence of write failures in several types of flip-flop cells is presented in reference (Khalid et al. 2015). An analysis based on the statistical characterization of WNMs is using transistor-level Monte Carlo simulations to evaluate the write failure probability as a result of an input voltage change in flip-flop cells.

8.3.4 Impact of Time Zero Variability and BTI on FinFET Devices

Time zero variation or prestress variations of the device are the main unease in scaled technology, and detailed study of time zero variability performance of the device helps to improve the circuit performance. Researchers evaluated that device degradation occurs due to time zero PV and BTI stress conditions and also studied the variations in a threshold voltage of a planar 10-nm FinFET system on chip, 16-nm FinFET and 20-nm FinFET device. The following points are studied in reference (Mukhopadhyay et al. 2018):

- An impact of BTI and time zero V_T variations on V_T.
- Evaluate NBTI and positive bias temperature instability (PBTI), and their statistical performance is compared.
- SNM degradation in SRAM cell due to bias temperature instability.
- At last, bit- and chip-level high-temperature operating life (HTOL) test results are studied.

8.3.5 *Impact of PVT Variations on FinFET Under Different Sizing Techniques*

Different transistor arrangements can cause variations in gate variability. We can find a most suitable topology which increases the robustness of cells regarding PV. Any obviously occurring variations in the attributes of transistor like length, width and oxide thickness during the fabrication of IC are related to PV issue. The best approach for reducing the PV issue is to utilize the network which has transistors in series and as far as possible to the output (Zimpeck et al. 2018). Researchers investigate the impact of variation on power consumption and performance for various transistor sizing approaches applied to circuits in FinFET technologies and evaluate PVT variations separately. Temperature and voltage variations are united to get an insight into their contributions. Results are beneficial to describe the variability effect in the initial design steps to choose suitable transistor sizing technique for an application (Zimpeck et al. 2016).

8.3.6 *Energy-Efficient Compressor Based on FinFETs*

Multiplier is one of the important required arithmetic blocks in digital signal processing (DSP) applications and also the major energy and time-consuming block for an enormous variety of applications. So, we can improve the efficiency of these circuits by introducing FinFET in it. A designer has introduced a new energy-efficient 4:2 compressor that has less transistors, smaller areas and superior energy efficiency. This compressor provides an improvement in terms of energy efficiency and has less area overhead than the previous design (Arasteh et al. 2018).

8.3.7 *Impact of Multiple Parallel Fins on FinFET*

The unique structure and geometry of FinFET as compared to CMOS makes FinFET more reliable and efficient than CMOS. 3D fin structure of FinFET contains the current conduction between source and drain. Multiple parallel fins can be fabricated between source and drain that increases the channel width. The number of fins on FinFET shows a great impact on circuit performances and reliability. Multiple parallel fins can be used to increase the total drive current but in this case, FinFET suffers from severe degradation. A multiple parallel fin structure of FinFET reduces the inversion charge by creating charge repulsion between the fins. This increases the coupling effect between fins, and HCI is always an issue in a deep submicron region (Yeh et al. 2018).

8.3.8 Multicore Power, Area and Timing (McPAT)-PVT: Modeling Framework for FinFET Under PVT Variations

FinFETs have become an appropriate transistor to replace the conventional MOS transistor due to their better scalability, efficiency and a better SCE control. FinFETs have some lithographic, fabrication and environmental limitations which lead to PVT variations in FinFET IC. So, due to these variations, delay and leakage are introduced into the FinFET ICs. McPAT-PVT is an integrated framework that is considered for analysis of delay, power and PVT variations of FinFET devices. This framework consists of FinFET logic, design library and memory cells to represent circuit-level characteristics and PVT variations. Both SG and ASG FinFET-based processors are modeled by McPAT-PVT. ASG mode implementation provides the same performance but, it increases the area and more beneficial with temperature variations (Tang et al. 2015).

8.3.9 FinFET Performance Under Various Design Strategies

Designers varied the source and drain junction placement, punch-through stop implant and gate work function to investigate the new design approaches for 10-nm FinFET technology to satisfy low power and extremely low power requirements and to know the impact of I_{OFF}, gate capacitance, transconductance and intrinsic frequency (Walke et al. 2017). Research extraction and analysis of external resistance have become important in modern CMOS technology. By adding some assumptions in shift-and-ratio methods, it can be explored for use in short channel devices and also find application in FinFET devices (Zhang et al. 2018). A transistor with reduced size and fin gate has led to important change and add a set of constructive layout design rules. Additional layers and 3D structure of FinFET changed the parameters of a parasitic element, so a comparative analysis of 28-nm planar and 7-nm FinFET CMOS is performed (Ilin et al. 2018). A FinFET with a modified drain extension exhibits a better analog and radio frequency (RF) behaviors. We can boost the cutoff frequency of power FinFET from 30 to 53 GHz by changing the drain extension from narrow fin to a planar layout. Researchers investigated the analog and radio frequency parameters of power FinFETs with diverse drain extension structures for microwave applications.

Researchers replace the bipolar junction transistor diodes with FinFET diodes in some cases and evaluate the device output without degradation. Minimum voltage headroom and less power dissipation are two benefits of using the FinFET diodes in subthreshold operation (Prilenski and Mukund 2018). New self-aligned double-gate silicon on insulator (SOI) structure FinFET is proposed as a nano-MOS device. This proposed structure suppresses SCE, even with 17-nm gate length, provides a proper V_T for ultra-thin body and reduces the parasitic resistance (Hisamoto et al. 2000).

The author has explored the FinFET's best suitability for low-power applications in very short gate-length future technologies (Rosner 2003). This paper shows that the proposed FinFETs offer low-power output for the state-of-the-art bulk MOSFETs, even with relaxed gate oxide thickness. A new method of estimating the leakage is being studied in Gu et al. (2008), and the results show that the effect of the quantization of the width on the estimate of the statistical leakage is important for FinFET devices. This approach can reliably determine the statistical characteristics of the leakage current under process variation.

Designers often try to create an innovative design and structures to remove the disadvantages related to FinFET including gate buckling, fin bottom erosion, structural instability and less uniformity between fin shapes. Inverted T (IT) FinFET is an innovative design that can be used to increase a drive current with limited size (Yu 2002; Mathew 2005). IT FinFET is more beneficial than SOI FinFET because it requires wider fin width and less fin height as compared to SOI FinFET. IT FinFET is a mechanically stable structure and reduces the random dopant fluctuation and fin bottom erosion, but suffers from high OFF state current. Fin width and ultra-thin body height parameters can be used to optimize the performance of the IT FinFET, and outcomes manifest that fin width should be less than 10 nm for better immunity against SCE (Yu et al. 2018).

In the new era of VLSI technology, the compact size of devices is the primary requirement and maintaining the good performance of devices with compact size is one of the biggest challenges for research designers. Nowadays, FinFET is the most promising transistor and it is the most competent device to substitute the MOS transistors because of its outstanding controllability of the SCE, great insulation, high driving efficiency and reduced leakage current for both high-speed and low-power applications. But, some scaling challenges faced by FinFET devices and improvement are required for the betterment of VLSI technology. The impact of PVT variation is mainly on the nanotechnologies and degrades the performance of FinFET, so relevant methods and techniques are needed to improve the FinFET technology.

8.4 Results and Discussion

With the advancement of technology, further scaling of MOS transistors is a challenging task for research designers. FinFET is one of the best alternatives to be used for the scaling process. The main reason for FinFET's success is its excellent SCE controllability compared to a conventional planar system. The fin like geometry of FinFETs, where the regions of depletion enter the body region from the gates, indicates that there are no free charge carriers available, making it possible to suppress SCE. Furthermore, FinFET technology dominates because it offers great isolation, less current leakage and higher driving capability.

Nonetheless, FinFET technologies face many scaling challenges. For example, fin engineering (channel length, fine thickness, oxide thickness and balancing height) is important to minimize I_{OFF} and maximize I_{ON}. PVT variation also exacerbates circuit

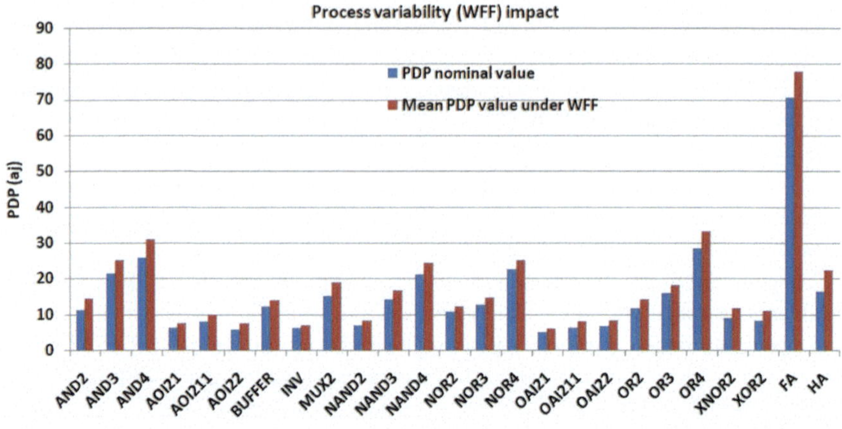

Fig. 8.6 Nominal PDP outcomes for standard cell gates under the WFF compared to mean values

degradation which makes the circuit inadequate for its initial purpose. A PVT variation causes a severe effect on the delay, leakage and performance of FinFET devices. Any variations in temperature affect the leakage current that leads to increases in energy–delay product (EDP) by up to 4X and 7X for full VDD operation and near-threshold voltage schemes, respectively (Turi and Delgado-Frias 2017). Researchers studied the 8T FinFET SRAM cell that reveals there are up to 42% of variations in EDP due to supply voltage variation. Temperature variability influences leakage current and the increase of up to 32X, and from the literature we noticed that a low-power inverter scheme is the highest rated 8T FinFET SRAM scheme (Turi and Delgado-Frias 2017). Fabrication of FinFET is a critical step for improvement in the performance of the device in deep submicron regime. Small variations during fabrication completely alter the circuit behavior, so we can conclude that nanoscale devices are becoming very sensitive to process variations. Figure 8.6 shows nominal PDP outcomes for standard cell gates under the WFF compared to mean values.

AND4, half adder and full adder standard cells exhibit more sensitivity due to WFF variations and show deviations of 19.76, 10.33 and 35.36% above the nominal PDP value, respectively. INV, NAND2 and AOI21 standard cells are less sensitive to WFF deviations. Figure 8.7 indicates the differences in power, PDP and timing due to voltage fluctuations (Zimpeck et al. 2015). Supply voltage variations play a very crucial part in the performance of FinFET. Figure 8.7 shows the total power, timing and PDP values for a voltage range from 0.9 to 0.3 V.

NAND4, AND4 and NOR3 standard cells show about 70% of PDP reduction by using FinFET devices. The main drawback of voltage variations is timing violations. The total power consumption parameter is mainly affected by temperature variations that can increase power consumption 5X higher than the nominal value in case of high temperature (Zimpeck et al. 2015). We can examine that WFF can considerably influence leakage current of the FinFET from the above results.

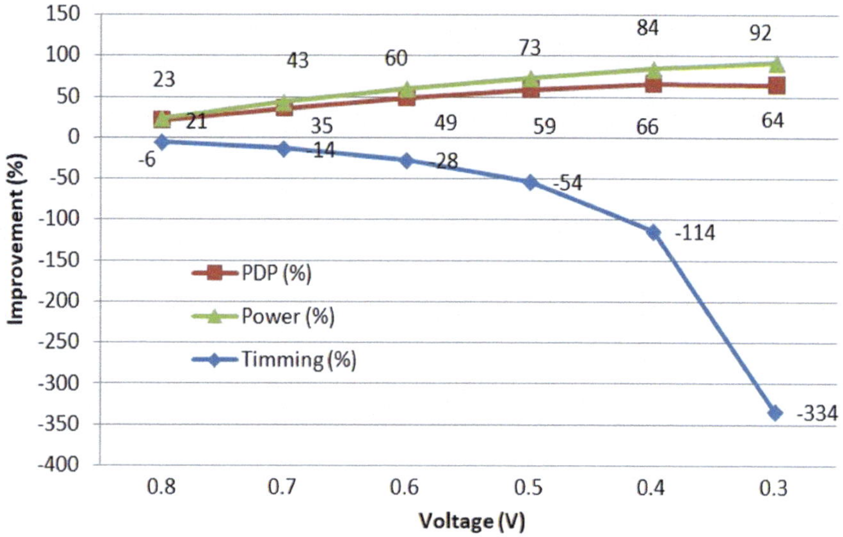

Fig. 8.7 Differences in timing, power and PDP due to voltage fluctuations

Diverse transistor arrangements for the similar logic function can reveal the different electrical and physical characteristics under PVT variations. To mitigate the impact of PVT variations, complex cells can be implemented in various transistor arrangements that can provide the most suitable topology for evaluation. Different transistor arrangements show a distinct impact on gate variability and concluded that far topology is best for OAI211 and OAI221 complex gates (Zimpeck et al. 2018). PDP determines the impact of process variability on complex cell by evaluating the delay and power of various circuits under the influence of WFF variations. Close arrangement is better for remaining complex gate. Far topology having three or more inputs provides better performance but causes the power penalty, i.e., increasing the power consumption mean value. Table 8.1 shows the mean and standard deviation of power consumption, worst-case delay and PDP (Zimpeck et al. 2018).

Impact of PVT variations on different transistor sizing techniques is also scrutinized in the previous literature in which transistor sizing techniques like optimized transistor sizing (OTS), logical effort (LE) and minimum transistor sizing (MTS) are largely utilized. LE-based technique cells exhibit the highest deviation in PDP. On the other hand, the OTS-based technique cells represent higher nominal values. Voltage variations mostly influence the OTS worst cases that cause maximum energy consumption. The impact of temperature variations is very less in OTS-based technique cells. LE technique shows the largest deviation. It is also important to consider environmental variation when choosing the appropriate approach for defining the correct transistor sizes for standard cell libraries, considering variability (Zimpeck et al. 2016).

Table 8.1 Mean and standard deviation of the power consumption, worst-case delay and PDP

Metrics	AOI21		OAI21		AOI211		OAI211		AOI221		OAI221	
	Close	Far	Close	Far	Close	Far	Close	Far	Close	Far	Close	Far
Delay (ps)	4.2	6.4	4.4	6.4	9.3	10	9.4	8	11.6	12	9.9	10.7
σ/μ (%)	34.3	33.8	32.7	33.4	33.6	34.3	31.6	35	35	35.6	34.4	32.2
Power (nW)	274.2	297.8	255.6	270.8	278.5	306.4	302.8	308	308	328.5	393.4	307.5
σ/μ (%)	24	22.1	26	24.2	27.8	25.4	29.7	27.3	28.9	27.3	31.2	29.8
PDP (a_j)	2.3	2.6	2.1	2.2	3.4	4	3	3.2	4.7	5.2	3.9	4.1
σ/μ (%)	27	26	26.7	27.8	29.7	28.1	30.5	31.4	30.3	28.2	31.8	31.4

The process variability will introduce a power deviation of up to 100 percent. RSNM shows about 20% variation under PV which is the worst case dramatically reduced cell noise robustness (Almeida et al. 2018). The author introduced FinFET CAM architecture focused on parasitic aware nature in Bhattacharya et al. (2015). All asymmetric gate work function shorted gate (ALL-ASG) bit cell was more superior to all shorted gate (ALL SG) and core ASG bit cells in terms of DC metrics. Leakage power assumes slightly greater significance with decreasing mismatch probability. When the BJT diode is replaced with FinFET diode, it shows the less voltage headroom and power dissipation (Prilenski and Mukund 2018). But, the biggest disadvantage of a FinFET diode is enlarged vulnerability to PV. The traditional method of estimating the leakage will greatly underrate the average leakage current by 43%, while the approach makes less error than 5% (Gu et al. 2008). The PV remains the main source of power and timing deviation in new technologies. CAD tools may play a significant role in assessing the impact of variability and reliability. Various tools like Cadence Virtuoso, Synopsys and ELDO simulator are the best way to implement any design philosophy (Alam 2008).

8.5 Conclusion

Scaling challenges of MOSFET technology such as SCE, an aging effect and a variability effect are becoming a barrier in the progress of VLSI technology; therefore, an appropriate alternative is the best way for evolution in VLSI technology. FinFET is the best option for substituting MOS technology because of better SCE controllability, lower leakage, perfect isolation and high driving capability. In this chapter, we outline various challenges faced by MOSFET technology and various factors which explicate the superiority of a FinFET as compared to MOS transistor. Researchers adopted various methodologies to mitigate the impact of PVT variations, but the

PVT variation is still a dominant factor in FinFET devices, especially in deep submicron regimes. Nowadays, for better performance of FinFET, various techniques are necessitated to mitigate the impact of PVT variations.

References

Alam M (2008) Reliability- and process-variation aware design of integrated circuits. Microelectron Reliab 48(8–9):1114–1122

Almeida RB, Marques CM, Butzen PF, Silva FRG, Reis RAL Meinhardt C (2018) Analysis of 6 T SRAM cell in sub-45 nm CMOS and FinFET technologies. Microelectron Reliab 88–90:196–202

Arasteh A, Moaiyeri MH, Taheri MR, Navi K, Bagherzadeh N (2018) An energy and area efficient 4:2 compressor based on FinFETs. Integration 60:224–231

Arulvani M, Mohamed Ismail M (2018) Low power FinFET content addressable memory design for 5G communication networks. Comput Electr Eng 72:606–613

Bagheriye L, Toofan S, Saeidi R, Moradi F (2018) Highly stable, low power FinFET SRAM cells with exploiting dynamic back-gate biasing. Integration. ISSN 0167-9260

Ban Y, Choi C, Shin H, Lee J, Kang Y, Paik W (2014) Analysis of dynamic voltage drop with PVT variation in FinFET designs. In: 2014 International SoC design conference (ISOCC), 2014

Bhattacharya D, Bhoj AN, Jha NK (2015) Design of efficient content addressable memories in high-performance FinFET technology. IEEE Trans Very Large Scale Integr (VLSI) Syst 23(5):963–967

Chen B, Chen K, Chiu C, Huang G, Chen H, Chen C, Hsueh F, Chang EY (2018) Analog and RF characteristics of power FinFET transistors with different drain-extension designs. IEEE Trans Electron Devices 65(10):4225–4231

Colinge J-P (2008) The SOI MOSFET: from single gate to multigate. In: FinFETs and other multi-gate transistors. Springer, US, pp 1–48

Ezz-Eldin R, El-Moursy MA, Hamed HFA (2015) Analysis and design of networks-on-chip under high process variation. Springer International Publishing, 2015 [Online]. Available: http://dx.doi.org/10.1007/978-3-319-25766-2

Gu J, Keane J, Sapatnekar S, Kim CH (2008) Statistical leakage estimation of double gate FinFET devices considering the width quantization property. IEEE Trans Very Large Scale Integr (VLSI) Syst 16(2):206–209

Gupta A, Mathur R, Nizamuddin M (2019) Design, simulation and comparative analysis of a novel FinFET based astable multivibrator. Int J Electron Commun 100:163–171

Hisamoto D, Lee W-C, Kedzierski J, Takeuchi H, Asano K, Kuo C, Anderson E, King T-J, Bokor J, Hu C (2000) FinFET-a self-aligned double-gate MOSFET scalable to 20 nm. IEEE Trans Electron Devices 47(12):2320–2325

Ilin S, Ryzhova D, Korshunov A (2018) Comparative analysis of standard cells performance for 7 nm FinFET and 28 nm CMOS technologies with considering for parasitic elements. In: 2018 IEEE conference of Russian Young researchers in electrical and electronic engineering (EIConRus), Moscow, 2018, pp 1360–1363

Islam A, Hasan M (2012) A technique to mitigate impact of process, voltage and temperature variations on design metrics of SRAM Cell. Microelectron Reliab 52(2):405–411

Jacob AP, Xie R, Sung MG, Liebmann L, Lee RTP, Taylor B (2017) Scaling challenges for advanced CMOS devices. Int J High Speed Electron Syst 26(01n02):1740001

Khalid U, Mastrandrea A, Olivieri M (2015) Effect of NBTI/PBTI aging and process variations on write failures in MOSFET and FinFET flip-flops. Microelectron Reliab 55(12):2614–2626

Khoshavi N, Ashraf RA, DeMara RF, Kiamehr S, Oboril F, Tahoori MB (2017) Contemporary CMOS aging mitigation techniques: survey, taxonomy, and methods. Integration 59:10–22

Kushwah CB, Vishvakarma SK, Dwivedi D (2016) A 20 nm robust single-ended boost-less 7T FinFET sub-threshold SRAM cell under process–voltage–temperature variations. Microelectron J 51:75–88

Lee C-Y, Jha NK (2014) FinCANON: a PVT-aware integrated delay and power modeling framework for FinFET-based caches and on-chip networks. IEEE Trans Very Large Scale Integr (VLSI) Syst 22(5):1150–1163

Lee C-W et al (2010) Performance estimation of junctionless multigate transistors. Solid-State Electron 54(2):97–103

Mahapatra S et al (2013) A comparative study of different physics-based NBTI models. IEEE Trans Electron Devices 60(3):901–916

Mathew L et al (2005) Inverted T channel FET (ITFET)—fabrication and characteristics of vertical-horizontal thin body multi-gate multi-orientation devices ITFET SRAM bit-cell operation. A novel technology for 45 nm and beyond CMOS. In IEDM technical digest, Washington, DC, USA, pp 713–716

Mukhopadhyay S, Lee Y-H, Lee J-H (2018) Time-zero-variability and BTI impact on advanced FinFET device and circuit reliability. Microelectron Reliab 81:226–231

Pal S, Bhattacharya A, Islam A (2014) Comparative study of CMOS- and FinFET-based 10T SRAM cell in subthreshold regime. In: 2014 IEEE international conference on advanced communications, control and computing technologies, 2014

Pham D, Larson L, Yang J-W (2006) FINFET device junction formation challenges. In: 2006 international workshop on junction technology, 2006

Prilenski L, Mukund PR (2018) A sub 1-volt subthreshold bandgap reference at the 14 nm FinFET node. Microelectron J 79:17–23

Rosner W et al (2003) Nanoscale finFETs for low power applications. Int Semicond Device Res Symp

Sharma VK, Pattanaik M (2014a) Techniques for low leakage nanoscale VLSI circuits: a comparative study. J Circuits, Syst Comput 23(5):1450061

Sharma VK, Pattanaik M (2014b) Process, voltage and temperature variations aware low leakage approach for nanoscale CMOS circuits. J Low Power Electron 10(1):45–52

Sinha S, Yeric G, Chandra G, Cline B, Cao Y (2012) Exploring sub-20 nm FinFET design with predictive technology models. In: Proceedings of the 49th annual design automation conference on—DAC'12, 2012

Sneha G, Krishna BH, Kumar CA (2017) Design of 7T FinFET based SRAM cell design for nanometer regime. In: 2017 International conference on inventive systems and control (ICISC), 2017

Taghipour S, Asli RN (2017) Aging comparative analysis of high-performance FinFET and CMOS flip-flops. Microelectron Reliab 69:52–59

Tang A, Yang Y, Lee C, Jha NK (2015) McPAT-PVT: delay and power modeling framework for FinFET processor architectures under PVT variations. IEEE Trans Very Large Scale Integr (VLSI) Systems 23(9):1616–1627

Tawfik SA, Kursun V (2008) Portfolio of FinFET memories: innovative techniques for an emerging technology. In: 2008 international SoC design conference, 2008

Tawfik SA, Liu Z, Kursun V (2007) Independent-gate and tied-gate FinFET SRAM circuits: design guidelines for reduced area and enhanced stability. In: 2007 International conference on microelectronics, 2007

Turi MA, Delgado-Frias JG (2017) Full-VDD and near-threshold performance of 8T FinFET SRAM cells. Integration 57:169–183

Upasani DE, Shrote SB, Deshpande PS (2010) Standby leakage reduction in nanoscale CMOS VLSI circuits. Int J Comput Appl 7(5)

Walke A, Schlenvogt G, Kurinec S (2017) Design strategies for ultra-low power 10 nm FinFETs. Solid-State Electron 136:75–80

Yang Y, Jha NK (2013) Fin Prin: analysis and optimization of FinFET logic circuits under PVT variations. In: 2013 26th international conference on VLSI design and 2013 12th international conference on embedded systems, 2013

Yang Y, Jha NK (2014) FinPrin: FinFET logic circuit analysis and optimization under PVT variations. IEEE Trans Very Large Scale Integr (VLSI) Syst 22(12):2462–2475

Yeh W, Zhang W, Chen P, Yang Y (2018) The impact of fin number on device performance and reliability for multi-fin tri-gate n- and p-type FinFET. IEEE Trans Device Mater Reliab 18(4):555–560

Yu B (2002) Fabrication of a field effect transistor with an upside down T-shaped semiconductor pillar in SOI technology

Yu E, Heo K, Cho S (2018) Characterization and optimization of inverted-T FinFET under nanoscale dimensions. IEEE Trans Electron Devices 65(8):3521–3527

Zhang C, Liu Z, Miao X, Yamashita T (2018) FinFET external resistance analysis by extended shift-and-ratio method. IEEE Trans Electron Devices 65(8):3127–3130

Zimpeck AL, Meinhardt C, da Luz Reis RA (2015) Impact of PVT variability on 20 nm FinFET standard cells. Microelectron Reliab 55:1379–1383

Zimpeck AL, Meinhardt C, Posser G, Reis R (2016) FinFET cells with different transistor sizing techniques against PVT variations. In: 2016 IEEE international symposium on circuits and systems (ISCAS), Montreal, QC, pp 45–48

Zimpeck AL, Meinhardt C, Artola L, Hubert G, Kastensmidt FL, Reis RAL (2018) Impact of different transistor arrangements on gate variability. Microelectron Reliab 88–90:111–115

Part III
Emerging Technologies for Integrated Circuits

Chapter 9
Gallium Nitride—Emerging Future Technology for Low-Power Nanoscale IC Design

Sahil Sankhyan, Tarun Chaudhary, Gargi Khanna, and Rajeevan Chandel

Abstract The development of the silicon (Si)-based deep submicron devices has promised significant improvement in the quality of life, including new technologies for the treatment of diseases and greater efficiency for storing and processing the computer data. It is a well-known fact that electronics industry has undoubtedly benefited from the Si-based technology that uses much lower power and offers cost-effective circuits and devices due to mass fabrication. But is it feasible for Si technology to improve and revive the electronics industry, speed up its growth, and enable rapid development of portable and compact products? An additional aspect which needs to be established is the choice of the right innovative materials and devices that will allow the electronics industry to grow and develop new low-power systems, along with the possible potential of renovating this industry. Various researchers throughout the world are evaluating distinct and effective methodologies to solve this problem, and gallium nitride (GaN) technology has come out as one of the major breakthroughs and innovations. This chapter mainly focuses on the basics of advanced materials beyond Si and germanium (Ge) which can be used for the fabrication of various electronic devices such as transistors, gates, oscillators, and amplifiers. It addresses the advantages and disadvantages associated with the usage of these materials for modern electronic devices and low-power VLSI circuits.

Keywords CMOS technology · Si · Power · GaN · Transistors · SCEs

S. Sankhyan · T. Chaudhary (✉)
Electronics and Communication Engineering, Dr. B. R. Ambedkar NIT Jalandhar, Jalandhar 144011, India
e-mail: chaudharyt@nitj.ac.in

S. Sankhyan
e-mail: sahilsankhyan@gmail.com

G. Khanna · R. Chandel
Electronics and Communication Engineering, NIT Hamirpur, Hamirpur 177005, Himachal Pradesh, India
e-mail: gargi@nith.ac.in

R. Chandel
e-mail: rchandel@nith.ac.in

R. Dhiman and R. Chandel (eds.), *Nanoscale VLSI*,
Energy Systems in Electrical Engineering,
https://doi.org/10.1007/978-981-15-7937-0_9

9.1 Introduction

Researchers are facing several hard concerns due to continuous gate-length downscaling of currently available metal–oxide–semiconductor (MOS) devices. Device scaling has led to increase in leakage current and short-channel effects (SCEs) with continuous reduction in gate electrostatic control over the channel. Nowadays, device engineers are in a dilemma that how to further boost up the device performance using the conventionally opted scaling technique and to maintain the reliability of the circuits through this scaling process. As it has been observed that scaling drift cannot continue indefinitely, so to solve this obstacle, engineers must turn to new revolutionary device materials and structures for high-speed, low-power applications in order to retain greater efficiency of VLSI circuits. Attributable to rapid growth of portable systems and limitations of battery technology appliances, design of power-sensitive devices with much smaller size, low power consumption, and high density to incorporate multiple functions in upcoming electronic devices has also become essential (Pop 2010; Shakouri 2004). In this context, scientists and researchers have been continuously designing and monitoring the capability of various new device architectures based on Si since the past few decades. Few such technologies are mentioned below:

9.1.1 FinFET

A fin field-effect transistor (FinFET) technology was introduced for relentless increase in the levels of integration due to refined lithography techniques being used in its fabrication. FinFET is the structure that grows above the substrate and looks like a fin. The "Fin" increases the control of gate in controlling the current, and in this way, they perform much effectively than a traditional planar transistor for the same area (Yang et al. 2004). The gate surrounds the "Fin" and gives it more control over the channel as it has sufficient length to control. This formation of the gate gives upgraded electrostatic control on the channel area and supports reduction in leakage current levels and therefore, overcome short-channel effect problems as well. Figure 9.1a shows the basic structure of FinFET (Mishra et al. 2011).

9.1.2 CNTFET

Single-layer carbon atom with rolled-up sheets of cylindrical molecules constitutes carbon nanotubes (CNTs). They can be a single wall (SWCNT) with less than one nanometer (nm) of diameter or multi-wall (MWCNT) as shown in Fig. 9.1b. Their development in recent years reflects the effect of revolutionary nanomaterials, particularly in biomedical imagining, biosensing, and functional nanocomposites design

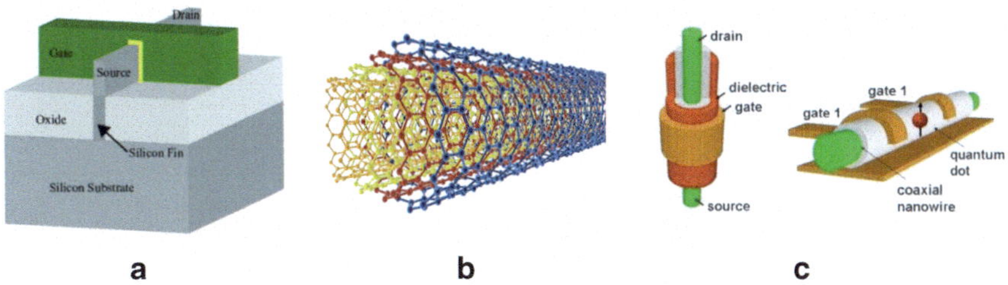

Fig. 9.1 **a** Device design of FinFET. **b** Molecular structure of MWCNT. **c** Device design of Si nanowire

(Dang et al. 2006). Despite various desirable characteristics of CNTs, many hurdles which need to nonetheless be conquering earlier than devices built with this generation are feasible. Most of these troubles surround the fabrication of the CNTs. Despite several advantages of the CNTs like the lightweight and small size, it is very difficult to work with CNTs. Also, the process to produce CNTs is very expensive (Singh et al. 2017).

9.1.3 Semiconducting Nanowires

Like the CNTs, nanowires (NWs) can also be used as interconnects to propagate signals in the electronic system as well as used as an active device. CNTs can act only device or wire at a time while NWs can behave as a device and an interconnect simultaneously. NWs are made up of semiconductor materials like silicon and germanium having a very small diameter up to 3 nm as presented in Figs. 9.1c, 9.2. Nanowire devices are now emerging as a class of ultrasensitive, powerful, and general electrical sensors to detect biological and chemical species directly. The sensitivity of the nanowire devices is an issue, and large-scale fabrication of nanowire devices is expensive. As the nanowire devices are made up of silicon materials, they cannot work

Fig. 9.2 Nanowire device

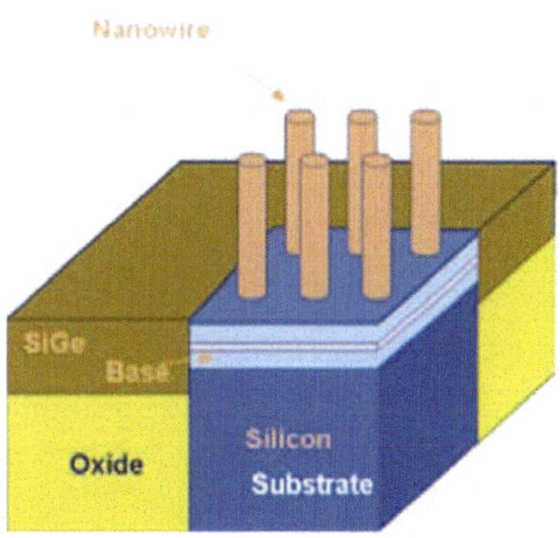

properly on the high power, voltage, current, frequency, and temperature (Amato and Rurali 2016; Huang et al. 2007).

9.1.4 Issues with Silicon Technology

In modern electronics, silicon-based devices are undoubtedly the main platform due to their low cost and vast experience base for chemical treatment on silicon oxide. However, silicon sensors usually cannot function at high temperatures and under server stress conditions and are degraded in a chemically corrosive environment. Si needs a thick crystalline layer as it is very brittle in nature and provides limited substrate options. Consequently, it has been observed that the process of fabrication for silicon-based devices is more expensive compared to the others (Singh 2006).

9.2 Gallium Nitride Technology

Silicon (Si) has been in demand in the field of power devices over a considerable period, due to its availability and abundant knowledge available of its material properties. However, Si devices are facing some operational limits based on intrinsic material properties. Thus, a new material needs to be investigated for the fabrication of power semiconductors (Qian et al. 2004).

Gallium nitride (GaN) is one such material that is on the rise to replace silicon. GaN a group III-V compound exhibits basic material properties which contribute to smaller devices resulting in reduction of parasitics size, fewer components count, higher frequency of operation, and lower switching losses. It is anticipated that GaN is a power semiconductor of next generation with much faster switching speed as compared to Si, a higher breakdown strength, improved thermal conductivity, and lesser on-resistance (Xing et al. 2001). Hence, power devices grounded on wide-bandgap GaN material can considerably outpace the conventional silicon power chips and can offer advantages of both Si-based MOSFET device and above-mentioned other technologies also (Wang et al. 2019). These materials possess high switching frequencies and operating temperatures comparable to that of silicon and thus require lower cooling requirements, smaller heat sinks, and transition from liquid-cooled to air-cooled and removing fans. This semiconductor has distinctive characteristics which favors the development of effective optoelectronic devices adding up with high-temperature and high-power applications. As GaN has eco-friendly inertial, thus these devices need to find vast practical applications in commercial markets and in defense arena (Mohammad et al. 1995). To sum up, GaN can offer the advantages of both MOSFETs and IGBTs for high-frequency, low-loss, high-voltage applications.

Figure 9.3 and Table 9.1 present some of the material properties which are experimentally derived and differ among various reference sources. The wide-bandgap (WBG) semiconductors are placed in terms of increasing bandgap with

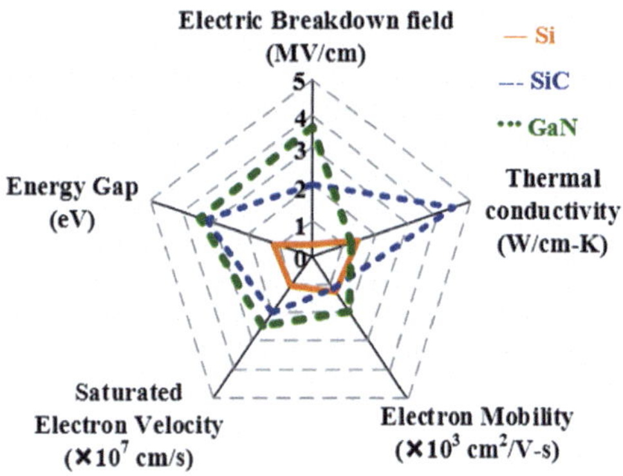

Fig. 9.3 Comparative evaluation of GaN, SiC and Si intended for power semiconductor applications (Chow 2014, 2015)

Table 9.1 Properties of Si and WBG semiconductors (Ozpineci and Tolbert 2003; Wang et al. 2015)

Property	Si	GaAs	6HSiC	4HSiC	GaN	Diamond	AlN
Bandgap, E_g (eV)	1.12	1.42	3.00	3.26	3.44	5.45	6.20
Electric breakdown field, E_c (MV/cm)	0.3	0.4	2.5	2.0	3.8	10.0	12.0
Electron mobility, μ_n (cm^2/V s)	1500	8500	500 80	1000	1250	2200	300
Saturated electron drift velocity, v_s (× 10^7 cm/s)	1.0	1.2	2.0	2.0	2.5	2.7	1.7
Dielectric constant, ε_r	11.8	13.1	9.7	10.0	9.5	5.5	8.5
Thermal conductivity, λ (W/cm K)	1.5	0.46	4.9	4.9	1.3	22	2.85

6H SiC exhibits anisotropy, therefore having different values of mobility in two different planes

respect to Si, in Table 9.1. Furthermore, it can be vividly observed that WBG semiconductors offer advantage over Si. The high breakdown field in WBG semiconductors permits the optimization of devices with slimmer drift regions, therefore resulting lower specific on-resistance in power devices.

GaN allows a small-scale die size to attain enough current capacity, and consequently lower input and output capacitances. Greater saturation velocity and reduced capacitances facilitate faster switching transients. Overall, it is summarized that these WBG semiconductors with improved material properties result in a device with lower on-resistance and switching losses than a Si material-based device with similar current capability and operating voltage (Chow 2014, 2015; Ozpineci and Tolbert 2003; Wang et al. 2015).

Basic structure of GaN-based heterostructure field-effect transistor (GaN HFET) is shown in Fig. 9.4. AlGaN/GaN heterojunction is the principle feature of this structure. As can be seen clearly from the figure, there is an interface between the

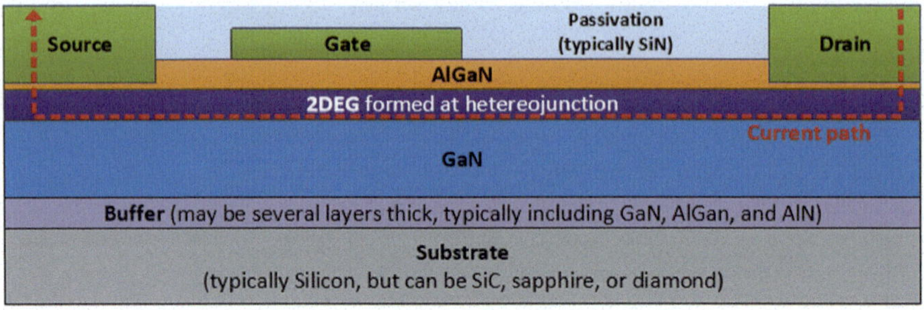

Fig. 9.4 Depletion-mode lateral GaN HFET basic structure (Jones et al. 2014)

layers of ALGaN and GaN, which is a "two-dimensional electron gas" (2DEG) a layer of high-mobility electrons and forms due to the crystal polarity and is also enhanced by piezoelectric crystal strain that results from lattice mismatch between AlGaN and GaN. This 2DEG structures a local channel for the current path between source and drain. Typically, Si is incorporated here as the substrate material; however, other materials, for instance, sapphire, SiC, and diamond, can also be used. Now, for the deposition of GaN layer on the substrate, a buffer layer has to be deposited which can directly provide the strain relief in between the foreign material and GaN. This buffer, however, frequently includes a number of thin layers of AlGaN, AlN, and GaN (Jones et al. 2014).

9.3 Device Design and Analysis of GaN FET and Silicon-Based FET

Above-mentioned properties of this upcoming material for developing new design technologies have motivated the authors to design simulate and analyze the characteristics of GaNFET and its comparison with its silicon-based counterpart. Both the devices are designed using Silvaco Atlas version 5.0.10.R (2020) electronic design automation (EDA) tool. Length of the device is kept 2 μm, metal gate as platinum, with gate length 0.5 μm is used with air as medium. The substrate doping concentration is $10^{17}/cm^3$; the source and drain doping concentrations are ranging from 10^{18} to $10^{19}/cm^3$. The oxide thickness is 1.2 nm with SiO_2 as insulating material as shown in Fig. 9.5a, b.

The SiO_2 dielectric layer restrains the virtual gate effect on the electric field distribution and dominantly reduces the current leakage on surface. This will on the other hand increase the electric field strength near the edge of the gate on the drain side, and therefore, Schottky gate leakage current is elevated. GaN devices display an excellent ability to accomplish breakdown voltage of several hundred volts. Together with the field plate mechanism, the large reduction in peak electric field at the edge of the gate is favorable for high-voltage application. GaN devices

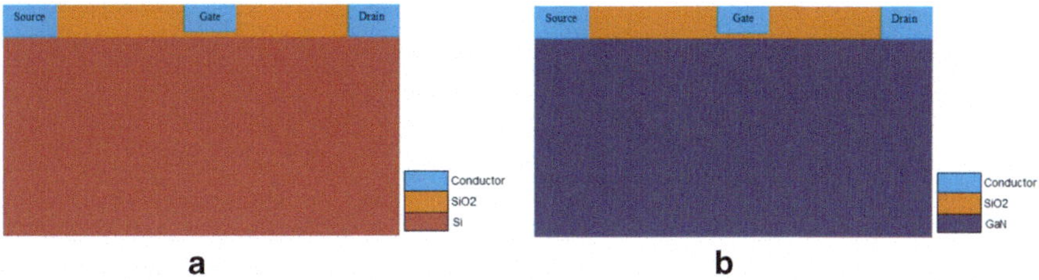

Fig. 9.5 **a** Structure of Si FET. **b** Structure of GaN FET

are high-electron-mobility field-effect transistor. In which two layers of different field and polarization field are grown on each other. Due to this, surface charge suspension in the polarization field heterointerface is generated. When the induced charge is positive, the electron tends to compensate for the induced charge resulting from the creation of the channel.

Threshold voltage (V_{th}) relationship for GaN-based transistor is given as (Charfeddine et al. 2012):

$$V_{\text{th}} = \text{Ø}^{\text{b}}_{\text{eff}} - \Delta E_{\text{c}} - \frac{qN_{\text{s}}d_{\text{AlGaN}^2}}{2.\varepsilon_{\text{AlGaN}}} - \sigma\frac{d_{\text{AlGaN}}}{\varepsilon_{\text{AlGaN}}} \quad (1)$$

where $\text{Ø}^{\text{b}}_{\text{eff}}$ is representing the barrier height of the Schottky gate, ΔE_{c} is conduction band discontinuity at the interface of UID-AlGaN and the GaN layers. Doping concentration of n-ALGaN is given by $\frac{qN_{\text{s}}d_{\text{AlGaN}2}}{2.\varepsilon_{\text{AlGaN}}}$, and the charge density at the interface induced due to polarization effect is given by σ.

Energy band diagram of the device obtained from Silvaco is shown in Fig. 9.6a, b. The figure depicts the mechanism of carrier flow between the conduction bands for GaNFET and SiFET.

For GaNFET with an increase in small drain–source voltage, i.e., if $V_{\text{ds}} > 0$ is applied, then a drain current which is proportional to the amount of V_{ds} applied, will

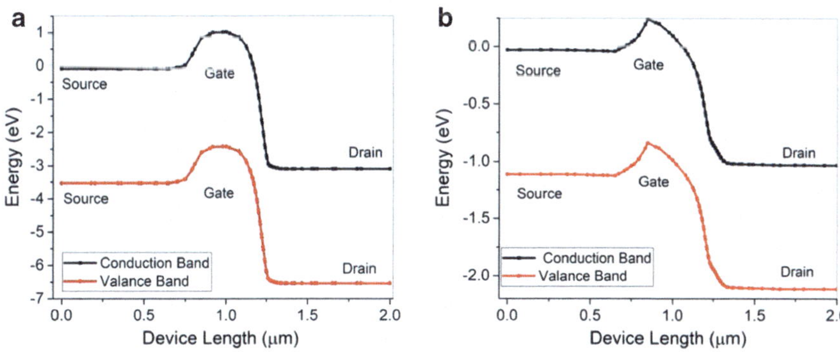

Fig. 9.6 **a** Energy band diagram of GaNFET. **b** Energy band diagram of silicon FET

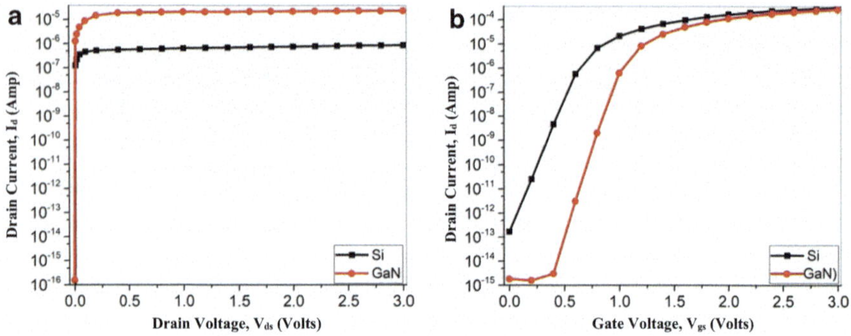

Fig. 9.7 **a** $I_d - V_{ds}$ characteristics of GaNFET and SiFET. **b** $I_d - V_{gs}$ characteristics of GaNFET and SiFET

flow between the source and drain through the conducting channel. In this mode, the device operates in linear mode or in linear region of operation. With further increase in applied drain voltage, the channel will form a continuous current path between source and drain. In this mode, the device operates in saturation mode or in saturation region of operation. Figure 9.7a represents the output (i.e., $I_d - V_{ds}$) characteristics of the devices. The drain current–gate voltage ($I_d - V_{gs}$) characteristics for GaNFET and Si-based FET operating in linear mode are presented in Fig. 9.7b. From Fig. 9.7a, b, it can be clearly inferred that GaNFET exhibits improved on current, as well as lower off-current, respectively. This amplifies the benefits of GaNFETs for developing high-performance circuits with reduced power consumption.

9.4 Conclusions

The work in this chapter investigates big hand solution for scalable high-performance and low-power devices for future ICs. Firstly, the performance parameters of various available devices and transistors have been overviewed and their potentials to overcome different challenges faced by electronics industry are studied. To boost up system performance, using different device geometries and materials, has been investigated. Further, their capabilities for small energy consumption, low leakage power, and reduced short-channel effects in nanoscale devices are analyzed. Secondly, it is examined that GaN technology offers a deep and profound physical insight of behavior of the device and is also found to be very effective in delivering an effective design space expedition for future nanoscale integrated circuits. Further, a comparative analysis of GaN-based and silicon-based FETs has been carried out in terms of on- and off-currents. It has been observed that GaNFET provides low I_{off} and improved I_{on} as compared to its silicon-based counterpart. The present work shall be highly beneficial for VLSI designers and particularly for next-generation low-power and high-performance integrated circuit design.

References

Amato M, Rurali R (2016) Surface physics of semiconducting nanowires. Prog Surf Sci 91(1):1–28

Charfeddine M, Belmabrouk H, Ali Zaidi M, Maaref H (2012) 2-D Theoretical model for current-voltage characteristics in AlGaN/GaN HEMTs. J Mod Phys 3:881–886

Chow TP (2014) Progress in high voltage SiC and GaN power switching devices. In: Proc Mater Sci Forum, pp 1077–1082

Chow TP (2015) Wide bandgap semiconductor power devices for energy efficient systems. IEEE workshop on wide bandgap power devices and applications, pp 402–405

Dang T, Anghel L, Leveugle R (2006) CNTFET basics and simulation. IEEE Int Conf Design Test Integr Syst Nanoscale Technol:28–33

Huang J, Momenzadeh M, Lombardi F (2007) An overview of nanoscale devices and circuits. IEEE Des Test Comput 24(4):304–311

Jones EA, Wang F, Ozpineci B (2014) Application-based review of GaN HFETs. In: Proceedings IEEE workshop on wide bandgap power devices and applications, pp 24–29

Mishra P, Muttreja A, Jha NK (2011) FinFET circuit design. Nanoelectronics circuit design springer, pp 23–54

Mohammad SN, Salvador AA, Morkoc H (1995) Emerging gallium nitride-based devices. Proc IEEE 83(10):1306–1355

Ozpineci B, Tolbert LM (2003) Comparison of wide-bandgap semiconductors for power electronics applications. Oak ridge national laboratory report

Pop E (2010) Energy dissipation and transport in nanoscale devices. Nano Res 3(3):147–169

Qian F, Li Y, Gradecak S, Wang D, Barrelet CJ, Lieber CM (2004) Gallium nitride-based nanowire radial heterostructures for nanophotonics. Nano Lett 4(10):1975–1979

Shakouri A (2004) Nanoscale devices for solid state refrigeration and power generation. IEEE Semicond Therm Meas Manag Symp:1–9

Silvaco Atlas version 5.0.10.R. (2020) https://www.silvaco.com/products

Singh R (2006) Reliability and performance limitations in SiC power devices. Microelectron Reliab 46(5):713–730

Singh A, Khosla M, Raj B (2017) Design and analysis of electrostatic doped Schottky barrier CNTFET based low power SRAM. AEU-Int J Electron Commun 80:67–72

Wang F, Zhang Z, Ericsen T, Raju R, Burgos R, Boroycvich D (2015) Advances in power conversion and drives for shipboard systems. Proc IEEE 103(12):2285–2311

Wang B, Dong S, Jiang S, He C, Hu J, Ye H, Ding X (2019) A comparative study on switching performance of GaN and Si power devices for bipolar complementary modulated converter legs. Energies 12:1146–1159

Xing H, Keller S, Wu YF, McCarthy L, Smorchkova IP, Buttari D, Coffie R, Green DS, Parish G, Heikman S, Shen L (2001) Gallium nitride based transistors. J Phys Condens Matter 13(32):7139

Yang FL, Lee DH, Chen HY, Chang CY, Liu SD, Huang CC, Chung TX, Chen HW, Huang CC, Liu YH, Wu CC (2004) 5 nm-gate nanowire FinFET. In: Digest of technical symposium on VLSI technology, pp 196–197

Chapter 10
A Low-Power Hybrid VS-CNTFET-CMOS Ring Voltage-Controlled Oscillator Using Current Starved Power Switching Technology

Ashish Raman, Vikas Kumar Malav, Ravi Ranjan, and R. K. Sarin

Abstract In analog and digital circuit, voltage controlled oscillator (VCO) plays a very important role in electronic circuits such as phase locked loop (PLL), radio frequency integrated circuits (RFICs), analog to digital converter (ADC) and other circuits. (Sun and KwasniewskI in IEEE J Solid State Circuits 36:910–916, 2001; Razavi in IEEE J Solid State Circuits 32:730–735, 1997; Jovanovic and Stojcev in Int J Electron 93:167–175, 2006; Hajimiri et al. in IEEE J Solid State Circuits 34:790–804, 1999; Jovanovic et al. in Sci Publ State Univ Novi Pazar 2:1–9, 2010). The VCO is an electronic circuit, which produces the frequency signal depending on its input voltage. VCO is voltage to frequency converter. VCO provides a better linear relationship among the variable control voltage and tuning oscillation frequency, which is a concern in many applications. In ring oscillator, the number of stages in the standard structure indicates the multiphase output in broad operating frequency (Jovanovic et al. in Sci Publ State Univ Novi Pazar 2:1–9, 2010). In this chapter, we have focused on the designing of stable frequency and low-power hybrid VS-CNTFET-CMOS VCO ring oscillator, which generate better linearity as compared to conventional CMOS design. Due to higher electron mobility and excellent transportation of career of CNTFET, it is used in many analog and radio frequency (RF

A. Raman
Assistant Professor, ECE Department, Dr. B.R. Ambedkar, National Institute of Technology, Jalandhar 144011, India

V. K. Malav
M.Tech, ECE Department, Dr. B.R. Ambedkar, National Institute of Technology, Jalandhar 144011, India

R. Ranjan (✉)
Pursuing PhD, ECE Department, Dr. B.R. Ambedkar, National Institute of Technology, Jalandhar 144011, India
e-mail: ranjan.ravi80@gmail.com

R. K. Sarin
Professor, ECE Department, Dr. B.R. Ambedkar, National Institute of Technology, Jalandhar 144011, India

R. Dhiman and R. Chandel (eds.), *Nanoscale VLSI*,
Energy Systems in Electrical Engineering,
https://doi.org/10.1007/978-981-15-7937-0_10

frequency) (Yang et al. in Appl Phys Lett 88:113507, 2006; Appenzeller et al. in 1:184–189, 2002; Akinwande et al. in IEEE Trans Nanotechnol 7(5):636–639, 2008; Cho et al. in MTL Annu Res Rep, 2007; Chakraborty et al. in IEEE TransCircuits Syst IRegul Pap 54(11):2480–2488, 2007). Recently, CNTFETs are a most popular device for RF applications. A chemical sensing application utilizing hybrid CMOS-CNTFET approach is reported in (Rahane and Kureshi in Int J Appl Eng Res 12:1969–1973, 2017). A low power and linear voltage controlled oscillator using hybrid CMOS-CNFET is used for RF application (Rahane and Kureshi in Int J Appl Eng Res 12:1969–1973, 2017). In this chapter, we designed five-stage "Hybrid VS-CNEFT-CMOS RVCO using Current Starved Power Switching Technology". The design VCO operates at the low supply voltage. This design does not only increase the frequency and current but also reduces the power dissipation and RMS jitter. We have used P-CNTFET and N-CNTFET in the place of convention PMOS and NMOS, respectively, which have small switching time, less threshold voltage (V_{th}) and less power consumption.

Keywords Carbon nanotube FET · Jitter · Ring voltage-controlled oscillator · Virtual source

10.1 Introduction

In 1960, the first MOSFET was demonstrated by Dawon Kahng and Atalla after resolving the problem of a surface state by growing an oxide insulator on Si. These devices are well suited for controlled switching between ON state and OFF state and are therefore useful in digital circuits. After three decades, the MOSFETs were being used in manufacturing of IC's (Integrated Circuits) popularity due to the use of silicon di oxide (SiO_2) as an insulator which provides good isolation and good control of gate terminal on conduction current (Kahng 1976). As per Moore's law, the number of transistors doubled on a chip in almost every two years. As the technologies are continuously scaled down to achieve high-speed MOSFET device by reducing the channel length, the power consumption has become the major problem in electronics devices (Moore 1998). When the number of transistors increases in unit area, the leakage or thermal power affects device's battery life, which is undesirable in electronic equipment (Borkar 2003). Figure 10.1 shows that the active power consumption has a quadratic relationship with a V_{dd} (supply voltage) as given in Eq. (10.1).

$$\text{Active power} = C * (V_{dd})^2 * f \tag{10.1}$$

Therefore, the active power consumption can be reduced by scaling down V_{dd}. Yet, V_{th} also needs to scale simultaneously with V_{dd}, which will effect on the leakage current of the device. Therefore, scaling down V_{dd} is hard to achieve.

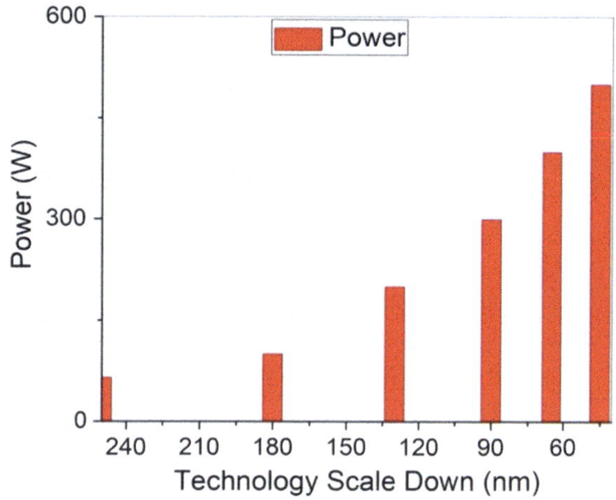

Fig. 10.1 Active power consumption increased when technology scaled down

10.1.1 MOSFET

The MOSFET consists of four terminals, i.e. source, drain, gate terminal and substrate/body which is normally connected to the source terminal of the device (Razavi 2002). The structure is symmetric with respect to source and drain. The terminal, which provides charge carriers, is termed as source, whereas the terminal, which collects the charge carriers, is termed as drain. The source and the drain may exchange their role on the basis of the variation in the voltages applied to the three terminals of the device (Fig. 10.2).

10.1.1.1 Operation of N-MOSFET

- The gate voltage value V_{gs} (gate-source voltage) at which the transistor is "turned ON" is termed as V_{th}.

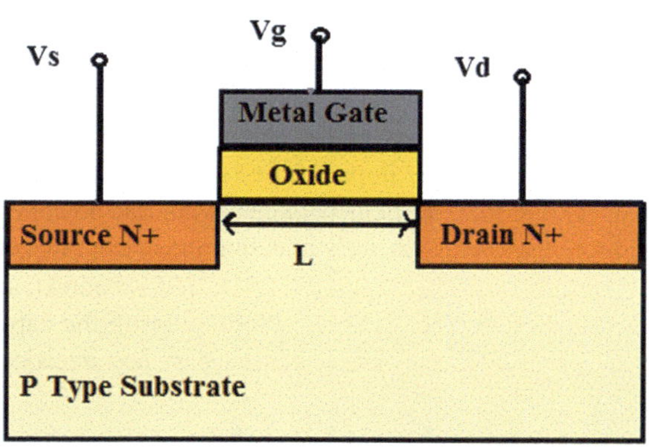

Fig. 10.2 Device structure of N-MOSFET

- When V_{gs} is increasing, but still below or equal to V_{th}, holes get repelled from gate area, and a depletion region is formed. Therefore, due to the absences of charges, no current flows from source to drain.
- When $V_{gs} \geq V_{th}$ and V_{ds} (drain-source voltage) $\leq (V_{gs} - V_{th})$, device operates in triode region, it is also known as a linear region in which channel forms between source and drain region, and current flows from drain to source.
- When $V_{gs} \geq V_{th}$ and $V_{ds} \ll 2(V_{gs} - V_{th})$, device operates in the deep triode region, i.e. the source to drain path is represented as a linear resistor, and it is controlled by overdrive voltage, i.e. $V_{gs} - V_{th}$.
- When $V_{gs} \geq V_{th}$ and $V_{ds} > (V_{gs} - V_{th})$, device operates in the saturation region, where drain current is strongly controlled by gate voltage, and it became a weak function of V_{ds}.

10.1.2 Limitation of MOSFET

As we are moving forward with Moore's law, the packing density, speed and power dissipation improve with scaled down the technology. To keep device functioning properly, various parameters of the device should be scaled. Unfortunately, power supply voltages are not scaled simultaneously with another device dimension practically. Due to such device scaling, electric field increases, and device do suffer from various second-order effects or SCEs like hot carrier effect (HCE), drain induced barrier lowering (DIBL), mobility degradation, subthreshold current, subthreshold swing and leakage current, etc.

10.1.2.1 HCE

In N-MOSFET device as electrons travel from source to drain, it gains kinetic energy and becomes "hot" electron with an increase in the electric field due to increase in drain bias voltage. But in case of the short channel device, the localized charge carriers have sufficient energy for trapping Si–SiO_2 interface or tunnelling into oxide region. Hence, effective charge carrier concentration is reduced in channel region, which reduces the drain current in the device. This is also harmful to the device because it reduces the lifetime of the device. It can be minimized by the halo (light) doping at the drain and heavy doping at source side in the MOSFET device. This structure is labelled as lightly doped drain (LDD) MOSFET. The LDD MOSFET structure reduces the electric field between drain and channel region, which subsequently reduces the charge carrier's injection into the oxide region (Razavi 2002).

10.1.2.2 DIBL

In small channel length MOSFETs due to the improper scaling of device, too low channel doping or if the source/drain junctions are too deep results in unexpected electrostatic interactions between source–drain which is acknowledged as DIBL (Razavi 2002). Simplistically, the DIBL is considered to correspond to the expansion of the drain depletion region which merges with the source depletion region, which results in punch-through breakdown among source and drain. To prevent the device from DIBL effect, the source/drain junction must be made shallow with a reduction in channel length. Secondly, the channel doping must be made sufficiently high to prevent the source junction control by drain.

10.1.2.3 Mobility Degradation

As drain current is dependent on the mobility of charge carrier, it is an important parameter. The mobility degradation occurs mainly due to two reasons as follows:

Lateral field effect: With increase in drain voltage, this effect increases. When carriers are travelling from source to drain, they suffer through microscopic roughness at the oxide-silicon interface and due to which scattering occurs and results in degradation of mobility.

Vertical field effect: This effect occurs due to application of V_g (gate voltage). Scattering phenomenon occur at the interface of oxide and substrate interface with increasing gate voltage, and due to this mobility of the devices will degrade.

10.1.2.4 Subthreshold Conduction

In ideal MOSFET devices, it is examined that drain current goes to zero as soon as V_{gs} reduces to V_{th}. But in reality, still some amount of drain conduction occur below threshold voltage due to the presence of weak inversion channel under the gate which allow the charges to flow from source to drain, and this conduction of drain current is known as subthreshold conduction. Due to this, turn off condition of a device is failed for the gate voltage below the threshold voltage, and it could be worse by DIBL effect.

10.1.2.5 Punch Through

When V_{ds} increases, the depletion region around the drain merges with source results in the formation of the single depletion region. Due to which, field below the gate becomes the strong function of V_{ds} and hence the drain current. This current is called as drain punch-through current, and the V_{pt} (punch-through voltage) is given by Eq. (10.2) (Razavi 2002).

$$V_{pt} = \frac{qN_aL^2}{2\varepsilon_s} \tag{10.2}$$

where q, N_a, L and ε_s are an electronic charge, doping concentration, channel length and dielectric constant of silicon, respectively.

10.1.2.6 Leakage Current

When the device parameter like supply voltage is scaled down for minimizing power dissipation in the MOSFET. We know that drain current is a function of overdrive voltage. Hence, the overdrive voltage ($V_{gs} - V_{th}$) should be as large as possible for achieving the maximum drain current. The overdrive voltage can be increased by scaling down the threshold voltage by a factor. The threshold voltage scaling down results in increasing the subthreshold leakage current exponentially. As the transistor is working in the weak inversion region (when $V_{gs} \leq V_{th}$), there is I_{sub} (subthreshold current) flows between the drain and source terminal and calculated by Eq. (10.3) (Semenov et al. 2003).

$$I_{sub} = I_0 e^{\left(\frac{V_{gs}-V_{th}}{\eta V_T}\right)}\left(1 - e^{-\frac{V_{ds}}{V_T}}\right) \tag{10.3}$$

$$I_0 = \frac{W\mu_0 C_{ox} V_T^2}{L}(\eta - 1) \tag{10.4}$$

$$V_T = \frac{kT}{q} \tag{10.5}$$

where, W, L, V_{gs}, V_{th}, V_{ds}, C_{ox}, V_T and μ_o are the width, length of the channel, gate to source biasing voltage, threshold voltage, oxide capacitance, thermal voltage and mobility of charge carriers, respectively.

10.1.3 Alternative Solution of MOSFET

The decrease in performance of the MOSFET device occurs, due to the occurrence of SHEs and second-order effects in a MOSFET device, as the technology scales down. Due to the high OFF state current in the device, OFF state condition is highly affected. The subthreshold slope of the MOS device is approximated to 60 mV per decade at room temperature, and it is highly dependent on the thermal voltage. The ON current to OFF current ratio is very less in this device due to the low ON current. The drain current depends on the drift and diffusion mechanism of this device due to which reduction in subthreshold slope is impossible. Therefore, we are moving towards the different devices in which different current conduction mechanism is

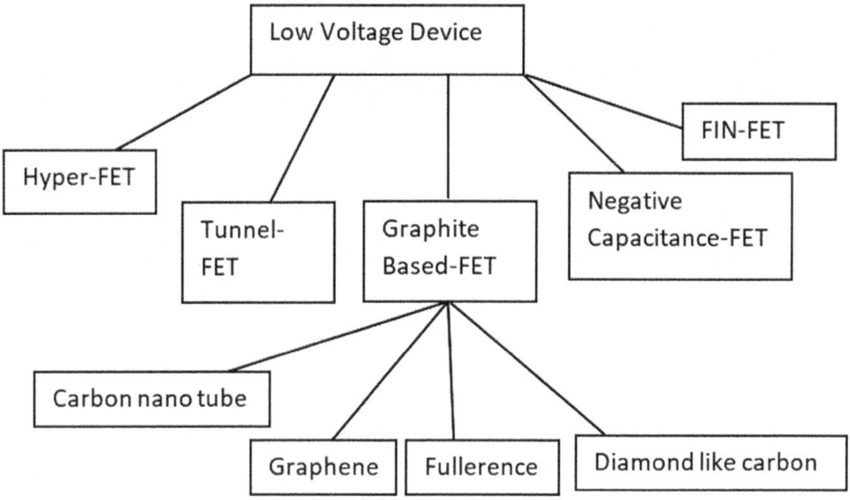

Fig. 10.3 Different low voltage device

used for the transport of majority and minority carriers from source to drain. These devices can be given as FINFET, carbon nanotube field effect transistor (CNTFET), nano wire, hyper-FET, tunnel field effect transistor (TFET), Negative capacitance FET (NCFET) (Fig. 10.3).

10.2 CNTFET

Carbon nanotubes (CNTs) were invented by Sumio Iijima in Japan in 1991. The biggest breakthrough in the invention of first carbon nanotube computer came in 1998 at Delft University of Technology when Dekker and colleagues made the first practical carbon nanotube transistor, and also, the first CNTFET was fabricated. Carbon nanotubes (CNTs) are cylindrical structures of graphene of Nano-scaled diameters wrapped up to form a tube (Ray Chowdhury and Roy 2005). The single sheet of graphite called graphene. Graphene is a two-dimensional carbon structure, which is held together by carbon–carbon bonds. These bonds provide extraordinary strength to graphene, making it stronger than steel. Carbon nanotubes are allotropes of carbon which belongs to the fullerenes family and are sheets of graphene rolled in the shape of a tube having a large length to diameter (132000000:1) ratio (lengths are in micro range and diameters in nm range).

Carbon nanotube can act as either a semiconductor or metallic depending on their chirality, i.e. angle of the atom arrangement along the tube (Ray Chowdhury and Roy 2005). Chirality vector is represented by the integer pair (n, m). The tube is metallic if $(n\text{-}m)$ is divisible by three, and the tube is semiconducting if $(n\text{-}m)$ is not divisible by three. The uniqueness of CNFETs lies in its ability that the threshold

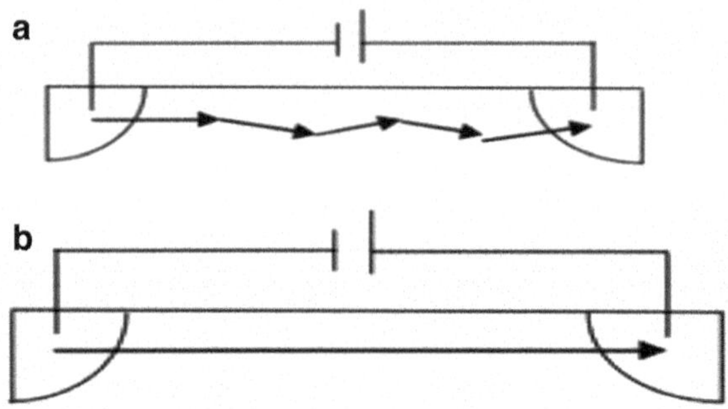

Fig. 10.4 **a** Diffusive electron transport, **b** ballistic transport

voltage (V_{th}) can be controlled either by varying the chirality vector or by changing the diameter of the carbon nanotube. In CNFET, ballistic transport mechanism is used. In nano-sized devices, scattering free/collision free charge transport is possible under appropriate conditions. This is called ballistic transport (Ray Chowdhury and Roy 2005) (Fig. 10.4).

10.2.1 *Parameters*

10.2.1.1 Chirality

The circumference of carbon nanotube can be shown in terms of a chiral vector, which joins two equivalent sites of the two-dimensional graphene sheets, i.e.

$$\overrightarrow{C_h} = n * \overrightarrow{b_1} + m * \overrightarrow{b_2} \tag{10.6}$$

where n and m are integers called chiral indices and b_1 and b_2 are the unit vectors of the hexagonal honeycomb lattice. By considering the indexes (n, m), a CNT can be determined whether it is a metallic or semiconducting. The nanotube is metallic if $n = m$ or $n - m = 3i$, where i is an integer otherwise, the tube is semiconducting (Zhou 2014). The structural parameters such as unit cell and its carbon atoms, diameter as well as size and shape of brillouin can be determined by the chiral vector of the tube and geometry of the graphene lattice.

Chiral vector is categorized into three types:

- Zigzag
- Armchair
- Chiral

If the cylinder axis has along x-axis (Fig. 10.5), the resulting tube called zigzag (n, 0) CNTFET. When the cylinder axis is in the y-direction, the tube formed is called an armchair (n, n) CNTFET. In case the axis of the cylinder is neither x- nor y-axis direction, the resulting nanotube is called chiral (n, m) CNTFET (Zhou 2014).

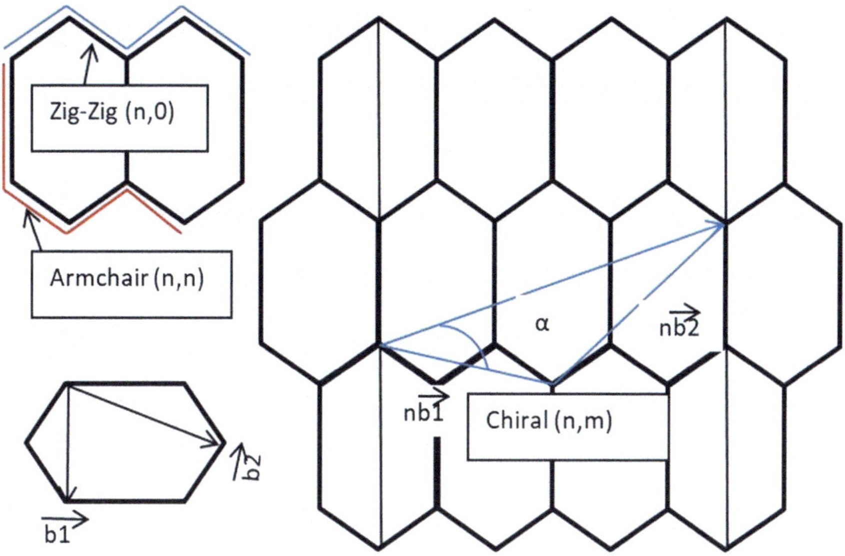

Fig. 10.5 Three different types of nanotubes

10.2.1.2 Diameter of CNTFET

The diameter of CNT is depending upon the chirality vector (*m*, *n*) as follows (Lin et al. 2009), i.e.

$$D_{\mathrm{cntfet}} = \frac{\sqrt{3} * A_{\mathrm{c-c}} * \sqrt{(n^2 + n * m + m^2)}}{\pi} \tag{10.7}$$

where, $A_{c\text{–}c} = 0.142$ nm is inter atomic distance between each carbon atom and its neighbours.

10.2.1.3 Threshold Voltage of CNTFET

The threshold voltage of a CNTFET (Lin et al. 2009) is given by

$$V_{\mathrm{th}} = \frac{\sqrt{3} * A_{\mathrm{c-c}} * V_{\pi}}{3 * e * D_{\mathrm{cntfet}}} \tag{10.8}$$

$$V_{\mathrm{th}} = \frac{0.43}{D_{\mathrm{cntfet}}(\mathrm{nm})} \tag{10.9}$$

where, $A_{\mathrm{c\text{–}c}}$ is the carbon to carbon atom distance (2.49 A^0), $V = 3.033$ eV bond energy of two atom, $D_{\mathrm{cntfet}} =$ Diameter of the CNT and e = Unit electron charge. As the threshold voltage of carbon nanotubes FET modifies, the diameter of the carbon nanotubes will also modify (Guo et al. 2004).

10.2.1.4 Chiral Angle of CNTFET

The direction of the chiral vector is measured by the chiral angle α. The chiral angle α can be calculated as following.

$$\cos\alpha = \frac{\frac{(n+m)}{2}}{\sqrt{(n^2 + n * m + m^2)}} \tag{10.10}$$

The changes in the chiral angle and the diameter cause the changes in the properties of the carbon nanotubes (Sinha et al. 2014).

10.2.1.5 Energy Band Gap of CNTFET

For a semiconducting carbon nanotube, the band gap can be varied by varying tube diameter.

The energy band gap of carbon nanotube FET is inversely proportional to the D_{cntfet} following as (Lin et al. 2009).

$$E_G = \frac{0.84}{D_{\text{cntfet}}(\text{nm})}\text{eV} \tag{10.11}$$

As previously mentioned and due to Eqs. (10.6) and (10.7), modification of CNTFET threshold voltage is possible only by changing the diameter of the nanotubes. So, CNTFETs are appropriate for implementing multiple threshold circuits. By changing the chiral vector indices, the nanotube diameter of transistor changes, and consequently, the threshold voltage of CNTFET sets simply.

10.2.2 Operation of CNTFET

The principle operation of carbon nanotube FET is almost same as silicon devices. It has three or four terminal devices where channel material is replaced with semiconductor carbon nanotube, which is bridging elements of source and drain contacts. The tube has heavily doped source and drain regions. Electrostatically, this device can be turned on/off via the gate. Since chirality vector is equivalent to $\overrightarrow{C_h} = n * \overrightarrow{b_1} + m * \overrightarrow{b_2}$, let $m = 0$ always in this formula, then for two CNTFETs having dissimilar chirality vectors (Lin et al. 2009).

$$\frac{V_{\text{th1}}}{V_{\text{th2}}} = \frac{D_{\text{cntfet1}}}{D_{\text{cntfet2}}} = \frac{n_2}{n_1} \tag{10.12}$$

Here, V_{th1} and V_{th2} are the respective threshold voltage for two CNFETs. The threshold voltages (V_{th}) can be varied by varying chirality of the CNT. The electronic structure of carbon nanotubes is purely dependent upon their physical structure (chirality and diameter) which is unique when compared to other materials (Lin et al. 2009). To achieve high performance characteristics, CNTs are best suited because the transport as well as ballistic or near-ballistic transport is determined with low voltage bias. It have mean free path for elastic scattering.

Several CNTs could be placed nearby each other under the transistor gate and set its width. The number of tubes which are placed under the transistor gate determines the width of a CNTFET transistor. The width also depends on the distance between two adjacent tubes which is called a pitch (Lin et al. 2009). Therefore, the width of a transistor is determined by the following equation.

$$W_{\text{gate}} = (W_{\text{min}}, N * \text{Pitch}) \tag{10.13}$$

where, N is the number of nanotubes that are placed under the transistor gate and W_{min} is the minimum width of the gate.

Carbon nanotubes, as novel materials with unique electronic characteristics, have been anticipated to be exploited to construct electronic devices for their better physical properties than those of conventional silicon, for example, longer mean free path, larger carrier mobility and higher transport current density.

10.2.3 *Types of CNTFETs*

On the basis of operation, CNTFETs are categorized into two types:

- Schottky barrier CNTFET (SB-CNTFET)
- MOSFET like CNTFET
- Virtual source CNTFET (VS-CNTFET)

10.2.3.1 Schottky Barrier CNTFET (SB-CNTFET)

In this type of CNFETs, a semiconducting CNT channel is connected directly to metallic source and drain contacts. As in ordinary contacts between semiconductors and metals, a Schottky barrier is formed at the interface. Charge carriers transport from the contacts to the channel by quantum mechanical tunnelling through the barriers (O'Connor et al. 2007). The tunnelling rate of charge carriers and the transistor current are controlled by the gate by changing the thickness of the Schottky barrier. This way of operation is different from the conventional transistors where the current switching is accomplished by modifying the channel conductance not the

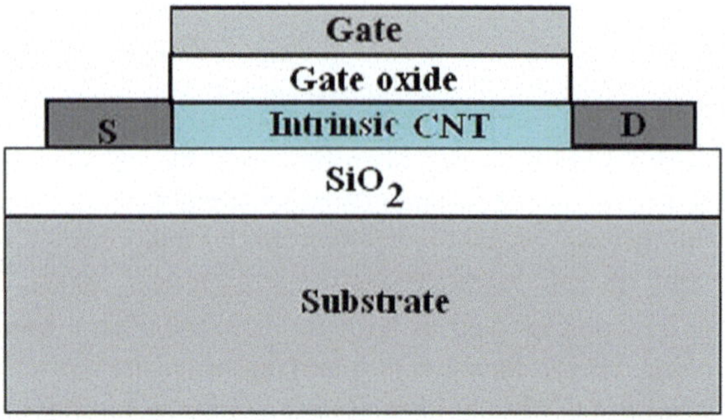

Fig. 10.6 Schottky barrier CNFET

contact resistance. Thus, the operation of the transistor is basically controlled by the electric field near the contact. Hence, transistor characteristics are affected by both the oxide thickness and the geometry of the metallic contact as proved by Heinze et al. (O'Connor et al. 2007) (Fig. 10.6).

10.2.3.2 MOSFET Like CNTFET

Here, heavily doped nanotubes sections as drain/source and intrinsic nanotubes section acting as channel exhibit substantially improved performance. It suppresses the ambipolar conduction found in SB carbon nanotubes FETs showing unipolar behaviour (O'Connor et al. 2007). The heavily doped semiconducting source and drain has a large band gap energy range for which no current is induced into the channel. The conductivity of the channel is modulated by the gate to source voltage (Yong-Bin 2011). The parasitic capacitance between the source and gate electrode is reduced offering faster operation. The leakage current is also very small in comparison with the SB-CNFETs (Fig. 10.7).

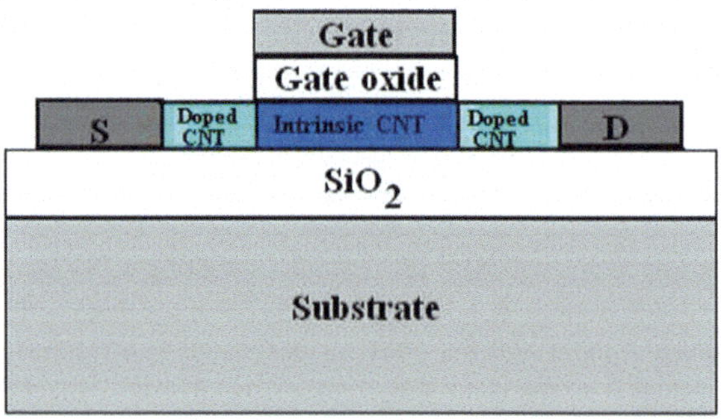

Fig. 10.7 MOSFET like CNFET

10.2.3.3 Virtual Source (VS-CNTFET)

The current voltage and capacitance voltage characteristics of MOSFET using carbon nanotube are described by semi-empirical model of VS-CNTFET. Tunnelling leakage currents, parasitic capacitance, parasitic resistance and scaling properties are included in the property of VS-CNTFET model. The product of carrier velocity and mobile charge density gives the drain to source current in VS-CNFET. The lateral electric field is small near the source in the ON state at the top of the energy barrier, and the gate voltage controls the potential of the top of energy barrier. The low-field carrier mobility (μ) and VS carrier velocity (V_{ox}) are few physical parameters of VS-CNFET (Lee et al. 2015).

10.3 Hybrid VS-CNTFET-CMOS Ring Voltage-Controlled Oscillator (RVCO)

10.3.1 Designed and Stages of Hybrid VS-CNTFET-CMOS RVCO

The hybrid VS-CNFET-CMOS RVCO consists of four stages such as input stage (transistor M1), current starved with power switching (top and bottom transistors), current starved circuitry (centre transistors) and ring oscillator.

10.3.1.1 Working of Hybrid VS-CNTFET-CMOS RVCO

The circuit of the hybrid VS-CNFET-CMOS RVCO using current starved with power switching technology is shown in Fig. 10.8. The working principle of ring oscillator and the hybrid VS-CNFET-CMOS RVCO is similar. CNTFET M1, operates as an amplifier, CNFETs M2 and M5 are operating as an inverter and providing delay in the circuit, while MOSFETs M3 and M4 are operating as a current starved. M3 and M4 are operating in the saturation region. The current sources ID1 and ID2 are equal to I1 and limit the current available to the inverter M2 and M5 (Zhou 2014).

10.3.1.2 Operating Frequency of Hybrid VS-CNTFET-CMOS RVCO

In order to calculate the operating frequency of proposed RVCO, the total time taken by capacitor C_{Total} to charge and discharge it which is seen by each inverter stage needs to be determined. The charging and discharging of the transistors take place only during the transitions, in inverter's triode region (Hwang et al. 2009; Kougianos and Mohanty 2009).

Therefore, C_{Total} can be written as follows:

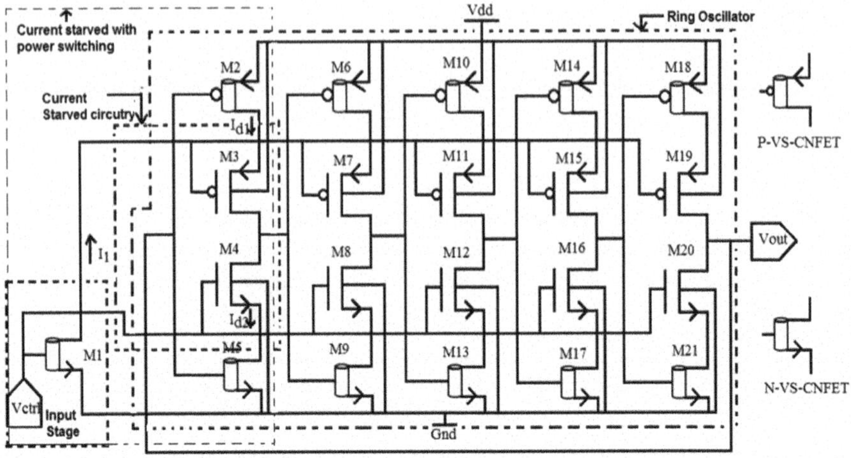

Fig. 10.8 Design and stages of hybrid VS-CNFET-CMOS RVCO

$$C_{\text{Total}} = C_{\text{out}} + C_{\text{in}} \tag{10.14}$$

where, C_{out} is output capacitance and C_{in} is input capacitance

$$C_{\text{out}} = C_{\text{ox}} * \left(W_{\text{P}} * L_{\text{p}} + W_{\text{Pc}} * L_{\text{pc}} + W_{\text{n}} * L_{\text{n}} + W_{\text{nc}} * L_{\text{nc}}\right) \tag{10.15}$$

$$C_{\text{in}} = C_{\text{ox}} * \frac{3}{2} * \left(W_{\text{P}} * L_{\text{p}} + W_{\text{Pc}} * L_{\text{pc}} + W_{\text{n}} * L_{\text{n}} + W_{\text{nc}} * L_{\text{nc}}\right) \tag{10.16}$$

From Eqs. 10.14, 10.15, and 10.16, the C_{Total} can be calculated as follows:

$$C_{\text{Total}} = C_{\text{ox}} * \frac{5}{2} * \left(W_{\text{P}} * L_{\text{p}} + W_{\text{Pc}} * L_{\text{pc}} + W_{\text{n}} * L_{\text{n}} + W_{\text{nc}} * L_{\text{nc}}\right) \tag{10.17}$$

where, C_{in} is input capacitance, C_{ox} is the gate oxide capacitance per unit area, C_{out} is output capacitance, W_{p}, W_{n} are the widths, and L_{p}, L_{n} are the lengths of the P-MOSFET and N-MOSFET transistors, W_{pc}, W_{nc} are the widths, and L_{pc}, L_{nc} are the lengths of the P-VS-CNFET and N-VS-CNFET transistors, respectively (Hwang et al. 2009; Kougianos and Mohanty 2009).

The operating frequency can be calculated by using following equation:

$$\text{Freq}_o = \frac{1}{2 * n * T_{\text{Total}}} \tag{10.18}$$

where

n odd number of inverters

T_{Total} total time taken by each stage of an inverter to charge or discharge the capacitance of it

$$T_{\text{Total}} = T_{\text{charge}} + T_{\text{discharge}} \tag{10.19}$$

where

T_{charge} Charging time from 0 to an inverter, V_{sp}
$T_{\text{discharge}}$ Discharging time from V_{dd} to V_{sp}

$$T_{\text{Total}} = C_{\text{Total}} * \frac{V_{\text{sp}}}{I_{\text{D1}}} + C_{\text{Total}} * \frac{V_{\text{dd}} - V_{\text{sp}}}{I_{\text{D2}}} \tag{10.20}$$

where, $I_{\text{D1}} = I_{\text{D2}} = I_{\text{D}}$. I_{D1}, I_{D2} and I_{D} are charging current, discharging current and inverter current, respectively (Tous et al. 2012).

$$T_{Total} = C_{Total} * \frac{V_{dd}}{I_D} \tag{10.21}$$

There from Eqs. 10.17, 10.18, 10.19, 10.20 and 10.21, the equation of operating frequency of the RVCO is as follows (Tous et al. 2012):

$$\text{Freq}_{\text{o}} = \frac{I_{\text{D}}}{2 * n * C_{\text{Total}} * V_{\text{dd}}} \tag{10.22}$$

The applied control voltage controlled the operating frequency which adjusts the current I_{D} following through each inverter stage.

10.3.1.3 Power Dissipation

The different components of power dissipation under the consideration of CNFET technology are as follows Eq. 10.23.

$$P_{\text{TPD}} = P_{\text{DPD}} + P_{\text{SCPD}} + P_{\text{SPD}} + P_{\text{GLPD}} \tag{10.23}$$

where are,

Dynamic Power Dissipation ($\mathbf{P_{DPD}}$)—whenever the transition of gate switches occurs from low to high (0 to 1) and from high to low (1 to 0) to charge or discharge the load capacitance, dynamic power occurs. The dynamic power dissipation can be given by Eq. 10.24 (Ben et al. 2010; Jan et al. 2003).

$$P_{\text{DPD}} = \alpha * C_{\text{Total}} * \text{Freq}_{\text{o}} * (V_{\text{dd}})^2 \tag{10.24}$$

Short-Circuit Power Dissipation (P_{SCPD})—The slope of the input waveform affects a direct path for a current between V_{dd} and ground during switching time period. During which, P-MOSFET and N-MOSFET transistor are conducting simultaneously. Therefore, it is not current to assume the zero rise and fall time of the input waveform. The short-circuit power dissipation can be given by Eq. 10.25 (Ben et al. 2010; Jan et al. 2003).

$$P_{SCPD} = 0.15 * P_{DPD} \quad (10.25)$$

Static Power Dissipation (P_{SPD})—Static power dissipation is given by the Eq. 10.26

$$P_{SPD} = i_{Off} * V_{dd} \quad (10.26)$$

where i_{Off} is the current which follows in the circuit when the switching activity absent (Tous et al. 2012; Ben et al. 2010).

Gate Leakage Power Dissipation (P_{GLPD})—The power dissipation due to gate leakage is occurring due to the tunnelling current through the gate oxide (Ben et al. 2010; Jan et al. 2003). It can be given by Eq. 10.27 as the following.

$$P_{GLPD} = i_G * V_{dd} \quad (10.27)$$

- **P_{TPD}—Total Power Dissipation**

10.3.1.4 Phase Noise

An ideal clock source would generate a pure sine or square wave. All signal power should be generated at the desired clock frequency. However, in actual, all clock signals have some degree of phase noise. The clock signal power spreads to adjacent frequencies due to the noise which results in noise sidebands. Phase noise is the representation of frequency domain of the clock noise. The phase noise is generally expressed in dBc/Hz and shows the amount of signal power at a given sideband or offset frequency from the ideal carrier frequency (Natesan 2003).

10.3.1.5 Jitter

The deviation in periodicity of practical clock signal from the actual periodicity reference clock signal is defined as the clock jitter (Natesan 2003).

Clock jitter is of three types:

Deterministic Jitter: It occurs due to process variation, design decisions like buffer size, length of a wire and other devices. As the name suggests, deterministic jitter is "controllable" and "predictable" (Natesan 2003).

Random Jitter: Random jitter is less predictable. It occurs due to interference between wires and circuitry modules and also due to capacitive coupling (Natesan 2003).

Source of Jitter: Instabilities in the oscillator electronics, thermal noise also external interferences which may occur due to power supply, ground and also due to output connection of the oscillator all these are the responsible sources of jitter. EMI radiation is the responsible parameter for occurrence of the deterministic jitter. Megan tic field from an EMI source such as RF signal sources, power supplies and AC power lines may affect a sensitive signal path (Yong-Bin 2011). Many sources occur responsible for random jitter such as mobility variation due to the thermal vibration of the semiconductor crystal structure which depends upon temperature of the materials. One of the sources of random jitter is nonuniform doping density in semiconductor process variation (Yu 2016).

10.3.1.6 Intrinsic Device Noise

It can be divided in two types

Thermal Noise: It occurs due to thermal agitation of charge carriers. The presence or absence of DC current does not affect thermal noise. Thermal noise and absolute temperature have the direct proportionality relationship. It is also called as white noise due to its flat power spectral density with frequency. Thermal noise dominates at high frequency (Yu 2016).

Flicker Noise: When electrons get trapped and released in gate oxide of MOSFET, flicker noise occurs. The 1/f spectral shape is of the power spectral density is due to the time contents involved in the trap and release mechanism (Yu 2016). Flicker noise dominates at low frequency.

10.4 Comparison Between Hybrid VS-CNTFET-CMOS RVCO with Conventional CMOS RVCO

10.4.1 Conventional CMOS RVCO

There are total 21 transistors are used in the design "conventional CMOS RVCO circuit". 10 P-MOSFET and 11 N-MOSFET are used to design the CMOS RVCO circuit (Table 10.1).

Table 10.1 Input parameters of CMOS RVCO for P-MOSFET and N-MOSFET

MOSFET		Length (nm)	Width (nm)
P-MOSFET	M2, M6, M10, M14, M18	45	500
	M3, M7, M11, M15, M19	45	120
N-MOSFET	M4, M8, M12, M16, M20	45	120
	M1, M5, M9, M13, M17, M21	45	500

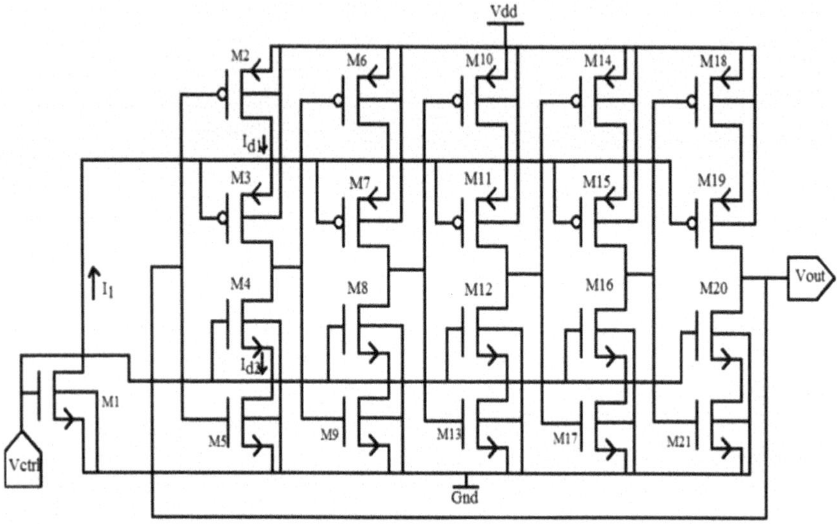

Fig. 10.9 Circuit of design of CMOS RVCO

10.4.2 Designed for Conventional CMOS RVCO

Figure 10.9 shows the circuit design of CMOS RVCO. The output voltage oscillation waveform has been shown in Fig. 10.10, which shows the voltage swing from 0 to 1 V. The input parameters are shown in Table 10.2.

10.4.3 Parameters of Conventional CMOS RVCO

10.4.3.1 RMS Jitter

After performing pss and pnoise analysis, the phase noise in design CMOS RVCO from the table, we can observe that as control voltage decreases, RMS jitter increases. We have achieved an RMS jitter of 867 fs for the control voltage at 1 V. The variation in RMS jitter with respect to control voltage is shown in Table 10.3.

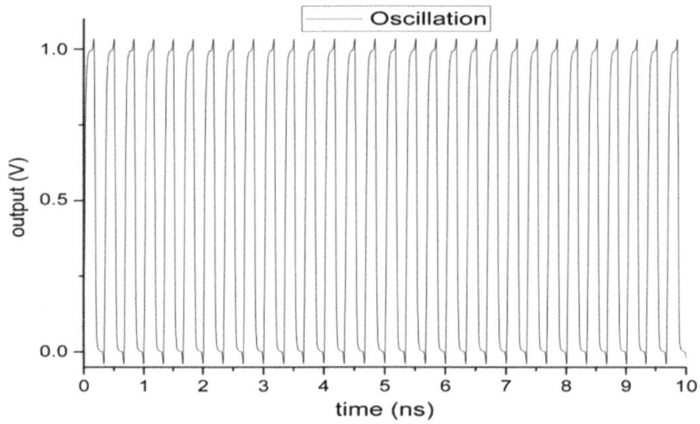

Fig. 10.10 Output voltage oscillation waveform of CMOS RVCO

Table 10.2 Input parameters of design of CMOS RVCO

Parameters	CMOS RVCO (V)
Voltage supply (V_{dd})	1
Control voltage (V_{ctrl})	1

Table 10.3 Variation in RMS jitter with respect to control voltage in Conventional CMOS RVCO

Control voltage (V)	RMS jitter
1	867 fs
0.9	1 ps
0.8	2.5 ps
0.7	6.7 ps

10.4.3.2 Operating Frequency

Figure 10.11 shows the operating frequency of CMOS RVCO. From the graph, we can say that the operating frequency of the proposed design gives approximately linear variation with receptive to the control voltage. As the control voltage varies from 0.7 to 1.0 V with the stepping of 0.1 V, operating frequencies vary from 1.345 to 2.99 GHz, respectively. In the design, we have achieved the operating frequency of 2.99 GHz for the control voltage of 1 V.

10.4.3.3 Power Dissipation

Figure 10.12 describes the power dissipation of CMOS RVCO. As the control voltage varies from 0.7 to 1.0 V with the stepping of 0.1 V, the power dissipation varies from

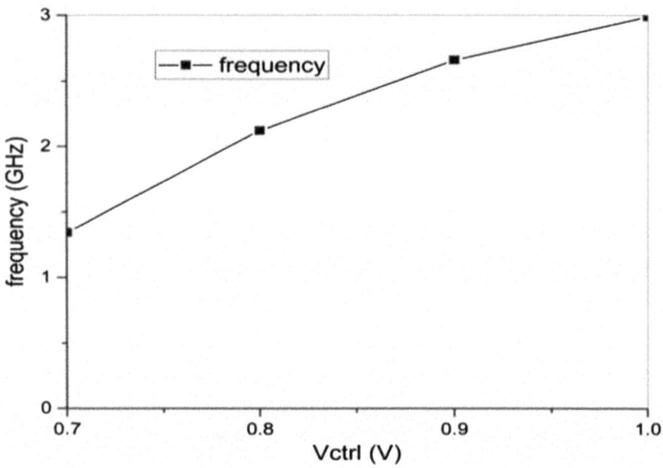

Fig. 10.11 Operating frequency of CMOS RVCO

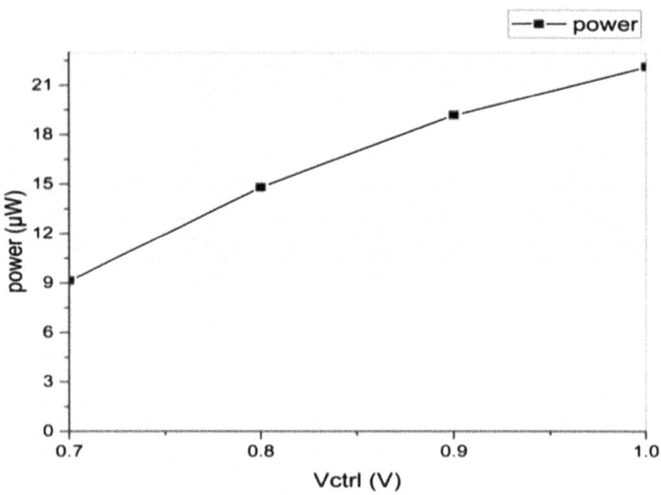

Fig. 10.12 Power dissipation of CMOS RVCO

9.15 to 22.15 μW, respectively. In the proposed design, we have achieved the power dissipation of 22.15 μW for the control voltage of 1 V.

10.4.3.4 Phase Noise

After performing pss and pnoise analysis, the phase noise in design CMOS RVCO is observed nearly −67.46 dBc/Hz at 1 MHz offset. Figure 10.13 describes the phase noise of CMOS RVCO.

10.5 Designed Input Parameters for P-VS-CNTFET, N-VS-CNTFET, P-MOSFET and N-MOSFET

Design input parameters for P-VS-CNFET and N-VS-CNFET are listed in Table 10.4. And design input parameters for P-MOSFET and N-MOSFET are shown in Table 10.5.

There are total 21 transistors are used in the proposed design "a low-power hybrid VS-CNFET-CMOS RVCO circuit". 5 P-MOSFET, 5 N-MOSFET, 5 P-VS-CNFET and 6 N-VS-CNFET are used to design the proposed hybrid RVCO circuit.

10.6 Designed for Basic Hybrid VS-CNTFET-CMOS Inverter

Figure 10.14 shows the circuit of the proposed design of basic hybrid VS-CNFET-CMOS inverter. The output voltage waveform has been shown inFig. 10.15, which represents that better voltage swings from 0 to 1 V. The basic input parameters details are listed in Table 10.6.

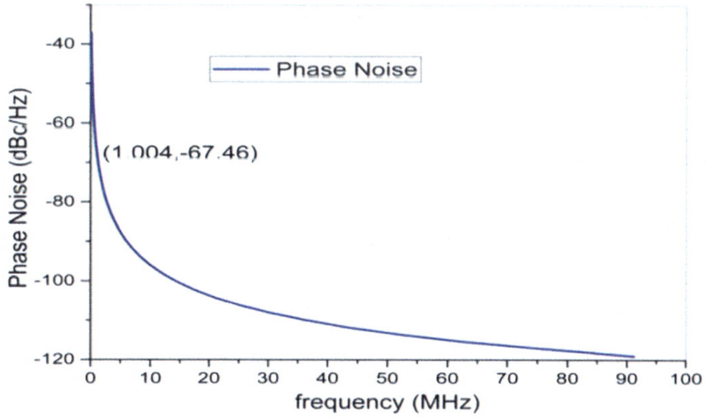

Fig. 10.13 Phase noise of CMOS RVCO

Table 10.4 Input parameters of proposed design hybrid RVCO for P-VS-CNFET and N-VS-CNFET

Parameters	P-VS-CNFET	N-VS-CNFET
Gate length (L_g)	45 nm	45 nm
Contact length (L_c)	11 nm	11 nm
S/D extension length	3 nm	3 nm
Device width (W_c)	500 nm	500 nm
Gate height (H_g)	15 nm	15 nm
Gate oxide thickness (t_{ox})	3 nm	3 nm
Gate oxide dielectric constant (k_{ox})	25	25
Diameter of CNFET (d)	1 nm	1 nm
Spacing between the VS-CNFETs (s)	400 nm	400 nm
Device structure (geo mod)	Gate-all-around (1)	Gate-all-around (1)
Flat band voltage (V_{fb})	0	0
Fermi level to the band edge at S/D (e_{fsd})	0.258 eV	0.258 eV

Table 10.5 Input parameters of proposed design hybrid RVCO for P-MOSFET and N-MOSFET

Parameters	P-MOSFET (nm)	N-MOSFET (nm)
Channel length (L)	45	45
Device width (W)	120	120

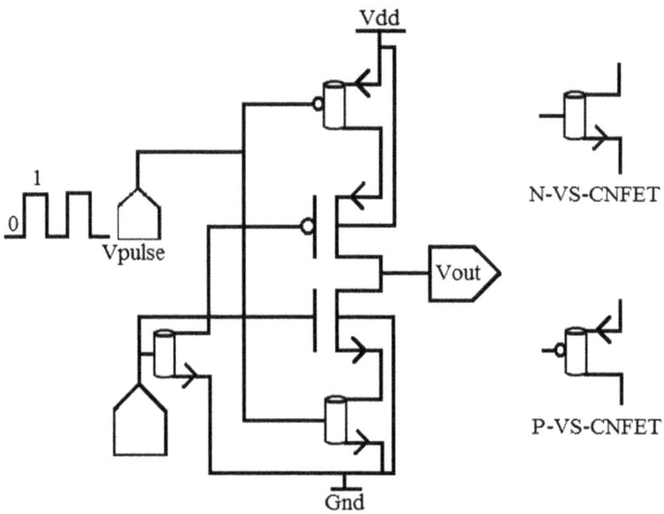

Fig. 10.14 Proposed design of basic hybrid VS-CNFET-CMOS inverter

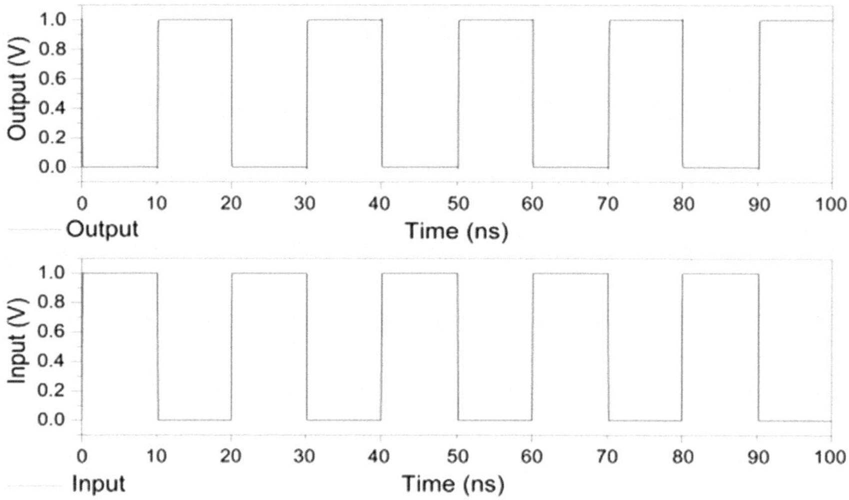

Fig. 10.15 Output voltage waveform of basic hybrid VS-CNFET-CMOS inverter

Table 10.6 Input parameters of proposed design hybrid RVCO for P-MOSFET and N-MOSFET

Parameters	Design of basic hybrid VS-CNFET-CMOS inverter (V)
Voltage supply (V_{dd})	1
Control voltage (V_{ctrl})	1
Voltage pulse (V_{pulse})	0 to 1

10.7 Designed for Hybrid VS-CNTFET-CMOS RVCO

Figure 10.16 shows the circuit design of hybrid VS-CNFET-CMOS RVCO. The output voltage oscillation waveform has been shown in Fig. 10.17, which shows the better voltage swing from 0 to 1 V. The input parameters are shown in Table 10.7.

10.7.1 Parameters of Hybrid VS-CNTFET- CMOS RVCO

10.7.1.1 RMS Jitter

We can calculate the standard deviation, peak-to-peak value and average clock period if number of clock period has been given. Peak-to-peak value, standard deviation also refer to peak-to-peak jitter, RMS jitter, respectively. After performing pss and pnoise analysis, the phase noise in proposed design hybrid VS-CNFET-CMOS RVCO from the table, we can observe that as control voltage decreases, RMS jitter increases. We have achieved an RMS jitter of 700 fs for the control voltage at 1 V. we have

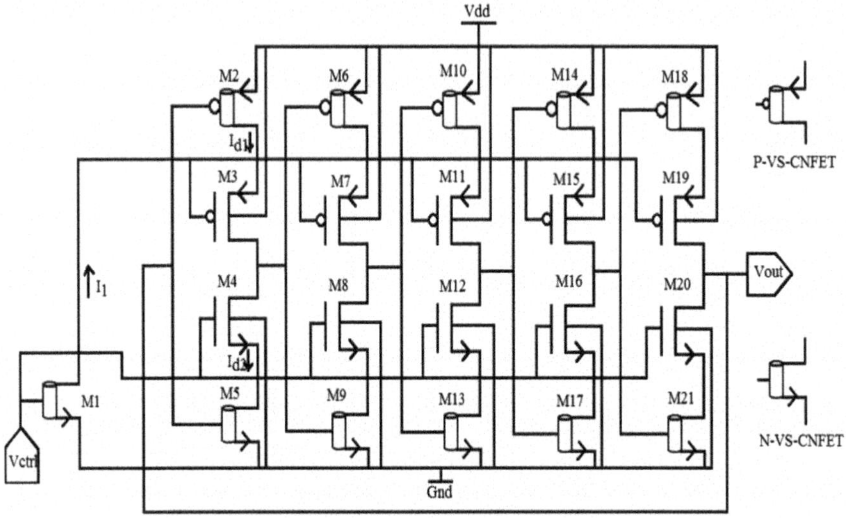

Fig. 10.16 Circuit of design of hybrid VS-CNFET-CMOS RVCO

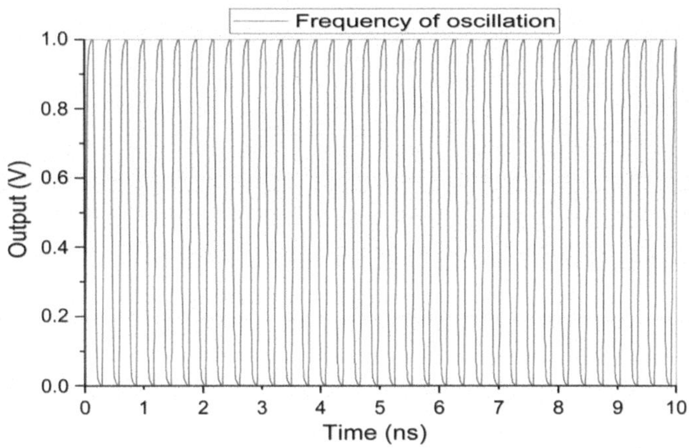

Fig. 10.17 Output voltage oscillation waveform hybrid VS-CNFET-CMOS RVCO

Table 10.7 Input parameters of design of hybrid VS-CNFET-CMOS RVCO

Parameters	Design of hybrid VS-CNFET-CMOS RVCO (V)
Voltage Supply (V_{dd})	1
Control Voltage (V_{ctrl})	1

Table 10.8 Variation in RMS jitter with respect to control voltage Hybrid VS-CNTFET-CMOS RVCO.

Control voltage (V)	RMS jitter
1	700 fs
0.9	1.1 ps
0.8	1.85 ps
0.7	4 ps
0.6	10.36 s

achieved the peak-to-peak RMS jitter of 5.2 ps. which is an advantage as compared to other RVCO. The variation in RMS jitter with respect to control voltage is shown in Table 10.8.

10.7.1.2 Operating Frequency

Figure 10.18 shows the operating frequency of hybrid VS-CNFET-CMOS RVCO. From the graph, we can say that the operating frequency of the proposed design gives approximately linear variation with receptive to the control voltage. As the control voltage varies from 0.6 to 1.0 V with the stepping of 0.1 V, operating frequencies vary from 0.908 to 3.43 GHz, respectively. In the proposed design, we have achieved the operating frequency of 3.43 GHz for the control voltage of 1 V. From Table 10.9, we can say that as the threshold voltage decreases, operating current increases, and hence, the operating frequency also increases.

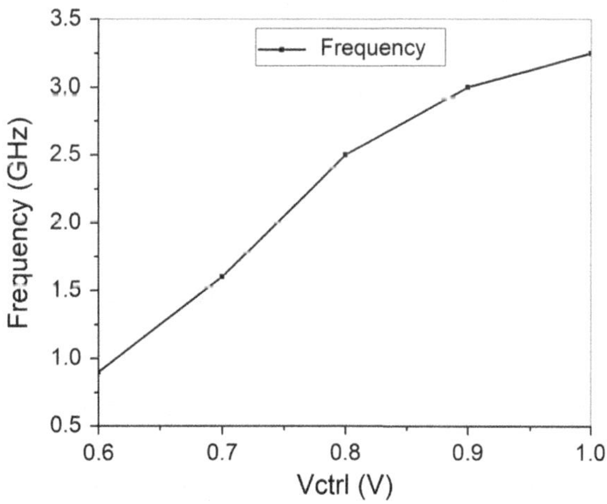

Fig. 10.18 Operating frequency of hybrid VS-CNTFET-CMOS RVCO

Table 10.9 Variation in operating frequency with respect to threshold voltage

Threshold voltage (V)	Frequency
0.426	2.97 GHz
0.355	3.43 GHz
0.284	3.73 GHz
0.236	3.84 Hz

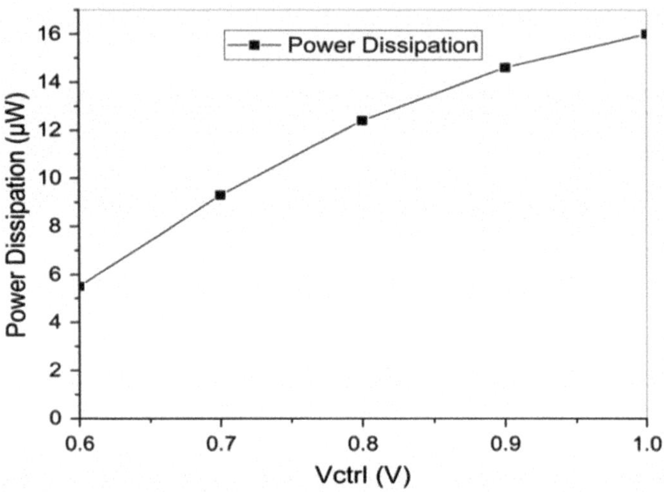

Fig. 10.19 Power dissipation of hybrid VS-CNFET-CMOS RVCO

10.7.1.3 Power Dissipation

Figure 10.19 describes the power dissipation of hybrid VS-CNFET-CMOS RVCO. As the control voltage varies from 0.6 to 1.0 V with the stepping of 0.1 V, the power dissipation varies from 5.5 to 15.97 μW, respectively. In the proposed design, we have achieved the power dissipation of 15.97 μW for the control voltage of 1 V.

10.7.1.4 Phase Noise

After performing pss and pnoise analysis, the phase noise in proposed design hybrid VS-CNFET-CMOS RVCO is observed nearly −69.57 dBc/Hz at 1 MHz offset. Figure 10.20 describes the phase noise of hybrid VS-CNFET-CMOS RVCO (Table 10.10).

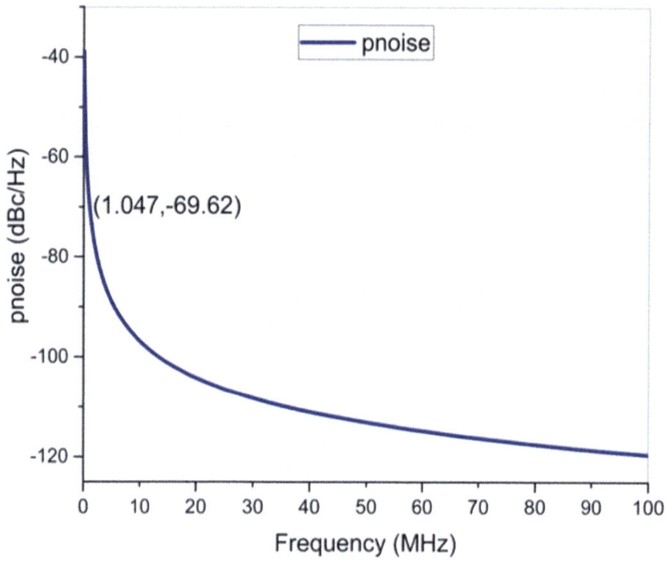

Fig. 10.20 Phase noise of hybrid VS-CNFET-CMOS RVCO

Table 10.10 Comparative analysis of different design circuits

Parameters	Hybrid VS-CNTFET-CMOS RVCO	CMOS RVCO	(Islam et al. 2017)	(Hwang et al. 2009)	(Raman and Sarin 2011)	(Chuang et al. 2004)
Voltage supply (V)	1	1	1	2	1.8	1.8
Technology	45 nm	45 nm	0.18 μm	0.35 μm	0.18 μm	0.18 μm
Operating frequency	0.908 to 3.43 GHz	1.345 to 2.99 GHz	4.52 to 6.02 GHz	1 to 25 MHz	0.958 to 4.43 GHz	440 to 1595 MHz
Phase noise	−69.57 dBc/Hz at 1 MHz offset	−67.46 dBc/Hz at 1 MHz offset	−76.27 dBc/Hz at 1 MHz offset	−42.1 dBc/Hz at 1 MHz offset	−94.51 dBc/Hz at 1 MHz offset	−93 dBc/Hz at 1 MHz Offset
Average power dissipation	15.97 μW	22.15 μW	0.295 mW	69 μW	0.226 mW	26 mW
RMS jitter	700 fs	867 fs	–	329 ps	–	–
Peak-to-peak jitter	5.2 ps	6.45 ps	–	–	–	–

10.8 Summary

A low-power hybrid VS-CNFET-CMOS RVCO using current starved power switching technology is designed, with 45 nm technology. The hybrid VS-CNFETCMOS RVCO shows better performance over the conventional CMOS RVCO using the same technology. The supply voltage (V_{dd}) equals 1 V is applied to both hybrid VS-CNFET-CMOS RVCO and conventional CMOS RVCO. The conventional CMOS RVCO is working on control voltage ranges from 0.7 to 1 V. The simulated results of conventional CMOS RVCO from the Cadence Virtuoso show that the operating frequency tuning range, power dissipation range, the phase noise and RMS jitter range are from 1.345 to 2.99 GHz, 9.15 to 22.15 μW, −61.32 dBc/Hz to −67.46 dBc/Hz at 1 MHz offset and 6.7 ps to 867 fs, respectively. The simulated peak-to-peak jitter is 6.45 ps at 1 V. The hybrid VS-CNFET-CMOS RVCO was optimized for low power dissipation by changing VS-CNFET parameters like intrinsic drive current, gate capacitance, gate width (W_c), gate length (L_g), VS-CNFET diameter (d), spacing between the VS-CNFETs (s) and gate oxide thickness (t_{ox}). The hybrid VS-CNFET-CMOS RVCO is working on control voltage ranges from 0.6 to 1 V. The simulated results from the Cadence Virtuoso show that the operating frequency of hybrid VS-CNFET-CMOS RVCO is inversely proportional to the threshold voltage (V_{th}) and average power dissipation of hybrid VSCNFET-CMOS RVCO is directly proportional to the threshold voltage (V_{th}). The simulated results of hybrid VS-CNFET-CMOS RVCO from the Cadence Virtuoso show that the operating frequency tuning range, power dissipation range, the phase noise and RMS jitter range are from 0.908 to 3.43 GHz, 5.5 to 15.97 μW, −58.89 dBc/Hz to −69.57 dBc/Hz at 1 MHz offset and 10.36 ps to 700 fs, respectively. The simulated peak-to-peak jitter is 5.2 ps at 1 V. The hybrid VS-CNFET-CMOS RVCO can be used for the applications like phase locked loops.

References

Akinwande D, Yasuda S, Paul B et al (2008) Monolithic integration of CMOS VLSI and carbon nanotubes for hybrid nanotechnology applications. IEEE Trans Nanotechnol 7(5):636–639

Appenzeller J, Knoch J, Martel R et al (2002) Carbon nanotube electronics. IEEE Trans Nanotechnol 1(4):184–189

Ben J, Haykel M, Mohanram K, De Micheli G (2010) Power consumption of logic circuits in ambipolar carbon nanotube technology. In: Proceedings of the conference on design, automation and test in Europe. European Design and Automation Association, pp 303–306

Borkar S (2003) Getting gigascale chips: challenges and opportunities in continuing moore's law. Queue 1(7):26

Chakraborty et al (2007) Hybridization of CMOS with CNT-based nano-electromechanical switch for low leakage and robust circuit design. IEEE Trans Circuits Syst I Regul Pap 54(11):2480–2488

Cho TS, Lee KJ, Pan T, Kong J, Chandrakasan AP (2007) Design and characterization of CNT-CMOS hybrid systems. MTL Annu Res Rep

Chuang Y-H, Jang S-L, Lee J-F, Lee S-H (2004) A low voltage 900 MHz voltage controlled ring oscillator with wide tuning range. In: The 2004 IEEE Asia-Pacific conference on circuits and systems. Proceedings, vol 1. IEEE, pp 301–304
Guo J, Ali J, Hongjai D, Lundstrom M (2004) Performance analysis and design optimization of near ballistic carbon nanotube field-effect transistors. In: IEEE international electron devices meeting, IEDM Technical Digest. IEEE, pp 703–706
Hajimiri A, Limotyrakis S, Lee T (1999) Jitter and phase noise in ring oscillators. IEEE J Solid State Circuits 34(6):790–804
Hwang Y-S, Kung C-M, Lin H-C, Chen J-J (2009) Low-sensitivity, low-bounce, high-linearity current-controlled oscillator suitable for single-supply mixed-mode instrumentation system. IEEE Trans Ultrason Ferroelectr Freq Control 56(2):254–262
Islam R, Suprotik ANK, Uddin SMZ, Amin MT (2017) Design and analysis of 3 stage ring oscillator based on MOS capacitance for wireless applications. In: International conference on electrical, computer and communication engineering (ECCE). IEEE, pp 723–727
Jan MR, Anantha C, Borivoje N (2003) Digital integrated circuits: a design perspective
Jovanovic G, Stojcev M (2006) Current starved delay element with symmetric load. Int J Electron 93(3):167–175
Jovanovic G, Stoj M, Stamenkovic Z (2010) A CMOS voltage controlled ring oscillator with improved frequency stability. Sci Publ State Univ Novi Pazar 2(1):1–9
Kahng D (1976) A historical perspective on the development of MOS transistors and related devices. IEEE Trans Electron Devices 23(7):655–657
Kougianos E, Mohanty SP (2009) Impact of gate-oxide tunnelling on mixed-signal design and simulation of a Nano-CMOS VCO. Microelectron J 40(1):95–103
Lee C-S, Pop E, Franklin AD, Haensch W, Philip Wong H-S (2015) A compact virtual-source model for carbon nanotube field-effect transistors in the sub-10-nm regime-part II extrinsic elements, performance assessment, and design optimization. arXiv preprint arXiv:1503.04398
Lin S, Kim Y, Lombardi F (2009) CNTFET-based design of ternary logic gates and arithmetic circuits. IEEE Trans Nanotechnol 10:217–225
Moore GE (1998) Cramming more components onto integrated circuits. Proc IEEE 86(1):82–85
Natesan P (2003) Comparison and analysis of jitter in CMOS ring oscillators
O'Connor I, Liu J, Gaffiot F (2007) CNTFET-based logic circuit design. IEEE Trans Nanotechnol 42(4):201–212
Rahane SB, Kureshi AK (2017) A low power and linear voltage controlled oscillator using hybrid CMOS-CNFET technology. Int J Appl Eng Res 12(9):1969–1973. ISSN 0973-4562
Raman A, Sarin RK (2011) 1P6M 0.18-[μm] low power CMOS ring oscillator for radio frequency applications. Int J Comput Theory Eng 3(6):770
Ray Chowdhury A, Roy K (2005) Carbon-nanotube-based voltage-mode multiple-valued logic design. IEEE Trans Nanotechnol 4(2):168–179
Razavi B (1997) A 2-GHz 1.6-mW phase-locked loop. IEEE J Solid State Circuits 32:730–735
Razavi B (2002) Design of analog CMOS. Integrated circuits. Professor of Electrical Engineering
Semenov Oleg, Vassighi Arman, Sachdev Manoj (2003) Leakage current in sub-quarter micron MOSFET: a perspective on stressed delta IDDQ testing. J Electron Test 19(3):341–352
Sinha SK, Singh P, Chaudhury S (2014) Effect of temperature and chiral vector on emerging CNTFET device. In: 2014 International conference on computing for sustainable global development (INDIACom). IEEE, pp 432–435
Sun L, Kwasniewskl TA (2001) A 1.25-GHz 0.35-m monolithic CMOS PLL based on a multiphase ring oscillator. IEEE J Solid State Circuits 36:910–916
Tous SI, Asghari E, Pourandoost R, Razeghi B (2012) A 0.4 V low frequency voltage-controlled ring oscillator Using DTMOS technique
Yang MH et al (2006) Advantages of top-gate, high-k dielectric carbon nanotube field-effect transistors. Appl Phys Lett 88(11):113507
Yong-Bin K (2011) Integrated circuit design based on carbon nanotube field effect transistor. Trans Electr Electron Mater 12(5):175–188

Yu RB (2016) Design, analysis and simulation of a jitter reduction circuit (JRC) system at 1 GHz
Zhou YS (2014) Laser-assisted nanofabrication of carbon nanostructures. Austin J Nanomed Nanotechnol 2(2):1–18

Chapter 11
Chip-Level Optical Interconnect in Electro-optics Platform

Sajal Agarwal and Y. K. Prajapati

Abstract Interconnects are the basic connections used to establish between two silicon-based chips or devices. Basically, the interconnect quality and size differ based on the physics on which these work, such as electrical and/or optical. With the increased demand of the high quality and speedup communication, it is essential to work on the various different aspects of the chip. Conventional interconnects used in electronics devices are basically electrical, and these are reaching to their limits. Since Moor's law suggests that the density of electrical component gets double in every 18 months, but with the increasing density of the components on/off chip, it is not possible to scale interconnect beyond a particular limit. Optical interconnects are feasible option which overcome the delay, loss, parasitic capacitance, etc., of electrical interconnect. In optical interconnects, nonlinear signals are transmitted through silicon-based waveguide through either two-dimensional or three-dimensional fabrication. Progression in nanotechnology made it viable to arrange light source; laser, medium; waveguide, and detector; and photodiode into a single silicon chip. However, there are still a lot of challenges to commercially implement dense optical interconnects to silicon chips, such as losses, packaging, and integration of two different technologies in single chip. It is observed that optical interconnect technology is not mature enough and needs a thorough analysis. On the designing aspect, there are a lot of features of optical interconnects which need to be addressed. Thus, this chapter is focused on optical interconnects for silicon on insulator (SOI) chips, analyzed the work which is already done and the basic challenges in this technology to make it practical.

S. Agarwal
Electronics and Communication Engineering Department, Jaypee Institute of Information Technology, Noida, Uttar Pradesh 201309, India
e-mail: sajal.agarwal@jiit.ac.in

Y. K. Prajapati (✉)
Electronics and Communication Engineering Department, Motilal Nehru National Institute of Technology Allahabad, Allahabad, Uttar Pradesh 211004, India
e-mail: yogendrapra@mnnit.ac.in

R. Dhiman and R. Chandel (eds.), *Nanoscale VLSI*,
Energy Systems in Electrical Engineering,
https://doi.org/10.1007/978-981-15-7937-0_11

Keywords Optical interconnect · Parasitic capacitance · Nanotechnology · Capacity · Power dissipation

11.1 Introduction

Growing need of automation and dependency on electronics make it necessary to improve the built quality of the products. However, not only electronics, communication scenario is also evolved from 1948 to 2019 (Leydesdorff 1994), i.e., mathematical theory given by Shannon to 2 Gbps speed using optical fiber (Sezgin et al. 2019), but the backbone of any communication system is electronics. Since silicon-based devices (electronics) reach their limit of miniaturization and carry the data, it is highly recommended to switch or search other elements for channeling the data. It is widely accepted fact that silicon-based electronics is the backbone of our present and future. Most of the VLSI systems are limited by the power dissipation and time delay using electrical interconnects. Initially, in the early 1900s (Miller 2000) it was analyzed that replacing all the electrical component with the optical logic devices certainly is not a good idea due to its higher energy consumption; however, the idea of using optics in communication is much viable for long distance. Till then, use of optical interconnects either in chips or in intra-chip level was not considered, and perhaps electrical wires were capable of carrying maximum power with good efficiency at that moment (Liu et al. 2018). Further, improvement in VLSI technology means, reduced size of individual component for higher density over a single chip. This reduction does not only improve the throughput, but introduce some loss components which severely affect the device capability, such as parasitic capacitance, cross talk, and intersymbol interference (ISI) (Lanzillo et al. 2018). However, it is separately researched by the optical community to employ optical concept to the chip level to carry the signals. If this integration could happen, this would result much faster and light weighted dense electronic device. This idea of integration of optical interconnects with the electronics components was proposed by Goodman et al. (1984). Proposed study pointed out numerous reasons why this integration is important. Nevertheless, in the absence of practicality at that time this idea was obsoleted by the scientific community. Although the VLSI industry has its gem semiconductor material, i.e., silicon, this material has its own limitations to be used in optical industry, such as its indirect band gap (Oh 2015). The major setback of optical incorporation in silicon devices is lack of optical materials and its compatibility with the electronic devices. In the early 1980s, III–V material-based optoelectronic devices were proposed, nonetheless compatibility with silicon-based devices and power consumption was not up to the mark, and thus, packaging was difficult and expensive. After the proposal of III–V semiconductor quantum well-based quantum-confined Stark effect optical modulator in 1984 (Miller 1984), high-speed modulation was made possible. This effect offers low energy devices and high yield and is very important for optical interconnect point of view. Utilization of optical interconnect within a chip provides a much more efficient way of data transfer. Data carried from one place to another can experience two types of interconnects: One is the device interconnect; connect different devices; second is chip interconnect; connect different components of same chip to one another. Based on the interconnect type, need for the interconnect element is changed. Current interconnection schemes and electrical components are becoming incapable of sustaining the increasing day-by-day demand (Triverio et al. 2007). Figure 11.1 displays different interconnection schemes for the computer mind, i.e., motherboard. Basic distance for the device interconnect within the motherboard for two different modules approximately 40–100 cm which is quite large and conventional interconnect introduces losses while carrying the data for this distance which lowers the overall device efficiency. However, optical interconnects have much lower losses and thus allow more data to carry (Schares 2006). If we think of an extremely general example of data channeling through conventional, i.e., electrical interconnect, suppose a channel having the highest capability of carrying the data with bit rate of 10 kbps and the data to be carried is of 16 kbps, what will happen to the data? Certainly, some of the data will be lost in transition.

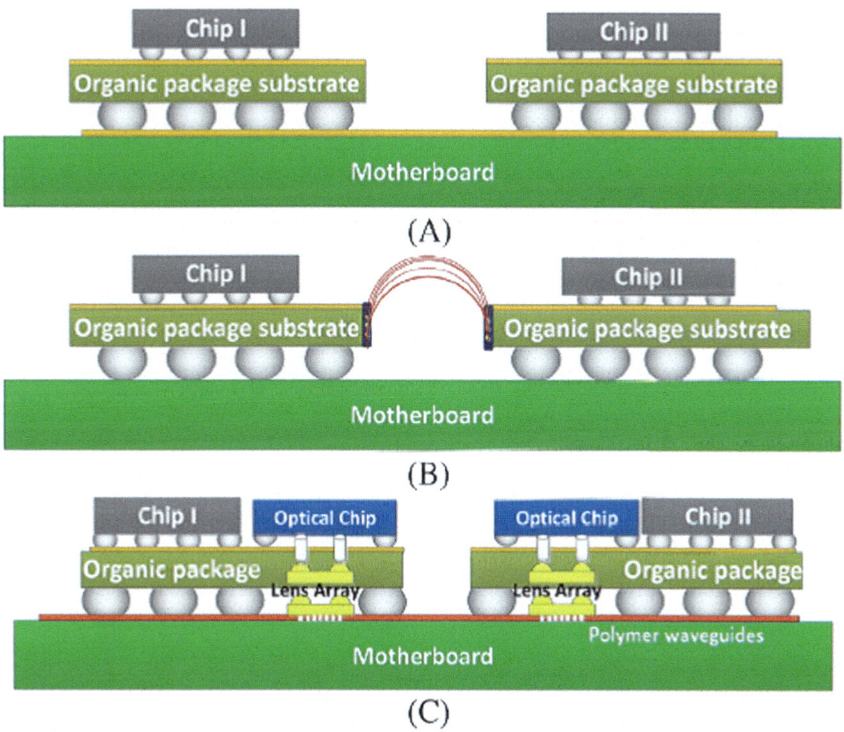

Fig. 11.1 Depiction of different interconnect schemes: **a** conventional, **b** flexing electrical interconnects, **c** optical interconnects (Zia et al. 2017)

Till now, short-distance optical interconnects have been employed and tested successfully. Recently, Intel launched the silicon photonics-based chip which is a new class of high-speed optical connectivity. There are numerous approaches proposed yet for the optical interconnects. However, there are major drawbacks of silicon-based optical interconnects, such as compatibility, efficiency, and cost-effectiveness. In this chapter different optical interconnects are discussed based on the methodology and practical challenges.

11.2 Methodology for Optical Interconnect Designing

Capacity (B) of the link can be given by a well-established relation between the link cross-sectional area (A) and length of the link (L) as:

$$B \leq B_0 \frac{A}{L^2} \tag{11.1}$$

where B_0 is constant for RC lines. From Eq. 11.1, it is observed that ratio A/L^2 made capacity independent of the wiring size, and this is the main reason that big or small wiring does not affect capacity. However, here no other loss is considered for simple calculations, like parasitic capacitance, cross talk, clock precision, etc. (Miller 2010) Interconnect energy is another major limiting factor in conventional interconnects due to its ecological effect on environment and carbon emission (Miller 2010). Consider the case of complementary metal–oxide–semiconductor (CMOS), capacitance of the gate oxide is nearly equal to the wire capacitance, and thus if transistor has to do a logical operation, energy dissipated is same for transistor capacitance and interconnect link. However, the result is transferred for the longer distance and exceeds the energy requirement than the operational energy. Except, interconnect energy; density, length of interconnect also affects the efficiency of the connect.

Size, power dissipation, and cost of optical transmitter and receiver are less important for larger distance. However, for short distances optical interconnects are utterly useful due to their integrity, timing, density, switching speed, etc. Optical fibers are widely known for the high bandwidth with very less density, though high-density exploitation is not easy since it involves high-speed transmitter and receiver multiplexing. Free-space optical interconnects are also an option of optical fiber interconnects, which are widely used in device-to-device optical interconnects. Since power dissipation is a major concern nowadays but at first glance it is not advisable to use optical interconnect because either short or long interconnect power required is almost same as conventional interconnect. This issue raised due to the transmitter and receiver designing. Quantum optical effect is a field which may overcome the above issue at transmitter side using either modulator or laser at low energy. These advantages of optical interconnects are very fascinating, but there are many requirements for optical interconnect realization which need to be addressed.

After observing various inherited properties of optical interconnect, it is seen that realization of the optical interconnect is not easy and it has many designing and practical implementation issues within the chip or silicon system as short-distance implementation. The major setback is the energy dissipation in the scale of few tens of picojoules per bit, while designing optical interconnect; backplane power should be limited to 1 picojoule/bit and approximately 100 fJ/bit for short connection of whole system (Miller 2009). Proper designing of the optical interconnect with source and receiver can make system enable to reduce dissipation loss using laser in place of light-emitting diode (LED). Lasers are advised to use because LED consumes more power for coupling between waveguide and detector. However, it is also directed to choose laser configuration wisely, such as vertical edge emitting laser cannot match with low energy target and those lasers may not be integrate-able with silicon for silicon on chip configuration.

Optical modulator is another option for optical interconnects with external optical source. This excess energy does not dissipate on chip as heat but clocked modulator implicitly reduces timing problem (Keeler et al. 2003). Since modulators are easily scalable and redundant to crystal defects, thus it is easy to practically design with silicon (Goossen et al. 1989). In recent years, silicon photonics made electronic manufacturing much efficient where electric, optoelectronic, and other components can be integrated into single chip. Since silicon photonics solves the density problem of waveguides on broad sense, but there are still significant challenges from integration point of view. Nano-photonics and other possible new approaches can be developed for the problem solution. Germanium-based modulator approach is also a viable option using quantum well, Franz–Keldysh effect, quantum-confined Stark effect (QCSE), etc. To take full advantage of the optical interconnects, it is essential to design the whole system such that board-to-board as well as chip-level interconnections should be optical with single-mode operation. Single-mode operation is desirable because this mode supports easy and efficient coupling with low power dissipation. Based on the various proposed approaches of optical interconnects, in this chapter different models are discussed with their classification as either board-to-board or chip-level interconnect.

11.3 Optical Interconnect

Optical interconnects can be short, ultrashort, and long haul based on the system. For computer-based system motherboards and backplanes, the optical interconnect distance may vary from few cm to tens of cm; however, optical interconnects can also be used to connect different data centers as long-haul connection, i.e., up to few kilometer-long links (Liu et al. 2010). Figure 11.2 shows the basic schematic circuit diagram of the optical interconnect. It is observed that the optical link cannot be applied readily to the system, and it requires a lot of components to carry data optically within the chip. Laser, optical modulator, and photodetectors are the main components required for the optical interconnect (Haurylau et al. 2006) as discussed

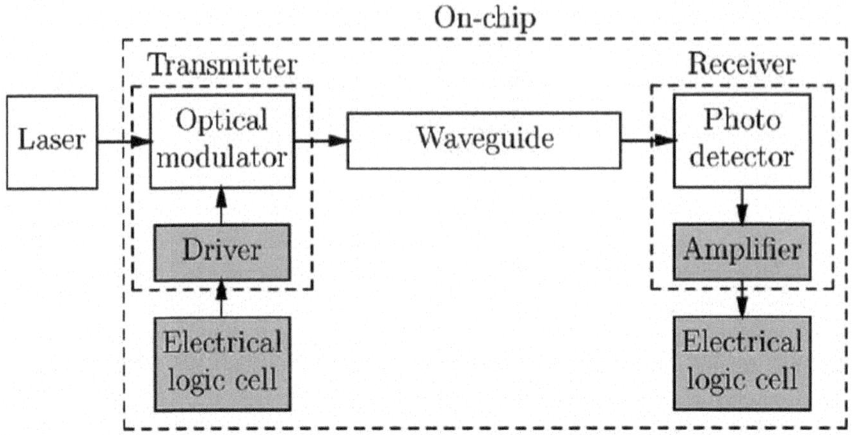

Fig. 11.2 Typical on-chip optical interconnect (Haurylau et al. 2006)

in the previous section. Waveguide used to carry the data can be either optical fiber or silicon on-chip waveguide.

Optical interconnect is the technology of choice by virtue of their unique bandwidth distance product that outperforms electrical links by far. Optical interconnect research goes way back and includes prototype demonstration of optical systems from the early 1990s. Although, simple optical devices are not the solution for the current technological advancement, due to their incapability to fit in small area and other losses. However, photonics introduction in practice improves the efficiency of optical components as well as interconnects. Photonics integration advances the capacity and throughput of the interconnect (Vlasov 2008). Moreover, integration of photonics made it more physically realizable and compatible to the silicon-based chips. As photonics technology is maturing rapidly, it is now finding its way to manifold applications. With increasing demand of cloud-based applications, rapid communication and huge channel capacity are necessary which can only be enabled by replacing the backbone of the communication carried from electrical connects to optical connects (Triverio et al. 2007). Optical interconnects can be of two types based on the interconnection length and the interlinked components as already explained in the previous section.

11.3.1 Board-to-Board Optical Interconnect

Connection from one board to other board within one device also utilizes the electrical interconnects conventionally. This type of connection does not only increase the loss budget of the link due to increased link length but also introduce parasitic capacitance of the circuit (Triverio et al. 2007). Generally, board-to-board connections are made

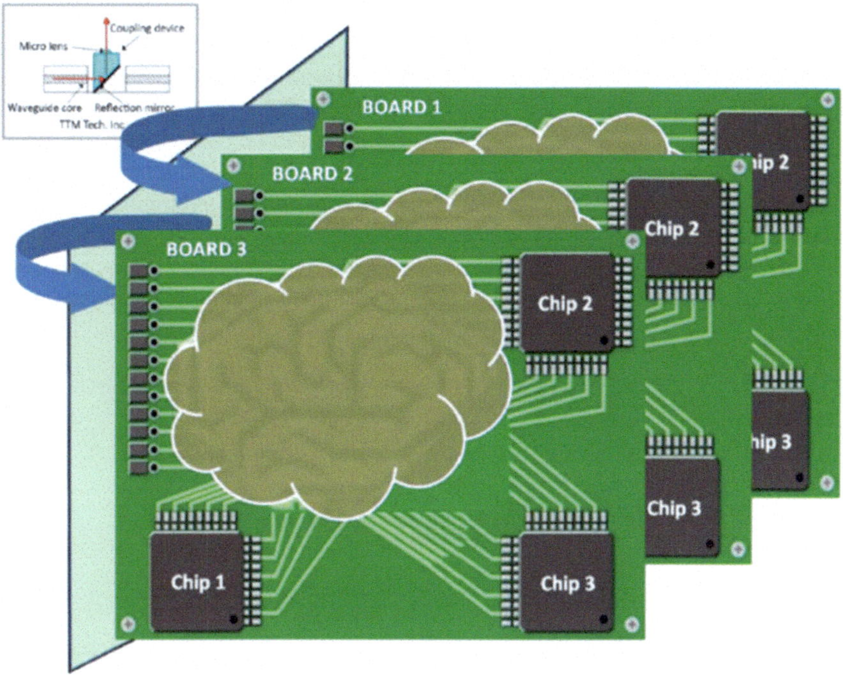

Fig. 11.3 Schematic diagram of board-to-board connection with vertical coupling (Tsiokos and Kanellos 2017)

using plug and play concept, which enables the connection of boards easily replaceable. Figure 11.3 displays that the electrical connections can be replaced with vertical optical interconnect using small mirror and lens structure. For optical interconnect, low loss mirrors with high optical bandwidth are necessary to accommodate power budgets and optical data from different sources (Tsiokos and Kanellos 2017).

Since 1994, various approaches are proposed for this job to be done with. In 1994, Hinton et al. (1994) proposed a relay system specially for optical systems which allow high-speed interconnection. In 1997, Boisset et al. (1997) incorporated an electronic backplane design to complete the practical use of previously proposed relay system. For the alternative approach, transmitters and receivers were dynamically aligned. For this, a number of different approaches were used, such as computer-controlled prisms (Boisset et al. 1995) and electrically controllable prism (Hirabayashi et al. 1997). After various studies, it is seen that holographic approach for the interconnection provides a better way to connect boards through free-space optical interconnect (Boisset et al. 1995; Hirabayashi et al. 1997). Source and detector are aligned to provide full space variant. However, misalignment introduces error which can be overcome through trial and error basis. In 2004, Dominic C. O'Brien et al. proposed programmable diffractive element to direct light from one board to the other board using electronic backplane (O'Brien et al. 2004). Various advantages of the holographic interconnect are correction of aberrations and curvature within the optical system, robust performance, switching between broadcast and fan-out interconnect. The proposed holographic system schematic diagram is given in Fig. 11.4. The presented model not only is programmable but also utilizes vertical cavity surface-emitting lasers with diffractive routing to be cheaper than the other proposed models.

In the proposed model, an array of collimated beams is used to eliminate the array of vertical cavity surface-emitting lasers. Collimated beams are used to illuminate reflective spatial light modulator via non-polarizing beam splitter. In this research, crossbar interconnect is used and it is seen that broadcasting of signal increases

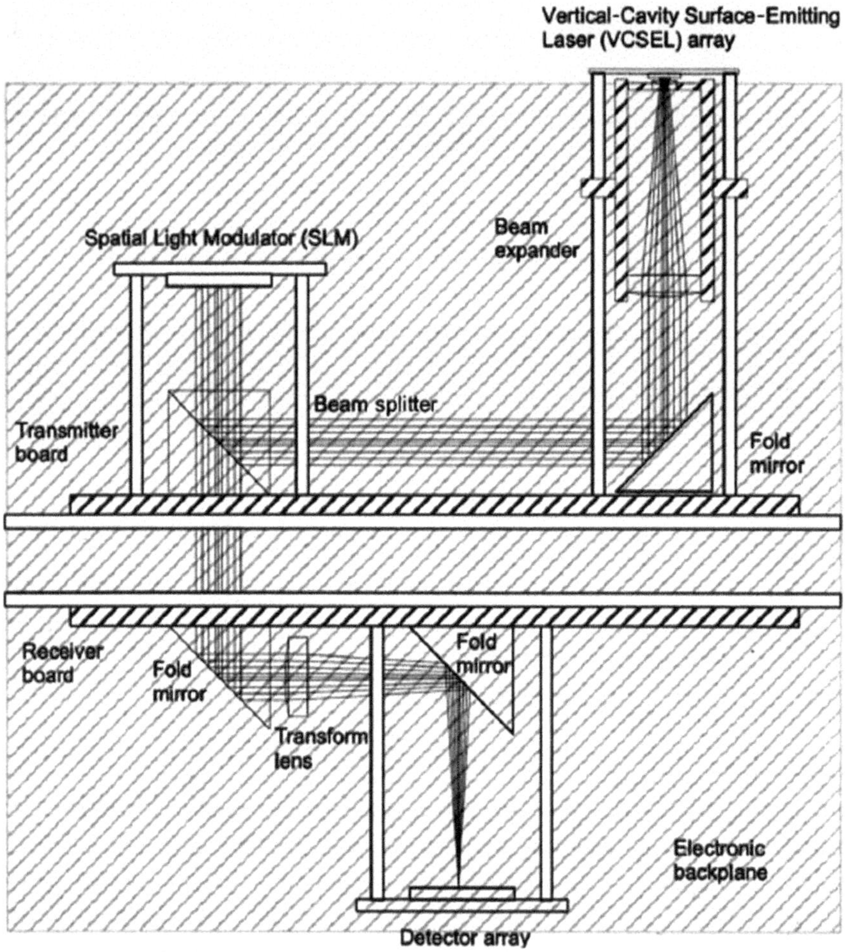

Fig. 11.4 Schematic diagram of programmable holographic interconnect (O'Brien et al. 2004)

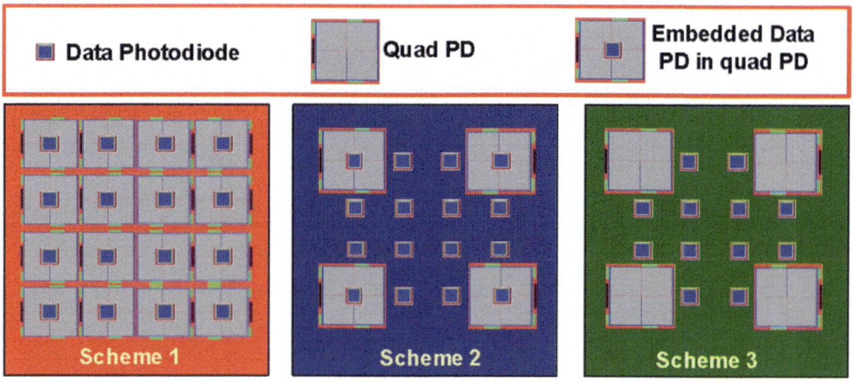

Fig. 11.5 Three schemes of proposed link having sixteen quad-photodiodes (QPDs), four QPD, and four QPD and twelve photodiodes (Wu 2009)

the complexity of the system. However, it is observed that the proposed system is extremely robust and offers an aligned system.

Feiyang Wu et al. proposed a new free-space interconnect in 2008 by designing the integrated receiver with high-speed interconnect in mechanically vibrational environment. In general, for free-space link, an array of laser beams is used to create link between the detection planes, i.e., photodetector (Wu 2009). This general linking scheme has three main modules: an array of laser, microelectromechanical system (MEMS) device for alignment, and an array of detector. Here, a monolithic integration of photodiode is proposed for positioning of beam for processing. Three different photodiode schemes are proposed by arranging four× four array of photodiode as shown in Fig. 11.5.

This study presented an integrated receiver for high-speed optical interconnect with beam tracking mechanism having small beam spot. The proposed algorithm not only is applicable to free-space optical interconnect but also enhances the performance of real-time vision-based tracking and automated robotic system. After 2009, Jin Hu et al. designed a protocol for the very shot reach optical interconnect system (Hu 2009). This is not the fabrication work however, is the software frame work to overcome the delay, power and interference for the optical interconnects ranges from centimeter to meters.

Figure 11.6 shows the system model, which displays interface between the upper protocol layers and lanes. Various blocks show lane initialization and generate protocol primitives that are inserted into the lane according to the received control signals. Using SERDES in field-programmable gate array (FPGA), single-lane protocol is realized as shown in Fig. 11.6. Along with this, it is already seen that board-to-board optical interconnections work on the free-space connection mostly because of its inherent advantages of scalability and density Fig. 11.7.

However, misalignment is the main bottleneck of this arrangement. This misalignment introduces cross talk, insertion losses, reliability, etc (Chou et al. 2009). Various techniques and designs were proposed to overcome these drawbacks such as Risley

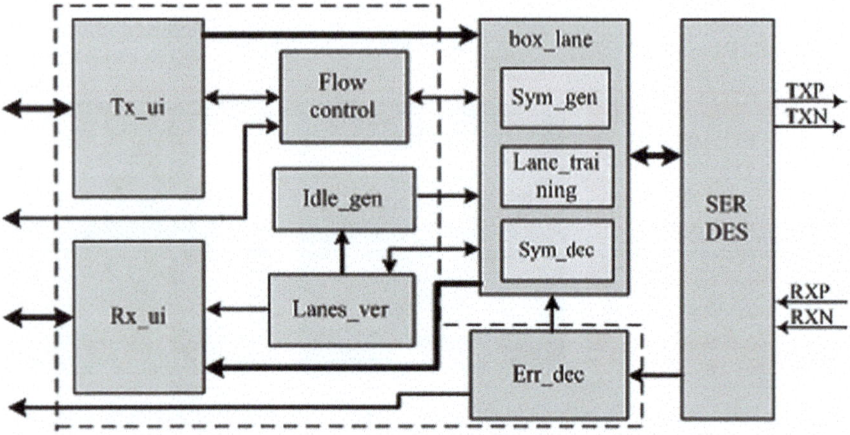

Fig. 11.6 Proposed protocol implementation diagram (Hu 2009)

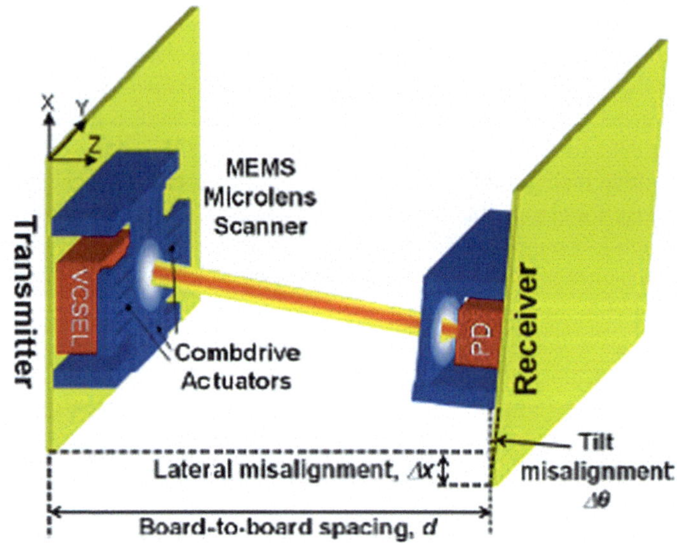

Fig. 11.7 Schematic diagram of free-space optical interconnect with misalignment by the angle and length using MEMS lens to correct the misalignment (Chou et al. 2009)

prism (Boisset et al. 1995), mechanical translational stages (Naruse et al. 2001), and liquid crystal spatial light modulators (Henderson 2006), and some of those are already explained earlier in this chapter. Among all the proposed designs, it is observed that MEMS technology is best due to its speed, low loss, etc. In 2009, an adaptive interconnect design was proposed by Chou et al. (2009) using MEMS micro-lens scanner with control loop to daze misalignment problem. This proposed system is used to correct the lateral and tilt misalignment between the two boards.

Beam scanning range is amplified by board-to-board distance, which compensates the lateral misalignment. Only one micro-lens was used in the proposed study for the one direction; however, it can be extended using large lens. This optical interconnect corrected the misalignment up to 40 dB with 700 MHz bandwidth. In 2017, Sen Lin et al. (Settaluri et al. 2015) proposed three-dimensional (3-D) optical transceiver using integrated photonics link in 300-nm CMOS foundry. Optical interconnect link was based on the dense wavelength-division multiplexing (DWDM) using 25 Gb/s channel using wafer-scale heterogeneous platform. From all the above studies, it is seen that optical interconnect between different chips or devices can successfully be established using free-space and medium-based optical connect.

11.3.2 On-Chip Optical Interconnect

On-chip optical interconnects have inherent property of high signal propagation efficiency, high speed, low loss, etc. Alike board-to-board interconnect, on-chip optical interconnect can be made of different materials and mostly used materials are silicon and polymer waveguide. Figure 11.8 shows a comparative analysis of propagation delay for different waveguide materials with varying interconnect length.

It is observed that delay is reduced for optical interconnect sufficiently; however, for the smaller interconnect length, delay is high but for the long interconnection optical interconnects are advantageous. Same as board-to-board optical interconnect technology, on-chip optical interconnect also has the same optical components for the interconnection establishment. However, the thing in this technology is all the optical components have to be placed or fabricated on a single chip. Thus, the size,

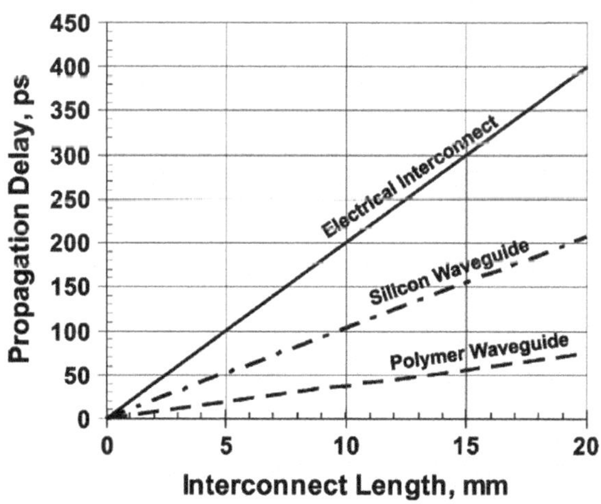

Fig. 11.8 Propagation delay for silicon, polymer, and electrical waveguide with different interconnect lengths (Haurylau et al. 2006)

power, delay, power–delay product, etc., are very important parameters which need to be addressed for on-chip interconnect. Modulator is the main component whose matching is most important, and Mach–Zehnder modulator is the most used one. With the current technological trend, it is important to observe whether the optical interconnect is capable to take place of electrical interconnect or not.

Since it is already discussed that delay and the area are the main bottleneck in optical interconnect implementation, a number of different models have been introduced over the years to overcome this problem. Figure 11.9 shows the advancement in the technological trend over the years in optical interconnect technology. Figure 11.10 displays the normalized critical length of the optical interconnect for different technological nodes. From the figure, it is depicted that the critical parameters are optimized for very small interconnect length which is CMOS compatible. However after significant progress for on-chip optical interconnect, there are still a number of issues unresolved. First is the large carbon footprint and power consumption of components, especially optical modulator. This drawback can be overcome by using advanced optical structures such as photonics crystal, photonics band gap, and resonator. However, these solutions have their own difficulty of fabrication effort.

Another problem is to produce sufficient optical power to continue the optical operation of signal transmission between different on-chip components. A simple optical operation may require 100–1000 detectors, and to drive such detectors huge optics will be required. This is a very big challenge to generate power in few watts and still reduce the power level below a particular value to avoid temperature rise and small carbon emission. Moreover, a set of integrated silicon compatible components needs to be developed to fully exploit the optical interconnect properties. Based on

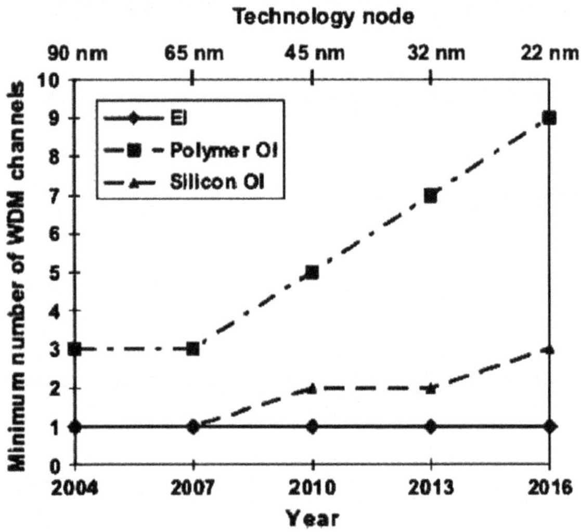

Fig. 11.9 Advancement in the optical interconnect technology for wavelength-division multiplexing (WDM) channels with year (Haurylau et al. 2006)

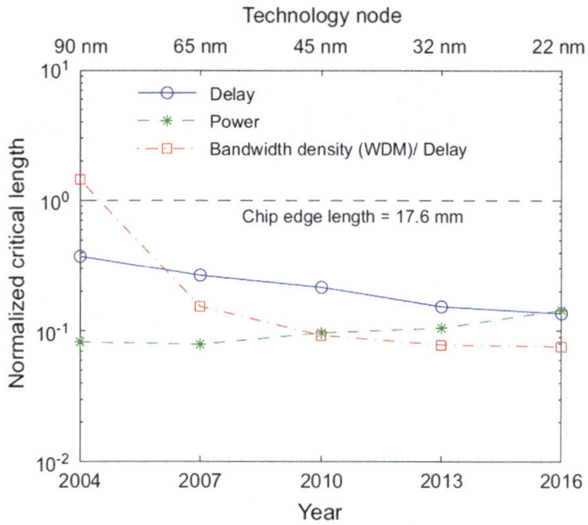

Fig. 11.10 Critical length for optical interconnect (Chen et al. 2007)

the challenges for on-chip optical interconnect implementation, various models have been proposed yet. In 2008, Jin Tae Kim et al. proposed a chip-to-chip optical interconnect using gold waveguide working of surface plasmon polariton (SPP) concept (Kim et al. 2008). For the purpose to serve, polymer-based long-range SPP waveguide having gold strips incorporated between transmission and receiver modules on the single board as shown in Fig. 11.11.

The proposed model comprised laser array chip, driver IC, photodetector array, etc. On analyzing the insertion loss and alignment tolerance characteristics of the model, it is observed that mode field diameter and loss are affected by the waveguide width characteristics of the interconnect and as the length of the interconnect increased insertion loss is also increased; however, smaller width caused high alignment losses. This shows that there should be a tradeoff between the length and

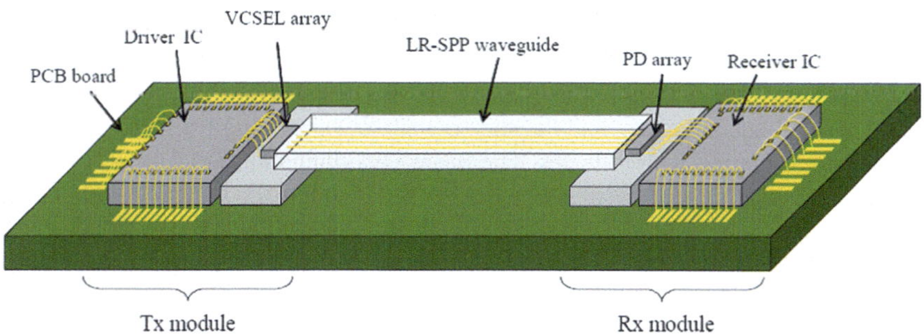

Fig. 11.11 Proposed SPP gold interconnect on a single board for chip-to-chip connection (Kim et al. 2008)

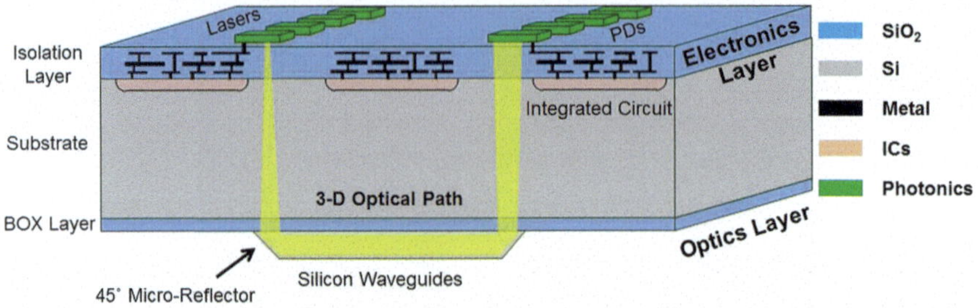

Fig. 11.12 Schematic diagram of on-chip optical interconnect with source and detector using 3-D guided wave path (Shen et al. 2014)

width of the waveguide and insertion loss and alignment loss, respectively. This model was proposed for 10 Gbps data transmission channel at 1.3 μm center wavelength. A breakthrough publication from Jacob S. Levy et al. published in 2008 reported a CMOS compatible multi-wavelength oscillator for on-chip interconnect (Levy et al. 2010). This oscillator was made of silicon nitride having nonlinear refractive index, and this material is chosen because of its compatibility with the CMOS industry and large band gap. Silicon nitride does not suffer from two photon absorptions and thus yields low loss waveguide in two different wavelength regions. However, for some particular thickness of silicon nitride, tensile stress is very high which causes deprived nonlinear optics due to delocalization from the material of interest. In this study, thicker film is used to confine the optical mode into silicon nitride layer with reduced modal area. Proposed integrated oscillator provides very narrow spaced linewidth sources with critical component to achieve high bandwidth wavelength-division multiplexed system for next-generation microprocessors.

In 2014, P. K. Shen et al. published an article for the chip-level implementation of optical interconnect using 3-D-guided wave path including laser and detector (Shen et al. 2014). Figure 11.12 shows the schematic diagram of the proposed on-chip interconnect.

The proposed design is assumed to be very useful for multi-core processor and memory-to-processor interface. 3-D-guided path provides large alignment tolerance and high coupling efficiency. This approach also simplifies the chip-level interconnect on silicon on-chip substrate. Figure 11.13 shows the simulation results for the proposed model based on ray-tracing method. The laser beam has been diverging at an angle in silicon substrate, and the shorter path and small angle allow most of the laser beam to couple with reflector. The beam is confined in the waveguide due to large refractive difference between the core and cladding made of silicon and silicon dioxide.

The simulated model also realized experimentally using chemical vapor deposition method. In this study, various geometrical parameters were also studied for the optimization of the model structure. It is demonstrated that the model has −2.19 dB optical transmission loss at 10 mA biased laser. It is also observed that the proposed model can be used for error free 10 Gbps data transmission at 9 mA biased

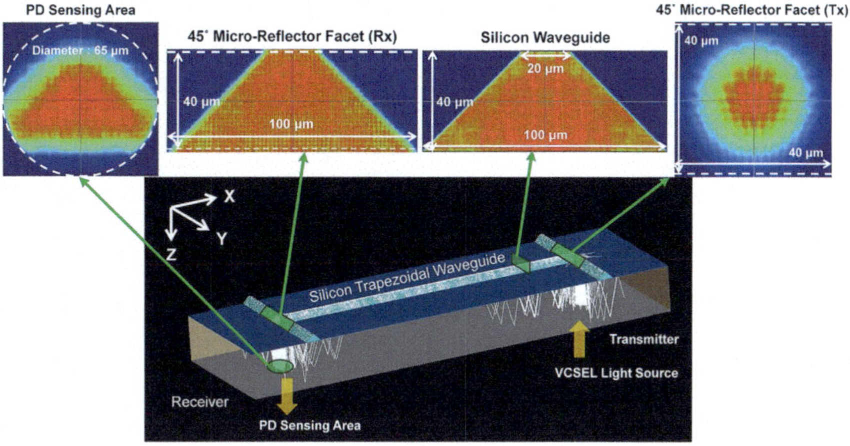

Fig. 11.13 Light propagation results for the proposed 3-D waveguide interconnect using ray tracing method (Shen et al. 2014)

laser. This study validates the physical concept of the optical interconnect for on-chip implementation and opens the new possibilities for the exploration.

In early 2017, silicon photonics bonded with the III–V group materials by adhesive or by molecular bonding. This integration made optical component to be attached with the electronic chips. However, this technology is mainly useful for the optical components to fabricate. Thus, through-silicon vias are proposed and used as 3-D optical interconnect (Hofmann et al. 2012). For this purpose, transistor polysilicon layer is to be combined with the optical waveguide for optical interconnect. Furthermore, stacking is another technology introduced which enables heterogeneous integration of different technologies to provide multi-functionality to a single chip. This technique is easily integratable with SOI and III–V technology for high performance. There are a number of studies proposed to utilize the above approach; along with that, integration is flip-chip bonding. Since, flip-chip has some disadvantages such alignment accuracy, complexity, and batch processing etc. Till now, IBM demonstrated optobus of two generations; initially, the throughput is 160 Gbps and recently 300 Gbps optobus is proposed. Thus, it is observed that various approaches have been proposed for on-chip optical interconnect with their own advantages and disadvantages. It is observed from the above review that optical interconnect reduces the memory latency by 35% but improved the power efficiency by 28% (Brunina et al. 2012). Unlike 3-D optical interconnect, stack interconnect suffers from the problem of thermal cross talk from high power logic die and thermal fluctuation of die. Figure 11.14 shows the 3-D stack of photonic-on-logic (Demir and Hardavellas 2015). This photonic stack is based on Intel Core i7 processor and consumes 5 W power. It is observed that the stack dissipates relatively low power and the maximum temperature of the chip is 93.35 °C; moreover, it is also observed that there is strong coupling between the two dice. Thus, this can be interpreted from the above study that there are still a number of challenges present in the field of optical interconnects

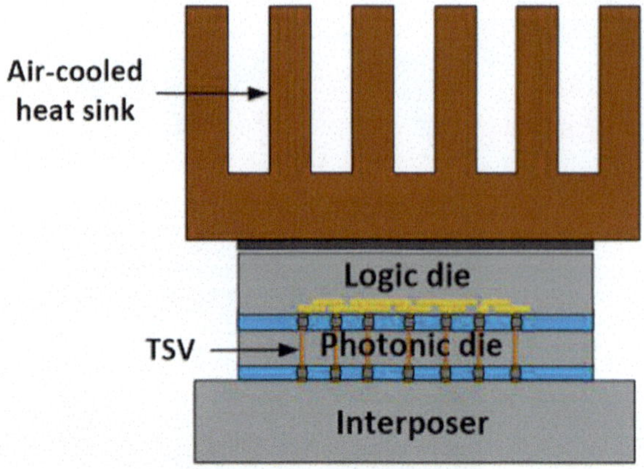

Fig. 11.14 Photonic-on-logic 3-D stack (Demir and Hardavellas 2015)

whether it is board-to-board, chip-to-chip, and/or on-chip interconnects. Nonetheless, advantages of using optical interconnect are much more than the drawbacks which encourage the researchers to work in this field to propose and explore new models and techniques for optical interconnects to couple with electronic chips.

Recently, Llewellyn et al. (2019) proposed a quantum teleportation-based silicon on-chip optical interconnect using multiqubit states. This study is the first of its kind which is physically realized till now having multiphoton sources which are nonlinear and linear multiqubit circuit interfaced naturally having low noise and controlled coherent system.

11.4 Challenges

Urgent need of faster, low power, and high bandwidth network for information transfer is very essential due to the dependency of every one of this era on Internet. For information transfer, processing, and storage, electronic medium is used whose efficiency is limited by electrical interconnect incompetence. Optical interconnect avoids the scaling issue of the wires and provides various other advantages. But this argument is valid for the long-distance communication (board to board or server to server); nevertheless, short interconnects are not easy to replace because these are cheap and consume less energy already. Noteworthy research and advancements took place in this field, but there are always possibilities for the improvement and integration with silicon technology.

Main challenges for this technology include packaging, integration of electrical and optical components, development of small components, single-mode optical sources, etc. Nanotechnology is the best technology which develops optical components compatible with the silicon electronics. However, novel optical multiplexing components are needed to be developed. Along with the other implementation and realization issues, temperature stability is of prime concern since optical components hold drastic effect on functioning.

Optical device bistability is also a very huge problem for practical implementation of optical logic systems to substantially small size. However, this challenge can be reduced by using self-electrooptic effect devices (SEEDs), and symmetric SEEDs are three terminal devices which made bistability possible using two beam powers. Along with this, there are design challenges also present, such as geometrical constraints and misalignment simulation. Laser is used to illuminate modulator with some particular intensity, and the size of the hologram should be big enough to reduce the overlapping of the laser pulses and subsidize the cross talk at acceptable level. From Fig. 11.15, it is seen that there are various detector planes which need to be aligned and symmetrical to diffraction order. Moreover, out-of-plane tilt causes light to diffract and spots misplacement. This misalignment is a major challenge for optical interconnect implementation because it limits the efficiency of the interconnect. Furthermore, misalignment situation is also a big challenge, since alignment of nano-components is very difficult, and ray-tracing package can be used and optimized

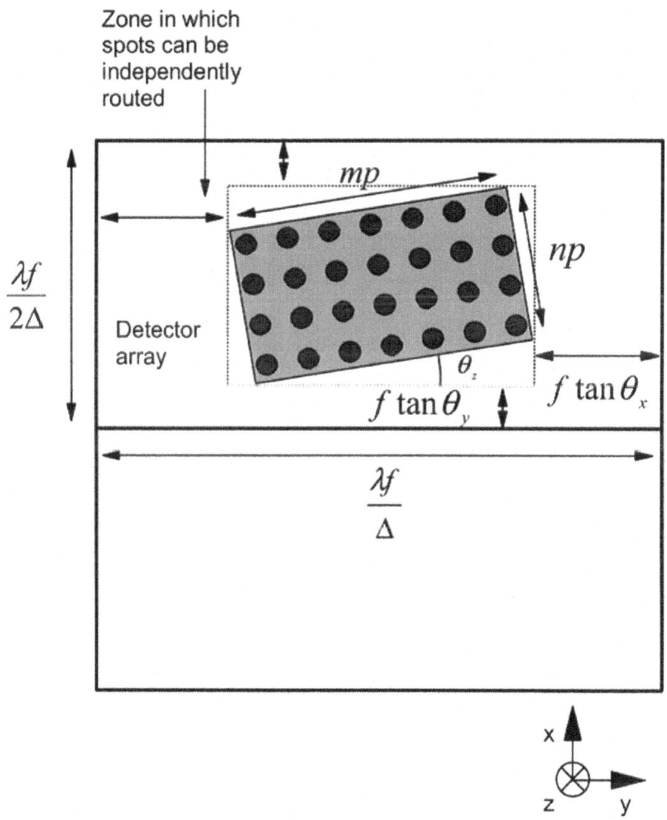

Fig. 11.15 Detector plane geometry (O'Brien et al. 2004)

using Monte Carlo technique. Measuring the actual position for perfect alignment has to be done very carefully. Link budget and cross talk are the parameters which are used to define the information-carrying capacity of optical interconnect. Hologram is used to direct the light toward the detector, and port is decided by standard direct binary search algorithm (Seldowitz et al. 1987) since this introduces only output cross talk and thus the illumination of desired port is necessary to improve the cross talk performance.

Link budget accounts all the power gains and losses in a communicating system. Mostly, losses are of two types: intrinsic and extrinsic, as optical switches are to be designed to operate with polarized light and beam splitter is used to direct the light beam toward the modulator.

Link budget severely affects the optical interconnect quality which can be overcome by improving modulator loss, imperfect phase modulation, reflectivity, alignment, etc. Except this, adaption of the optical system is also a challenge which is done in two steps: tracking and aberration correction. These corrections are to be done after the fabrication of the hologram using different positions of the component. However, till now there is no solid method to improve this drawback but research is continuously undergoing to eliminate this problem. Along with these challenges, there are various circuit-level and fabrication-level challenges which need to be addressed.

First of all, let us talk about the fabrication issues. Since detectors are very important, low input capacitance is essential for small and low power receiver circuit. And power dissipation is largest at the receiver circuit for optical interconnects; these need to be designed carefully. If the size of the detector is increased, more sensitive amplifier with large transistor is required. Since more sensitive amplifier means more stages which in turn increase the latency as well as power dissipation. Thus, small circuit leads to high voltage swipe and leads to better noise immunity and few amplifier stages. Design considerations should be taken care seriously for the silicon-based optical interconnects. While designing small area detector, one key problem is the large absorption length. Large absorption length leads to two problems; one is low efficiency, and second is diffusion will be long. Thus, the designing of the detector should be done carefully for CMOS optical interconnects. Threshold current is another design parameter of source, i.e., laser to be taken care. Mode and polarization, wavelength control, spot size, and power supply voltage are some other challenging parameters which need to be taken care at the time of the fabrication and implementation of the laser source for optical interconnect. One of the most serious issues is the high cost for practical implementation of dense optical interconnect. The listed design, fabrication, and placement issues are the most common and powerful points which are necessarily to be addressed; however, there are many other issues which are to be researched continuously to make optical interconnect more practical and useable with low cost and high efficiency.

11.5 Conclusion

The present chapter addresses details about a recent technology, i.e., optical interconnect in the silicon industry. Optical interconnects are increasingly high in demand in the current electronics and communication industry due to their numerous advantages. These interconnects are the replacement of the electrical interconnects basically used to connect chip to chip, board to board, component to component, etc. Need for optical interconnects arose because of the technological challenges faced by the electrical interconnects in modern era, such as parasitic capacitance, leakage current, and bandwidth limitation. Countless data to be accessed through cloud and storage of the same in data centers make it compulsory to exchange electrical connect to the optical one. This chapter summarizes the need of the optical interconnect and the methodology used to design and optimize the optical connect in various environments. It also comprises different approaches proposed yet for the board-to-board and silicon chip-level connects. It is discussed that optical interconnect incorporation within the current technologies will reduce the leakage current and parasitic capacitance generated by the thin and dense electrical interconnects. Moreover, it will also eliminate the bandwidth limitation as 100 Gbps data link can be produced using four 25 Gbps single optical channels which is sufficient to sustain the data traffic within a single data center such as small institute, and this does not only improve the bandwidth of the system but also reduce the required area in the chip which in turn optimize the system performance and reduce the realization cost. It can be concluded from the above discussion that the advancement in silicon photonics enables the researchers to further densify the chip components to accomplish advancement in the existing technologies. However, there are a number of bottlenecks still present in the implementation of the optical interconnect on commercial level, but the continuous research in this field will do the trick for the optical interconnects to incorporate with the help of silicon photonics for technological as well as economical developments.

References

Boisset GC, Robertson B, Hinton HS (1995) Design and construction of an active alignment demonstrator for a free-space optical interconnect. IEEE Photonics Technol Lett 7(6):676–678

Boisset GC, Ayliffe MH, Robertson B, Iyer R, Liu YS, Plant DV, Goodwill DJ, Kabal D, Pavlasek D (1997) Optomechanics for a four-stage hybrid-self-electro-optic-device-based free-space optical backplane. Appl Opt 36(29):7341–7346

Brunina D, Liu D, Bergman K (2012) An energy-efficient optically connected memory module for hybrid packet-and circuit-switched optical networks. IEEE J Sel Top Quantum Electron 19(2):3700407

Chen G, Chen H, Haurylau M, Nelson NA, Albonesi DH, Fauchet PM, Friedman EG (2007) Predictions of CMOS compatible on-chip optical interconnect. Integr, VLSI J 40(4):434–446

Chou J, Yu K, Horsley D, Yoxall B, Mathai S, Tan S, Wang S-Y, Wu MC (2009) Robust free space board-to-board optical interconnect with closed loop MEMS tracking. Appl Phys A 95(4):973

Demir Y, Hardavellas N (2015) Parka: thermally insulated nanophotonic interconnects. In: Proceedings of the 9th international symposium on networks-on-chip. ACM, p 1

Goodman JW, Leonberger FJ, Kung S-Y, Athale RA (1984) Optical interconnections for VLSI systems. Proc IEEE 72(7):850–866

Goossen KW, Boyd GD, Cunningham JE, Jan WY, Miller DAB, Chemla DS, Lum RM (1989) GaAs-AlGaAs multiquantum well reflection modulators grown on GaAs and silicon substrates. IEEE Photonics Technol Lett 1(10):304–306

Haurylau M, Chen G, Chen H, Zhang J, Nelson NA, Albonesi DH, Friedman EG, Fauchet PM (2006) On-chip optical interconnect roadmap: challenges and critical directions. IEEE J Sel Top Quantum Electron 12(6):1699–1705

Henderson, CJ, Leyva DG, Wilkinson TD (2006) Free space adaptive optical interconnect at 1.25 Gb/s, with beam steering using a ferroelectric liquid-crystal SLM. J Lightwave Technol 24(5):1989–1997

Hinton, HS, Cloonan TJ, McCormick FB, Lentine AL, Tooley FAP (1994) Free-space digital optical systems. Proc IEEE 82(11):1632–1649

Hirabayashi K, Yamamoto T, Hino S, Kohama Y, Tateno K (1997) Optical beam direction compensating system for board-to-board free space optical interconnection in high-capacity ATM switch. J Lightwave Technol 15(5):874–882

Hofmann WH, Moser P, Bimberg D (2012) Energy-efficient VCSELs for interconnects. IEEE Photonics J 4(2):652–656

Hu, J, Yang Y, Zhang Z, Zhao Y, Lin P (2009) Design and implementation of board-to-board optical interconnect protocol. In: 2009 symposium on photonics and optoelectronics. IEEE, pp 1–4

Keeler GA, Nelson BE, Agarwal D, Debaes C, Helman NC, Bhatnagar A, Miller DAB (2003) The benefits of ultrashort optical pulses in optically interconnected systems. IEEE J Sel Topics Quantum Electron 9(2):477–485

Kim JT, Ju JJ, Park S, Kim M, Park SK, Lee M-H (2008) Chip-to-chip optical interconnect using gold long-range surface plasmon polariton waveguides. Optics Express 16(17):13133–13138

Lanzillo NA, Restrepo OD, Bhosale PS, Cruz-Silva E, Yang C-C, Kim BY, Spooner T et al (2018) Electron scattering at interfaces in nano-scale vertical interconnects: a combined experimental and ab initio study. Appl Phys Lett 112(16):163107

Levy JS, Gondarenko A, Foster MA, Turner-Foster AC, Gaeta AL, Lipson M (2010) CMOS-compatible multiple-wavelength oscillator for on-chip optical interconnects. Nat Photonics 4(1):37

Leydesdorff L (1994) The evolution of communication systems. Int J Syst Res Inf Sci 6:219–230

Liu H, Lam CF, Johnson C (2010) Scaling optical interconnects in datacenter networks opportunities and challenges for WDM. In: 2010 18th IEEE symposium on high performance interconnects. IEEE, pp 113–116

Liu K, Fan H, Huang Y, Duan X, Wang Q, Ren X, Wei Q, Cai S (2018) A full-duplex working integrated optoelectronic device for optical interconnect. Opt Commun 414:102–105

Llewellyn D, Ding Y, Faruque II, Paesani S, Bacco D, Santagati R Qian Y-J et al (2019) Chip-to-chip quantum teleportation and multi-photon entanglement in silicon. Nat Phys 1–6

Miller, DAB, Chemla DS, Damen TC, Gossard AC, Wiegmann W, Wood TH, Burrus CA (1984) Band-edge electroabsorption in quantum well structures: the quantum-confined Stark effect. Phys Rev Lett 53(22):2173

Miller DAB (2000) Optical interconnects to silicon. IEEE J Sel Topics Quantum Electron 6(6):1312–1317

Miller DAB (2009) Device requirements for optical interconnects to silicon chips. Proc IEEE 97(7):1166–1185

Miller DAB (2010) Are optical transistors the logical next step? Nat Photonics 4(1):3

Naruse M, Yamamoto S, Ishikawa M (2001) Real-time active alignment demonstration for free-space optical interconnections. IEEE Photonics Technol Lett 13(11):1257–1259

O'Brien DC, Faulkner GE, Wilkinson TD, Robertson B, Gil Leyva D (2004) Design and analysis of an adaptive board-to-board dynamic holographic interconnect." Applied optics 43(16):3297–3305

Oh, YJ, Lee I-H, Kim S, Lee J, Chang KJ (2015) Dipole-allowed direct band gap silicon superlattices. Sci Rep 5:18086

Schares, L, Kash JA, Doany FE, Schow CL, Schuster C, Kuchta DM, Pepeljugoski PK et al (2006) Terabus: Terabit/second-class card-level optical interconnect technologies. IEEE J Sel Topics Quantum Electron 12(5):1032–1044

Seldowitz MA, Allebach JP, Sweeney DW (1987) Synthesis of digital holograms by direct binary search. Appl Opt 26(14):2788–2798

Settaluri KT, Lin S, Moazeni S, Timurdogan E, Sun C, Moresco M, Su Z et al (2015) Demonstration of an optical chip-to-chip link in a 3D integrated electronic-photonic platform. In: ESSCIRC conference 2015-41st European solid-state circuits conference (ESSCIRC). IEEE, pp 156–159

Sezgin IC, Gustavsson J, Lengyel T, Eriksson T, Simon He Z, Fager C (2019) Effect of VCSEL characteristics on ultra-high speed sigma-delta-over-fiber communication links. J Lightwave Technol 37(9):2109–2119

Shen P, Chen C-T, Chang C-H, Chiu C-Y, Li S-L, Chang C-C, Mount-Learn W (2014) Implementation of chip-level optical interconnect with laser and photodetector using SOI-based 3-D guided-wave path. IEEE Photonics J 6(6):1–10

Triverio P, Grivet-Talocia S, Nakhla MS, Canavero FG, Achar R (2007) Stability, causality, and passivity in electrical interconnect models. IEEE Trans Adv Packag 30(4):795–808

Tsiokos D, Kanellos GT (2017) Optical interconnects: fundamentals. In: Optical interconnects for data centers. Woodhead Publishing, pp 43–73

Vlasov, Y, Green WMJ, Xia F (2008) High-throughput silicon nanophotonic wavelength-insensitive switch for on-chip optical networks. Nat Photonics 2(4):242

Wu, F, Logeeswaran VJ, Saif Islam M, Horsley DA, Walmsley RG, Mathai S, Houng D, Tan MRT, Wang S-Y (2009) Integrated receiver architectures for board-to-board free-space optical interconnects. Appl Phys A 95(4):1079–1088

Zia M, Wan C, Zhang Y, Bakir M (2017) Electrical and photonic off-chip interconnection and system integration. In: Optical interconnects for data centers. Woodhead Publishing, pp 265–286

Chapter 12
Emerging Graphene FETs for Next-Generation Integrated Circuit Design

Yash Agrawal, Eti Maheshwari, Mekala Girish Kumar, and Rajeevan Chandel

Abstract Electronic devices are the basic building blocks in integrated circuits. Silicon-based devices are dominating the VLSI industry since decades. However, with miniaturization of the technology, quantum effects aggregate extensively at nano-dimensions, and silicon-based devices are harder to scale down than tens of nanometer. As a result, traditional silicon FETs at nano-era are becoming less significant. The rise of nano-era and recent research trends have shown that graphene and related materials (GRMs) are emerging as promising candidates for future devices. In this chapter, the physics governing the graphene material is discussed. Thereafter, analytical model of graphene FET (GFET) is presented. Further, advanced GFET is explored, and the high end novel GFET-based inverter and adder circuits are implemented using HSPICE. To investigate the GFET performance efficiency, a comparative analysis has also been made with respect to conventional SiFET devices. The technology node considered for SiFET is 22 nm for the various analyses presented in the chapter.

Keywords CNTFET · GFET · GRM · SiFET

Y. Agrawal (✉) · E. Maheshwari
VLSI and Embedded Systems Research Group, Dhirubhai Ambani Institute of Information and Communication Technology, Gandhinagar, Gujarat 382007, India
e-mail: mr.yashagrawal@gmail.com

E. Maheshwari
e-mail: maheshwari.eti787@gmail.com

M. G. Kumar
Electronics and Communication Engineering Department, Vidya Jyothi Institute of Technology, Hyderabad, Telangana 500075, India
e-mail: giri.frds@gmail.com

R. Chandel
Electronics and Communication Engineering Department, National Institute of Technology, Hamirpur, Himachal Pradesh 177005, India
e-mail: rajeevanchandel@gmail.com

R. Dhiman and R. Chandel (eds.), *Nanoscale VLSI*,
Energy Systems in Electrical Engineering,
https://doi.org/10.1007/978-981-15-7937-0_12

12.1 Introduction

Extensive scaling of feature size to accommodate higher transistor packing densities increases the functional capacity of the chip. This also reduces the total cost of fabrication as small sizes allow more chips on a single wafer. However, there is other side of the coin that brings in major hurdles with the current state-of-the-art technology. Progressive challenges at nano-dimensions are increasing power dissipation, short channel effects such as hot carrier injection, impact ionization, drain induced barrier lowering (DIBL) and other geometry effects (Islam 2015).

Scaling of technology demands lowering of supply voltages which in turn reduce switching ability of MOS and further impact the performance and reliability of miniaturized devices. Consequently, a new paradigm based on incorporation of futuristic and prospective materials is proposed by several researchers. Over the years, graphene has evolved that endeavors to integrate the complexities in a more fine-grained way. Graphene is an optimistic substitute for silicon FETs due to its excellent mechanical, electrical, thermal and physical properties.

Device characterization and analytical formulation are very essential for understanding its operation, designing and making application of it. A semi-empirical model for graphene FET has been shown in (Gelao et al. 2011). However, it is reported that this work is not suffice to characterize graphene FET operation for sub-threshold region of operation, and further work is needed in this area. Modeling of graphene transistors incorporating the quantum limits has been presented in (Thiele et al. 2010). This model's incorporation for circuit and system development is to be explored and not discussed in the paper. A parameterized SPICE-compatible compact model for GFET is presented in (Chen et al. 2015). Also, delay and power analysis are performed under process variation. However, the different analyses are performed for relatively simple circuits due to large computation time requirement in Monte-Carlo analysis. In (Marulanda et al. 2008), current–voltage characterization and model development of grapheme-based CNTFET are explored using numerical method. This model however could not explain profoundly the charge transport phenomena that are existing in graphene FET. A study of single layer and bilayer GFET has been detailed in (Anas 2016). It is investigated that bilayer GFET possesses better performance in terms of higher gate voltage control, better saturation region and lower sensitivity to low frequency noise than its counterpart single layer GFET. In (Aradhya et al. 2016), low power 8-bit ALU is designed using GFET. It is reported that further advancement in this can be attained using sub-threshold adiabatic logic. Several quantum effects governing CNTFETs and GFETs are discussed and presented in (Banadaki 2016; Tan et al. 2014; Wang 2014). Uniformly from all the reported work, it is investigated that prospective graphene FET is distinctly superior and effective than conventional SiFETs. Henceforth, comprehensively physics, modeling of GFET and its circuit realization have been systematically detailed and amalgamately presented in this chapter.

The chapter comprises four sections including the present introduction section. Section 12.2 details about the physics of graphene. In the next section, several different digital circuits based on graphene FETs and silicon FETs are designed. The performance comparison between the futuristic graphene and conventional silicon FETs has been made. Finally, conclusion is deduced in Sect. 12.4.

12.2 Physics of Graphene

Graphene is a sp^2 hybridized allotrope of carbon atom. The interactions between different hybridized orbitals result in three σ-bonds and one π-bond. Amongst the two, the strongest type of covalent bond is σ and responsible for providing high strength and mechanical properties to graphene. On the other hand, the electrons associated with the π-bond are delocalized and responsible for providing excellent electronic and optical properties to graphene (Philip Wong and Akinwande 2010).

To understand the behavior of electrons in graphene requires consideration of quantum mechanical wave nature of electrons and the periodic arrangement of atoms, i.e., the crystal structure and lattice. Since the electrons are treated as waves in quantum physics, Schrödinger's equation in its most basic form will be employed to solve for these properties. This can be described as (12.1) (Philip Wong and Akinwande 2010),

$$\frac{\mathrm{d}^2\psi}{\mathrm{d}x^2} + \frac{8\pi^2 m}{\hbar^2}(E - V)\psi = 0 \tag{12.1}$$

where ψ is the wave function, x is position, m is mass of charged particle, $\hbar$ is the reduced Planck's constant, E is the energy, and V is the potential. The relation between energy and wave vector gives the dispersion and band structure. To determine the band structure, time-dependent Schrödinger's equation is used

$$H\psi(k, r) = E(k)\psi(k, r) \tag{12.2}$$

where H is the Hamiltonian operator that operates on wave function ψ to produce allowed energy levels E. k is the wave vector, r corresponds to the parameter of the spherical polar coordinate system. The tight-binding approach is followed to characterize and model graphene FET behavior. For graphene FET, conduction and valence bands meet at Fermi energy level and the point where these touch is called Dirac point (denoted by K in Fig. 12.1), leaving a zero-band gap. At these K-points, the energy is dispersed around center of the K-point and can simply be expressed as a linear equation (Thiele et al. 2010):

$$E(k) = |\hbar v_F(k)| \tag{12.3}$$

where v_F is the Fermi velocity.

The band structure of graphene is in the linear dispersion form. The 3D plot denoting the band structure is shown in Fig. 12.2. The band structure is of conical

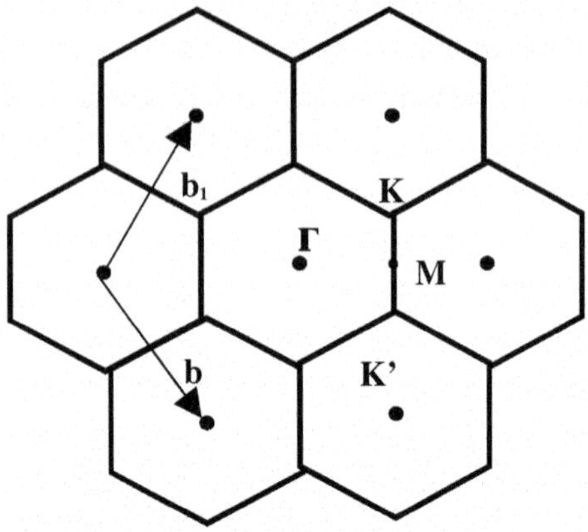

Fig. 12.1 Reciprocal lattice of graphene

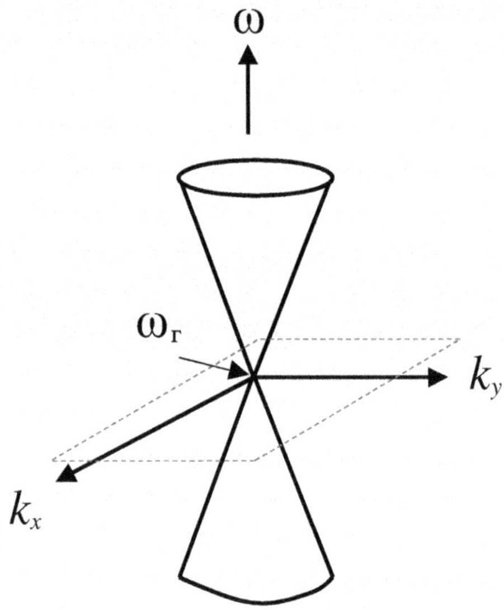

Fig. 12.2 Dirac cones

shape and referred to as Dirac cone. At Dirac points, electrons and holes behave as massless. It is reported that a band gap of several milli volts is necessary for digital logic implementation (Tan et al. 2014). Consequently, a thin layer of graphene sheet is patterned into several nano strips called graphene nano-ribbons (GNRs) in FET design. The small width of nano-ribbons leads to quantum confinement and restriction of electrons in dimension thereby inducing a band gap.

12.3 Modeling of Graphene Field-Effect Transistor (GFET)

In this section, the analytical drift-diffusion model and simulation model are discussed. The analytical model has been realized in MATLAB, while simulation model has been implemented in HSPICE.

12.3.1 Analytical Drift-Diffusion Model of GFET

The initial works on graphene metal-oxide-semiconductor field-effect transistor (MOSFET) were reported in early 2000s (Gelao et al. 2011; Thiele et al. 2010; Marulanda et al. 2008; Philip Wong and Akinwande 2010). In graphene field-effect transistors (GFETs), the channel material is graphene. To model the GFET, drift-diffusion model has been widely used (Chen et al. 2015; Banadaki 2016; Philip Wong and Akinwande 2010). The drift-diffusion model of a semiconductor is frequently used to describe semiconductor devices. The basic structure of GFET is shown in Fig. 12.3. Graphene is used as a semiconductor that creates a channel between the source and drain terminals. The channel is sandwiched by dielectrics that are present between the top and back gate terminals. C_e is the resulting capacitance between the top gate and the channel, while C_b denotes the capacitance between back gate and channel. C_q is the quantum capacitance which varies with channel charge density.

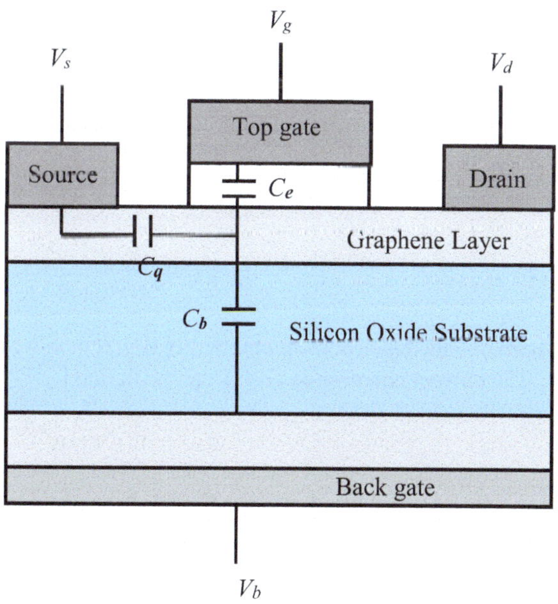

Fig. 12.3 Three basic structures of GFET with back gate

The potentials at the source, drain and gate terminals affect energy levels of both Dirac point E_d as well as the Fermi level E_F. The difference between E_F and E_d is of great importance as it determines the type of charge as well as the charge density in the channel.

The channel voltage can be obtained as

$$V_{ch} = -(E_f - E_d)/q \tag{12.4}$$

Electron and hole concentration inside the channel defines the quantum capacitance in GFET and can be determined as

$$p = \int_{-\infty}^{E_{cv}} D(E)[1 - f(E)]\mathrm{d}E \tag{12.5}$$

$$n = \int_{E_{cv}}^{\infty} D(E)[f(E)]\mathrm{d}E \tag{12.6}$$

where $D(E)$ is the density of states and $f(E)$ represents Fermi–Dirac integral.

Sheet charge (Q_{sh}) can be used to determine quantum capacitance (C_q). Q_{sh} can be computed as

$$Q_{sh} = q(p - n) \tag{12.7}$$

Quantum capacitance is the derivation of net sheet charge to channel potential and is obtained as

$$C_q = -\mathrm{d}Q_{sh}/\mathrm{d}V_{ch} \tag{12.8}$$

Under the condition $qV_{ch} \gg K_B T$, above expression can be simplified to

$$C_q = \frac{2q^2 q|V_{ch}|}{\pi(\hbar v_F)^2} \tag{12.9}$$

where q is the electronic charge, K_B is Boltzmann's constant.

The current equation in GFET can be defined as

$$I_{DS} = WQ(x)v_F \tag{12.10}$$

where W is the width of the graphene layer and $Q(x)$ is the electric charge density along the channel. v_F is the Fermi velocity and given as

$$v_F = \frac{\mu_0 F}{1 + \frac{F}{F_c}} \tag{12.11}$$

where μ_0 is the mobility, F is the electric field, F_c is the critical field. The net charge density is given by

$$Q(x) = -C_{top}\left[V_{g0} - V(x)\right] \tag{12.12}$$

Here, $V_{g0} = V_{gtop} - V_0$, where V_0 is the threshold voltage of the GFET and is defined as

$$V_0 = V_{gtop}^0 + \frac{C_{back}}{C_{top}}\left(V_{gback}^0 - V_{gback}\right) \tag{12.13}$$

V_{gtop}^0 and V_{gback}^0 in (12.13) are the top and back gate voltages of Dirac point, respectively.

12.3.2 HSPICE Simulation of the Model

The simulation model is necessary for readily implementing large circuit designs. First, the GFET model is developed. This is realized using two library files viz. 'param.lib' and 'gfet.lib'. The param.lib contains fixed device constants, and gfet.lib contains the derived mathematical equations that characterize the device working. The test-bench file calls the gfet.lib wherein GFET can be used as instance for realizing big circuits. The simulator performs the operation as per commands specified. The results are then analyzed on the output panel of the simulator. The complete flow diagram of the simulation model is shown in Fig. 12.4. This is performed in HSPICE electronic design automation tool.

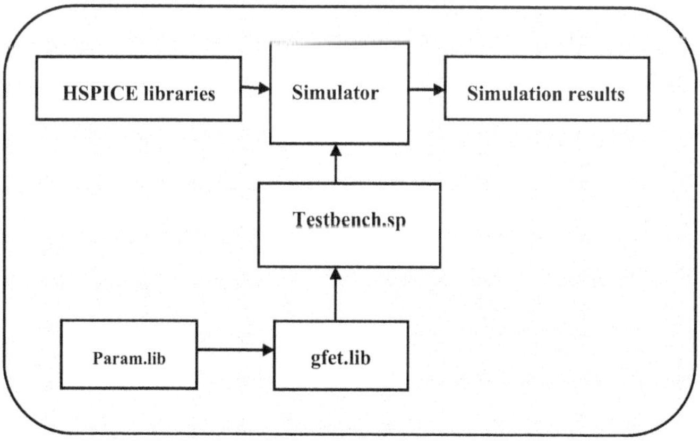

Fig. 12.4 Flowchart detailing steps for HSPICE circuit simulations

12.4 Results and Discussion

In this section, analyses corresponding to GFET have been presented. The performance analyses have been carried out using both the analytical and simulation models. Further, to validate the efficacy of the futuristic graphene FETs, it has been compared with its counterpart conventional silicon FET-based designs. Different circuit designs such as inverter, NAND, NOR gates and half adder have been implemented.

First, the current–voltage characteristics of graphene FET are determined using analytical model and validated with the HSPICE simulation model. This is shown in Fig. 12.5. The *y*-axis in the figure represents the drain current (I_{DS}), while *x*-axis delineates the output voltage (V_{DS}). V_{DS} is varied from 0 to −3 V. The bulk voltage is kept at −40 V. The top gate voltage is varied as −0.8 V, −1.3 V, −1.8 V, −2.3 V and −2.8 V. It is seen from the figure that the analytical and HSPICE simulation results are in close agreement with each other and hence can be effectively used for realizing large circuit designs.

Once, validating the GFET characteristics and correctness of the analytical and simulation models, different digital circuits are implemented. Figure 12.6 shows the schematic of inverter circuit using silicon and graphene FETs. The input and output waveforms of inverter circuit using both the FETs are shown in Fig. 12.7. It is analyzed that functionality wise, both the inverters using different FETs give correct output. However, performance wise, GFET has a higher edge over SiFET. It has been analyzed that delay and power dissipation in inverter circuit using GFET is much lesser as compared to SiFET. This can be seen in Figs. 12.10 and 12.11, respectively.

Next, different logic gates viz. NAND and NOR are realized. These are shown in Fig. 12.8. Figure 12.9 shows the waveform of half adder circuit. For all the simulation results, it is analyzed that circuits implemented using GFETs accurately produce the desired output results with added advantage of lower delay and power dissipation

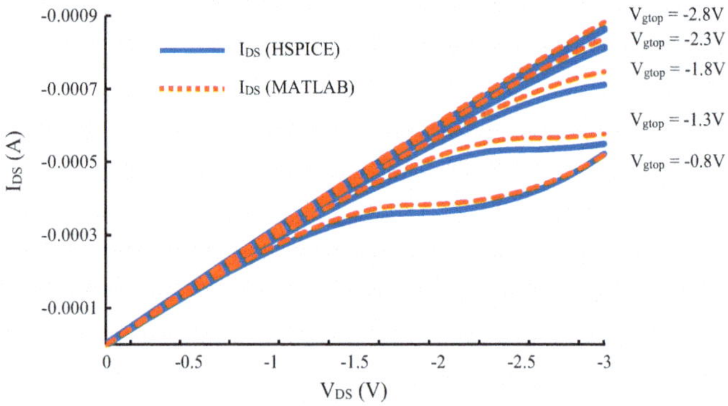

Fig. 12.5 Current–voltage characteristics of graphene FET

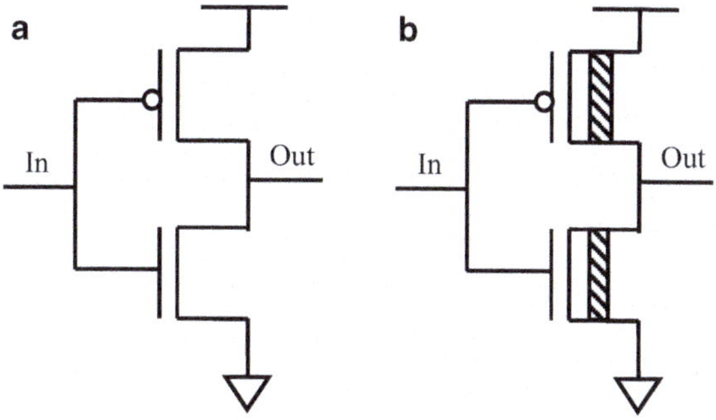

Fig. 12.6 **a** Circuit schematic for SiFET inverter. **b** Circuit schematic for GFET inverter

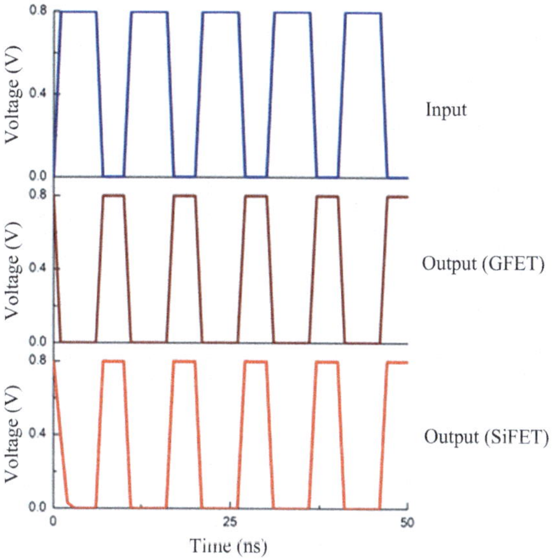

Fig. 12.7 Input and output characteristics of an inverter circuit using SiFET and GFET

(as shown in Figs. 12.10 and 12.11). Hence, it can be deduced that GFETs can be convincely incorporated for next-generation integrated circuit designs.

Figures 12.10 and 12.11 show delay and power comparison, respectively, in between circuits implemented using SiFET and GFET. In case of inverter circuit, it is investigated from both the figures that GFET has comparatively lesser delay than its counterpart SiFET-based inverter circuit. This is due to the fact that graphene exhibits ultra fast switching capabilities because of much higher mobility of charge carriers. The decrease in propagation delay in GFET is about 11% lesser than SiFET-based circuit and is 62% more power boosted. Considering NAND and NOR logic gates, from the delay and power calculations, it is envisaged that GFET-based NAND is nearly 22% more superior in terms of delay and 67.3% in terms of power, whereas

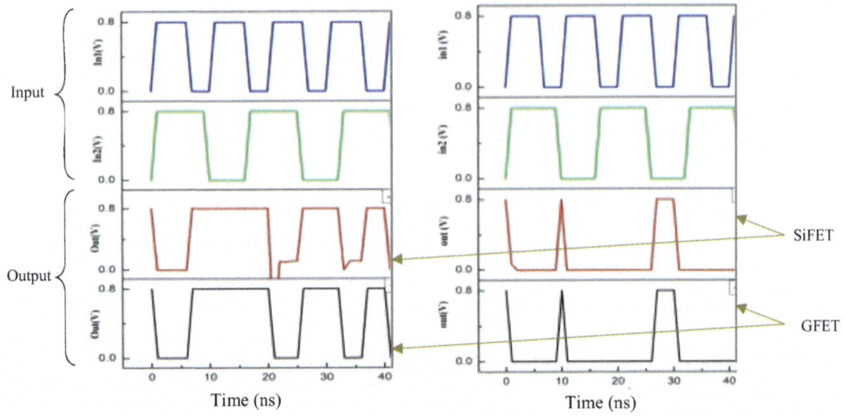

Fig. 12.8 Output waveforms for NAND and NOR gates using SiFET and GFET

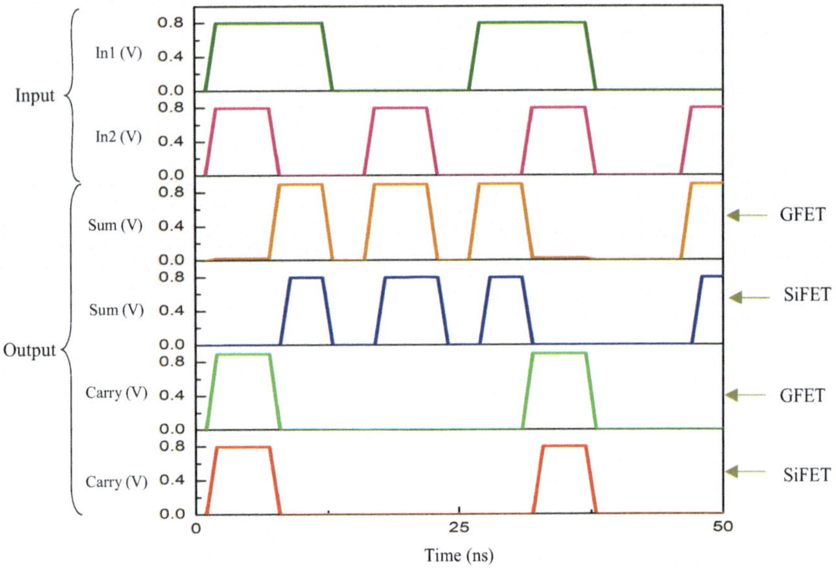

Fig. 12.9 Output waveforms for half adder using SiFET and GFET

GFET-based NOR is 28% more faster and consumes 47% lesser power than its SiFET counterparts. Similarly, for half adder circuit also, it is illustrated from both the figures that GFET-based circuit is faster as compared to SiFET. Hence, it is inferred that GFET-based devices are best suited for high speed applications, the reason being size of PMOS is greater than twice the NMOS which increases the parasitics at the output, whereas in GFET, the size of both pull-up and pull-down devices is same which reduces the efforts and makes it faster and better circuit as well

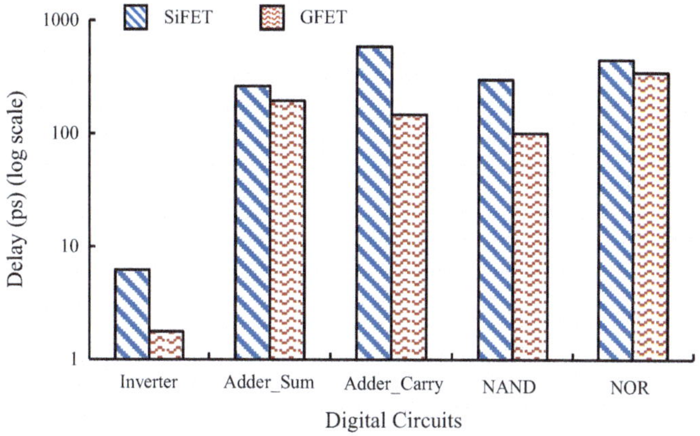

Fig. 12.10 Delay analysis using SiFET and GFET

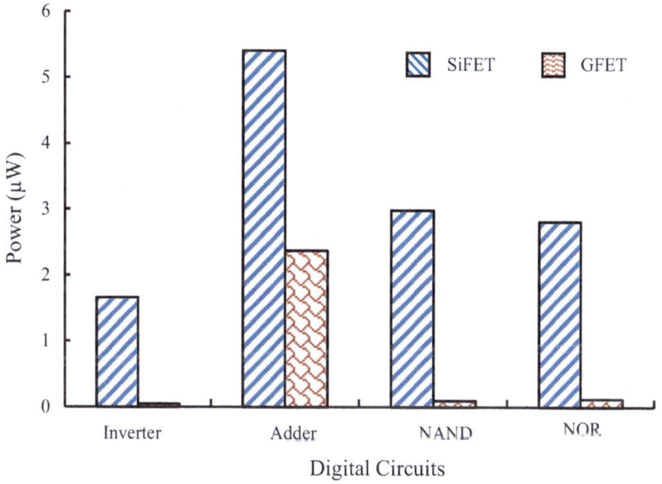

Fig. 12.11 Power analysis using SiFET and GFET

as reliable interconnects (Agrawal et al. 2017; Umoh et al. 2013; Patel et al. 2019; Patahde et al. 2018). Moreover, power dissipation in GFET-based circuits is comparatively much lower than SiFET. Hence, GFETs can be effectively incorporated for low power applications.

12.5 Conclusion

In this chapter, extensive physics, analytical modeling and simulation of graphene-based transistor and its realization for implementation of various gates and circuits have been shown. Graphene FET is a nano device and incorporates several quantum phenomena for its operation. The device is modeled using drift-diffusion model in MATLAB. The simulation of GFET device and its corresponding circuits are implemented using HSPICE. For validating the efficacy of the prospective GFETs for circuit and system design, it is compared with its counterpart conventional silicon FET. The results show that graphene FET can lead to better efficiency at device level. The remarkable properties inhibited by graphene make it superior than silicon-based transistors. It is envisaged that circuits using GFETs faithfully produce the correct output. Energetically with correctness, GFETs lead to lesser delay and power dissipation in the circuit. Thus, it can be promisingly concluded that graphene FETs are efficient and convincely can be incorporated for developing fast, low power and efficient circuits and systems in next generation integrated circuit designs.

References

Agrawal Y, Kumar MG, Chandel R (2017) A novel unified model for copper and MLGNR interconnects using voltage-and current-mode signaling schemes. IEEE Trans Electromagn Compat 59(1):217–227

Anas MM (2016) A study of single layer and bilayer GNRFET. Paper presented in IEEE UK Sim-AMSS 18th international conference on computer modelling and simulation, 2016

Aradhya H, Mahadikar M, Muniraj R, Suraj M, Moiz M, Madan H (2016) Design analysis and performance comparison of GNRFET based adiabatic 8-bit ALU. Paper presented in IEEE international conference on recent trends in electronics, information & communication technology (RTEICT), 2016

Banadaki YM (2016) Physical modeling of graphene nanoribbon field effect transistor using non-equilibrium green function approach for integrated circuit design. Ph.D. Thesis, Louisiana State University and Agricultural and Mechanical College, 2016

Chen Y, Sangai A, Rogachev A, Gholipour M (2015) A spice-compatible model of MOS-type graphene nano-ribbon field-effect transistors enabling gate and circuit-level delay and power analysis under process variation. IEEE Trans Nanotechnol 14(6):1068–1082

Gelao G, Marani R, Diana R, Perri A (2011) A semiempirical spice model for n-type conventional CNTFETs. IEEE Trans Nanotechnol 10(3):506–512

Islam A (2015) Technology scaling and its side effects. Paper presented at 19th international symposium on VLSI design and test (VDAT) at Nirmal University, Ahemdabad, 2015

Marulanda JM, Srivastava A, Yellampalli S (2008) Numerical modeling of the I–V characteristic of carbon nanotube field effect transistors (CNT-FETs). Paper presented in 40th IEEE Southeastern symposium on system theory, 2008

Patahde T, Shah U, Agrawal Y, Parekh R (2018) Preeminent buffer insertion technique for long advanced on-chip graphene interconnects. Paper presented in IEEE electrical design of advanced packaging and systems symposium (EDAPS) at Chandigarh, India, 2018

Patel N, Agrawal Y, Parekh R (2019) A literature review on next-generation graphene interconnects. World Sci J Circuits, Syst Comput 28(9):1930008

Philip Wong H-S, Akinwande D (2010) Carbon nanotube and graphene device physics. Cambridge University Press

Tan S, Tang L, Chen K (2014) Band gap opening in zigzag graphene nanoribbon modulated with magnetic atoms. Curr Appl Phys 14(11):1509–1513

Thiele S, Schaefer J, Schwierz F (2010) Modeling of graphene metal-oxide-semiconductor field-effect transistors with gapless large-area graphene channels. J Appl Phys 107(9):094505

Umoh I, Kazmierski T, Hashimi BA (2013) A dual-gate graphene FET model for circuit simulation—SPICE implementation. IEEE Trans Nanotechnol 12(3):427–435

Wang W et al (2014) Quantum transport simulations of CNTFETs: performance assessment and comparison study with GNRFETs. JSTS: J Semicond Technol Sci 14(5):615–624

Part IV
System Level Applications

Chapter 13
Power and Area-Efficient Architectural Design Methodology for Nanomagnetic Computation

Santhosh Sivasubramani, Sanghamitra Debroy, and Amit Acharyya

Abstract Magnetic quantum-dot cellular automata (MQCA)-based nanomagnetic logic computation started emerging to augment the CMOS-based traditional computing devices as Moore's law approaching towards its end. Computation performed using nanomagnets exhibits non-volatility and adheres to the thermodynamic law (second). The emerging advents in the field of artificial intelligence computing on edge with the constrained resources necessitate rebooting the computing paradigm beyond CMOS and more than Moore to cater for area and power efficiency. In this regard, digital logic arithmetic circuits should be revisited using this energy-efficient computing paradigm using nanomagnets. This chapter summarizes the undergoing research in the design of such arithmetic architecture development and its corresponding nanomagnetic implementation. Researchers have demonstrated the MQCA-based arithmetic architecture implementation using inplane nanomagnetic logic (iNML) utilizing the dipole coupling. Design methodologies presented in the literatures have exploited the shape (S), positional (P), shape & positional-based hybrid nanomagnetic anisotropies pertaining to the optimization in terms of required number of resources in terms of nanomagnets (NMs), clock cycles (CCs) and majority gates (MGs) which are the critical constraints leading to high speed, area and energy-efficient design. Subsequently, researchers have exploited physical analogy of the basic building block, i.e. the three inputs nanomagnetic majority logic gate for enhanced optimization in the nanomagnetic design. However, for higher integration densities and efficient area consumption, the scalability of the dipole coupling-based nanomagnetic devices is an important aspect which is eventually limited by its susceptibility to thermal fluctuations. In this regard, interlayer exchange-coupled (IEC) scheme has been demonstrated and has been shown to offer stronger interaction

S. Sivasubramani · S. Debroy · A. Acharyya (✉)
Advanced Embedded Systems and IC Design Laboratory, Department of Electrical Engineering, Indian Institute of Technology, Hyderabad, India
e-mail: amit_acharyya@ee.iith.ac.in

S. Sivasubramani
e-mail: ee15m16p100001@iith.ac.in

S. Debroy
e-mail: ee14resch12002@iith.ac.in

R. Dhiman and R. Chandel (eds.), *Nanoscale VLSI*,
Energy Systems in Electrical Engineering,
https://doi.org/10.1007/978-981-15-7937-0_13

between thin nanomagnets, resulting in greater scalability and better data retention at the deep sub-micron level, hence allowing magnetic interaction to be manipulated both in the vertical and lateral directions at the same time. In this regard, interlayer exchange-coupling scheme has been discussed as a possible solution to better scalability and data retention. Interlayer exchange-coupled system comprises of a non-magnetic metal layer (known as spacer layer) sandwiched between two ferromagnetic layers. The two ferromagnetic layers may be coupled ferromagnetically (FM) or antiferromagnetically (AFM), decided by the thickness and material (e.g. chromium, copper, ruthenium) of the spacer layer. On the other hand, perpendicular Nanomagnetic Logic (pNML) has involved a lot of interest for 3D architecture exploration. This chapter gives an overview on the emerging nanoscale architecture circuits, design and its implementation using nanomagnets. The implementation of nanomagnetic logic for data transmission in 3D IC has also been discussed, resulting in higher packing densities in 3D IC's.

Keywords NML · Magnetic computing · Nanomagnets · Architecture design · iNML · pNML · Inplane magnetization · Out-of-plane magnetization · Design methodology · Copper interconnects · Graphene interconnects

13.1 Introduction

The term "Rebooting Computing" IEEE (2016, 2020) was coined by Professor Peter Denning for rethinking/reimagining the learning-based education in computing. Professor Tom Conte has reinvented this term "rebooting computing" which becomes the motivation for the Future Directions working group in 2012, to rethink the computer processing capabilities, "from soup to nuts", including all aspects from device to user interface (IEEE 2016). With the increasing demand in the computing power, generated big data, emerging need for edge computing, scaling issues, foundry limitations, demise of Moore's prediction and many such obstacles necessitates the need for rebooting computing from the scratch. The term rebooting is defined as revisiting the series of known traditional computing paradigms to discard all previous problems leading to performance degradation and to restart alternative paradigms from its base. Extensive research has been performed in the areas of CMOS and approximate logic-based design methodologies. On another note, there exists ongoing research in the upcoming NML devices based on the principles of magnetic-QCA. CMOS possesses advantages in read/write and clocking circuitry, approximate logic design inherently possesses ultra-low-power operations with insignificant loss in data accuracy, and MQCA-based logic design incurs no static loss, no heat dissipation, low power consumption, non-volatility and is radiation hard. Research vision is to leverage the advantages of these technologies and to propose various arithmetic architectural design methodologies resulting in area and energy efficiency. This chapter explores such arithmetic architecture designs and their corresponding nanomagnetic implementation using theoretical modelling, micromagnetic simulations, analysis and validation considering the constraint of resources in terms of the number of nanomagnets, majority gates and clock cycles.

The main issue of scaling devices is the detrimental short channel effects and high circuit power densities imposing several challenges and thus scaling beyond is not practical (The International Roadmap for Devices and Systems 2016, 2017; The IRDS Roadmap 2018; Blank 2018; Moore 2018, 2019). The standby power (the power required to maintain data in a circuit) is speedily approaching to the power consumed while actual computation is performed (Blank 2018; Moore 2018, 2019). Figure 13.1 illustrates the big picture of Moore's law and its varied perspectives of the challenges associated with further scaling of CMOS by semiconductor industries, industrialists, academicians, government and research organizations (Blank 2018; Moore 2018, 2019). It can be observed that there exist quantum effects at 7/5 nm process nodes and beyond though, not only by technical limitations we are in

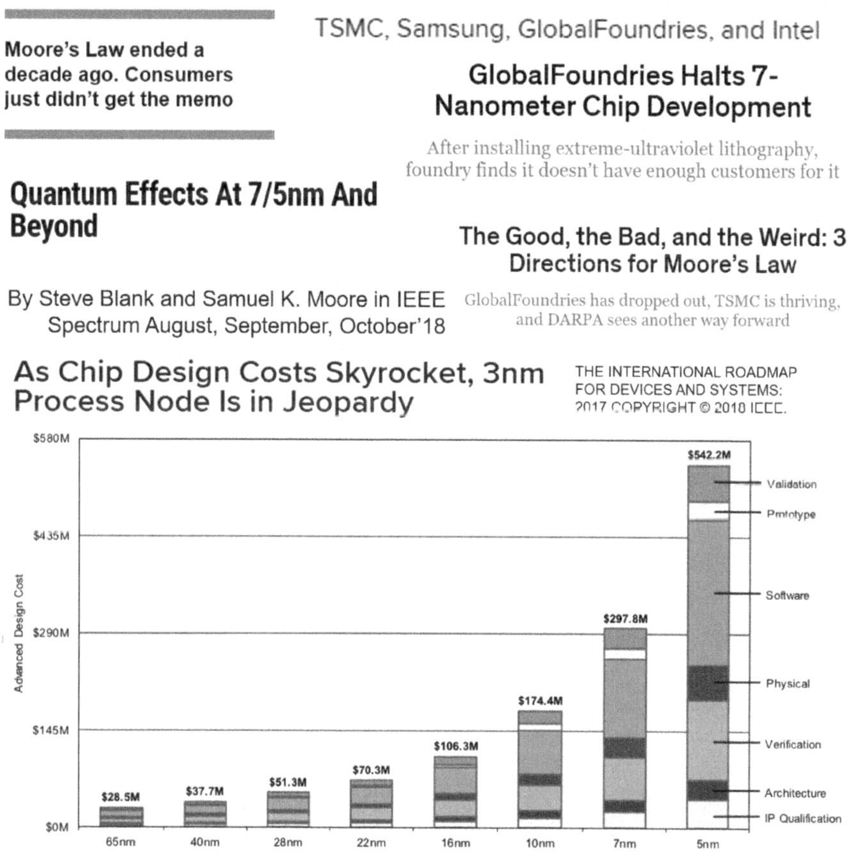

Fig. 13.1 More than Moore and few major challenges associated with further scaling down of transistors along with perspectives from different eminent visionaries (Blank 2018; Moore 2018, 2019; Lapadeus 2018)

search for More than Moore but in major by the skyrocket increase in cost for the chip design beyond 5 nm process nodes (Blank 2018; Moore 2018, 2019) (cf. Fig. 13.1). To surpass the existing problems and to mitigate this, there arises a significant need to think beyond known computing paradigms and to reboot computing with varied alternatives (Porod and Niemier 2015). One such potential candidate of consideration in rebooting computing is nanomagnet-based computing which is detailed below (Porod and Niemier 2015). International Roadmap for Devices and Systems (IRDS) working group on Beyond CMOS presents the taxonomy of options for the emerging logic devices (The International Roadmap for Devices and Systems 2016, 2017; The IRDS Roadmap 2018) as depicted in Fig. 13.2a, b. It can be inferred that the focused block on nanomagnetic logic is a potential candidate identified to complement beyond CMOS computing paradigm (IEEE 2016). The functional and dimensional scaling of CMOS devices is driving the information processing technology hooked on a broadening spectrum of novel applications resulting in increased performance and complexity (The International Roadmap for Devices and Systems 2016, 2017; The IRDS Roadmap 2018). However as the scaling of CMOS eventually approach fundamental limits, numerous new information processing devices and micro-architectures for both existing and new functions is required to be explored (Blank 2018; Moore 2018, 2019). This is driving interest in new devices for information processing and memory and new paradigms for system architecture (The International Roadmap for Devices and Systems 2016, 2017; The IRDS Roadmap 2018; Debroy et al. 2019). Therefore, the following discussion provides an IRDS perspective on emerging focus technologies and serves as a bridge between conventional CMOS and the realm of nanoelectronics beyond the end of CMOS scaling. The three major identified work areas as per the IRDS 2018 The IRDS Roadmap (2018) update under Beyond CMOS work group report are emerging materials, devices/process and

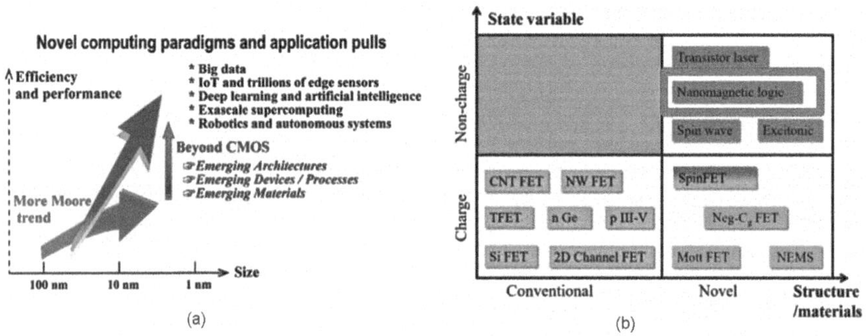

Fig. 13.2 **a** Illustrates the novel computing paradigms and appliation pulls as per International Roadmap for Devices and Systems (IRDS) 2017, 2018, Beyond CMOS working group (The International Roadmap for Devices and Systems 2017; The IRDS Roadmap 2018), **b** The emerging logic devices as identified by Beyond CMOS computing working group as part of the IRDS 2017, 2018. A bounding box is superimposed on "Nanomagnetic Logic (NML)" under the novel structure/materials in the non-charge category as illustrated (The International Roadmap for Devices and Systems 2017)

architectures (cf. Fig. 13.2a,b and Table 13.2 The International Roadmap for Devices and Systems 2017; The IRDS Roadmap 2018). Similarly, as identified in the IRDS 2017 The International Roadmap for Devices and Systems (2017) edition under the Emerging Research Materials work group category, researchers have explored the tunable intrinsic magnetism, electronic transport properties and interconnect applications of graphene and its role in nanomagnetic logic devices contributing towards the emerging materials and devices research (Debroy et al. 2019; Sivasubramani et al. 2018). As part of this chapter, we present the emerging architecture design methodology focusing on the MQCA-based NML complementing the work on emerging materials and devices as aforestated contributing towards the system development of Beyond CMOS, and more than Moore leading towards rebooting computing IEEE (2016). MQCA-based NML architecture designs overview will be presented in the subsequent section (Table 13.1).

Table 13.1 List of abbreviations

CMOS	Complementary metal oxide semiconductor
CA	Cellular automata
QCA	Quantum-dot cellular automata
MQCA	Magnetic quantum-dot cellular automata
NML	Nanomagnetic logic
iNML	Inplane nanomagnetic logic
pNML	Perpendicular Nanomagnetic Logic
MtM	More than Moore
IRDS	International Roadmap for Devices and Systems
NM	Nanomagnets
MG	Majority gate
CC	Clock cycles
S	Shape anisotropy
P	Positional anisotropy
SP	Shape and positional hybrid anisotropy
FMG or SMG	Ferromagnetically coupled fixed input majority gate
OOMMF	The object-oriented micromagnetic framework
RRN	Runtime reconfigurable nanomagnetic adder
ACN	Accurate nanomagnetic adder
APN	Approximate nanomagnetic logic
UMG	Universal majority gate
SLA	System level approach
RC	Rebooting computing

Table 13.2 Beyond CMOS difficult challenges and summary of issues and opportunities as identified in IRDS 2018 Beyond CMOS work group report (The IRDS Roadmap 2018)—Highlights focus on the architecture design & development of novel information processing paradigm

Difficult challenges The IRDS Roadmap 2018	Summary of issues and opportunities The IRDS Roadmap 2018
Continue functional scaling of information processing technology substantially beyond that attainable by ultimately scaled CMOS	Invent and reduce to practice a new information processing technology to replace CMOS as the performance driver
Bridge the gap between novel devices and unconventional architectures and computing paradigms	Identify suitable opportunities in unconventional architectures and computing paradigms that can utilize unique characteristics of novel devices

13.2 Nanomagnetic Computing-Prerequisites and Challenges

Information flow is controlled through the interaction of cell amongst the nearest neighbours in the quantum-dot cellular automata (QCA) (Berto et al. 2017; Porod et al. 1999). Contrasting the approach of the CMOS computing paradigm, the cell's state (Porod et al. 1999) stores the information. Nanomagnetic computing also exhibits non-volatility; ie., their magnetic states are retained when powered off. Its working principle (cf. Figure 13.3), state-of-the-art literature pertaining to architecture design methods, will be discussed in the subsequent subsections.

13.2.1 Why Computing with Nanomagnets?

- Nanomagnetic computing consumes the least amount of energy (Porod and Niemier 2015; Bhanja et al. 2016).
- Reductions in power consumption is possible in this evolving electron-free magnetic microprocessors (Porod and Niemier 2015).
- Nanomagnetic computing emerged as a promising candidate because the magnetic bits can be differentiated by direction (Porod and Niemier 2015; Bhanja et al. 2016).
- A new era of magnetronics based design could be our future.
- In this emerging magnetic computing, the logic devices would be intrinsically memory devices (Porod and Niemier 2015; Bhanja et al. 2016).
- Landauer limit–in any computer, each single bit operation must expend an absolute minimum amount of energy Porod and Niemier 2015; Bhanja et al. 2016).
- Nanomagnetic computing only took 15 millielectron volts of energy–to flip a magnetic bit from one state to another (Porod and Niemier 2015)

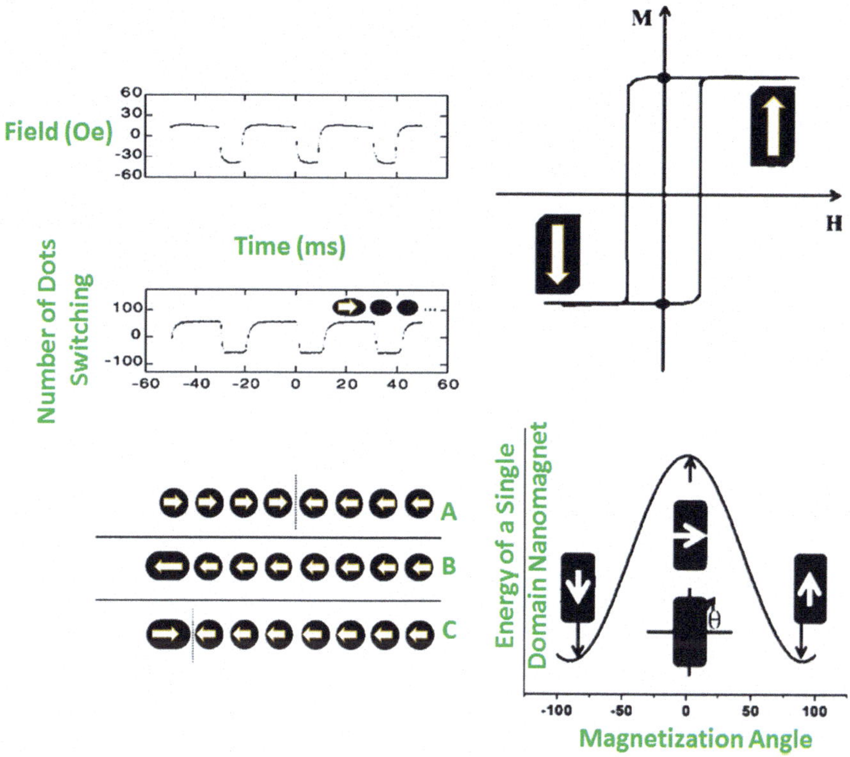

Fig. 13.3 Room Temperature Magnetic Quantum Cellular Automata. A schematic of the vector magnetization (arrows) in a number of dots. Two stable states of a single-domain nanomagnet (Madami et al. 2017). Hysteresis loop along easy axis, and energy barrier between the two states (Cowburn and Welland 2000; Cowburn 2002)

- Nanomagnetic computing devices exhibit non-volatility (Porod and Niemier 2015; Bhanja et al. 2016).
- The magnets retain their state even when power is off, and no time or energy is wasted in booting up (Porod and Niemier 2015; Bhanja et al. 2016).
- Energy Minimization Nature of Nanomagnets: "When collection of nanomagnetic discs are driven to an excited state and relaxed, they tend to couple magnetically with one another to minimize total magnetic energy of the system" (Bhanja et al. 2016).

13.2.2 Working Principle

The methodology in which the information is propagated amongst few number of nanomagnets is depicted in Fig. 13.4. The flux generated by the single-domain nano-

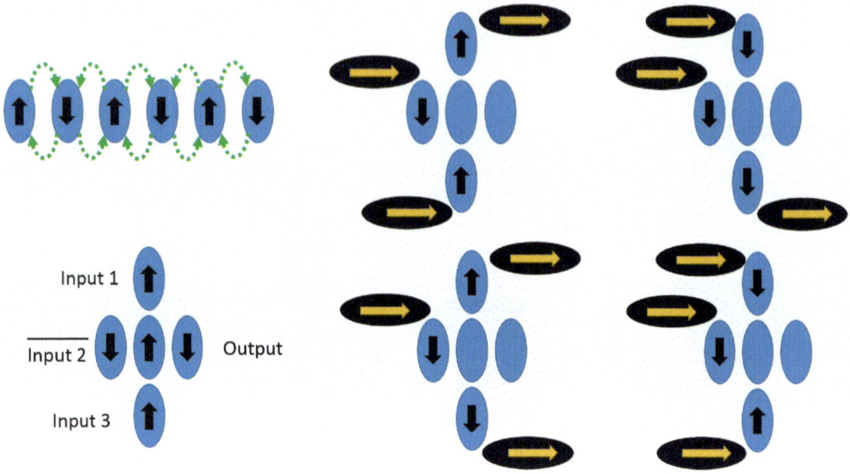

Fig. 13.4 Flux direction indication along with the data flow in few number of nanomagnets. Universal majority gate (UMG) implementation using nanomagnets. Operational illustration of the MQCA-based UMG (Csaba et al. 2002; Imre et al. 2006)

Table 13.3 Logic states summary–MQCA-based university majority logic gate for all input combinations. Two input programmable NAND and NOR gate can be obtained using upper/lower four rows of the table (Imre et al. 2006)

Logic State of		
Input Nanomagnets	Central Nanomagnet	Output Nanomagnets
000	0	1
010	0	1
110	1	0
100	0	1
001	0	1
011	1	0
111	1	0
101	1	0

magnet by the application of an external field tends to drive the neighbouring single domain nanomagnet, and this process is repeated till the last nanomagnet (Csaba et al. 2002). The major components of a NML device are (a) binary wire for ferromagnetic or antiferromagnetic data propogation (b) inverter (c) input, output and computing nanomagnets (Porod and Niemier 2015; Cowburn 2002; Bhanja et al. 2016). A binary wire is used for propagating information, and the majority logic gate is used for computing information (Imre et al. 2006). As shown in Table 13.3, to attain all the logic combinations, eight design configurations are used. The states of input 1 and 3 nanomagnets are connected as it is, and the state of the nanomagnet concerning input 2 is negated, i.e. inverted to perform the logic gate operations (Imre et al. 2006).

Figure 13.5 depicts various magnetic ensembles to understand the working principle of data propagation and computation using nanomagnets. Figure 13.6 depicts the MQCA-based NML UMG working principle (Orlov et al. 2008; Varga et al. 2010; Pulecio et al. 2011; Csaba et al. 2004). Figure 13.7 depicts the overview of the primary literatures. Figure 13.6 illustrates various methodologies to implement the MQCA-based nanomagnetic universal majority logic gate. Traditional oval-shaped driver nanomagnet, standalone input using slanted edge nanomagnet (Hesjedal and Phung 2010; Niemier et al. 2010; Dey et al. 2013) and 45 degree aligned oval-shaped driver nanomagnet-based implementations (Li et al. 2014; Gu et al. 2015) are shown (Fig. 13.8).

13.3 Nanomagnetic Logic Architecture Design an Overview

In view of the above, the nanomagnetic logic architecture design started emerging as a potential alternative to CMOS-based computing which faces challenges with Moore's law approaching towards its end. However, in contributing towards our vision aforementioned, there is a need for a design approach which focuses on:

- the proposal of a novel architecture for logic computation resulting in the reduction of the required number of resources (NMs, MGs & CCs)

Proposing novel architecture design methodology and optimization are the main research challenges identified to realize the arithmetic circuits using MQCA-based NML contributing towards rebooting computing. To address the aforestated issues, logic should be computed using novel efficient architectures, thus remains the motivation for exploring the MQCA-based nanomagnetic architecture designs for next-generation rebooting computing platform. The primary research challenge is the design and development of NML architectural design methodology which offers:

- **Area Efficiency**–with reduced design footprint;
- **High Speed**–with reduced number of MGs and CCs and
- **Energy Efficiency**–in combination with number of nanomagnets and majority gates.

This chapter presents the overview of such design methodologies which addresses these research challenges leading towards rebooting computing in the subsequent section.

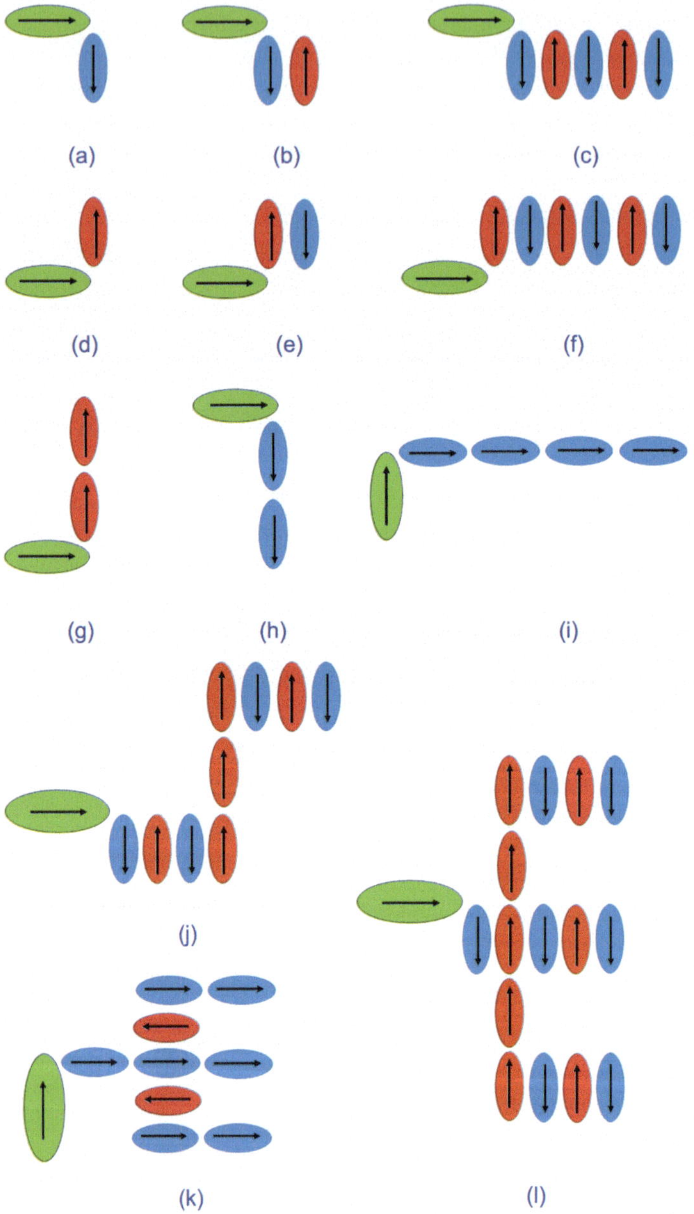

Fig. 13.5 MQCA-based NML working principle **a** magnetic ensemble 0 **b** antiferromagnetic magnet from state 0 to 1 **c** antiferromagnetic data propogation as buffer **d** magnetic ensemble 1 **e** state 1 to 0 **f** antiferromagnetic data propogation as inverter (horizontal) **g–i** ferromagnetic data propogation (vertical, horizontal) **j** single driver-multiple nanomagnets for both ferromagnetic and antiferromagnetic data propogation **k, l** fanout structures. [driver–green; 0–blue and 1–red]

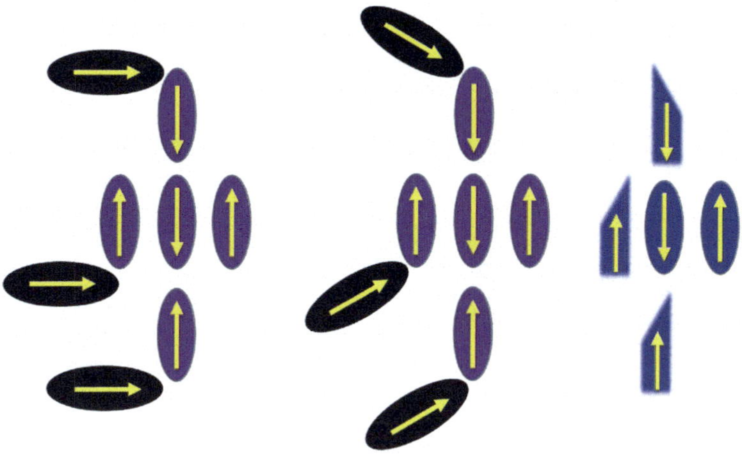

Fig. 13.6 Majority gate implementation using traditional elongated oval-shaped input driver nanomagnets, 45 degree aligned oval-shaped elongated driver input nanomagnets and the slanted edge standalone input nanomagnets

13.4 Inplane Nanomagnetic Logic and Perpendicular Nanomagnetic Logic-Based MQCA Architecture Design

This section introduces the readers to the nanomagnetic logic architecture design methodologies. The two ventricles for implementing nanomagnetic logic are by exploiting inplane magnetization and out-of-plane/ perpendicular magnetization referred as inplane nanomagnetic logic and perpendicular nanomagnetic logic, respectively. Such architecture design methodologies are designed using two known coupling enabling nanomagnetic interaction. Fringing field interactions between the non-magnetic neighbours are achieved using

- Dipole coupling
- Interlayer exchange coupling.

13.4.1 INML-dipole Coupling-Based Arithmetic Design

In this subsection, we present the nanomagnetic logic-based rebooting computing architecture design methodology by exploiting inplane magnetization of the dipole-coupled single domain nanomagnets. Here, we present the overview of two such designs, namely

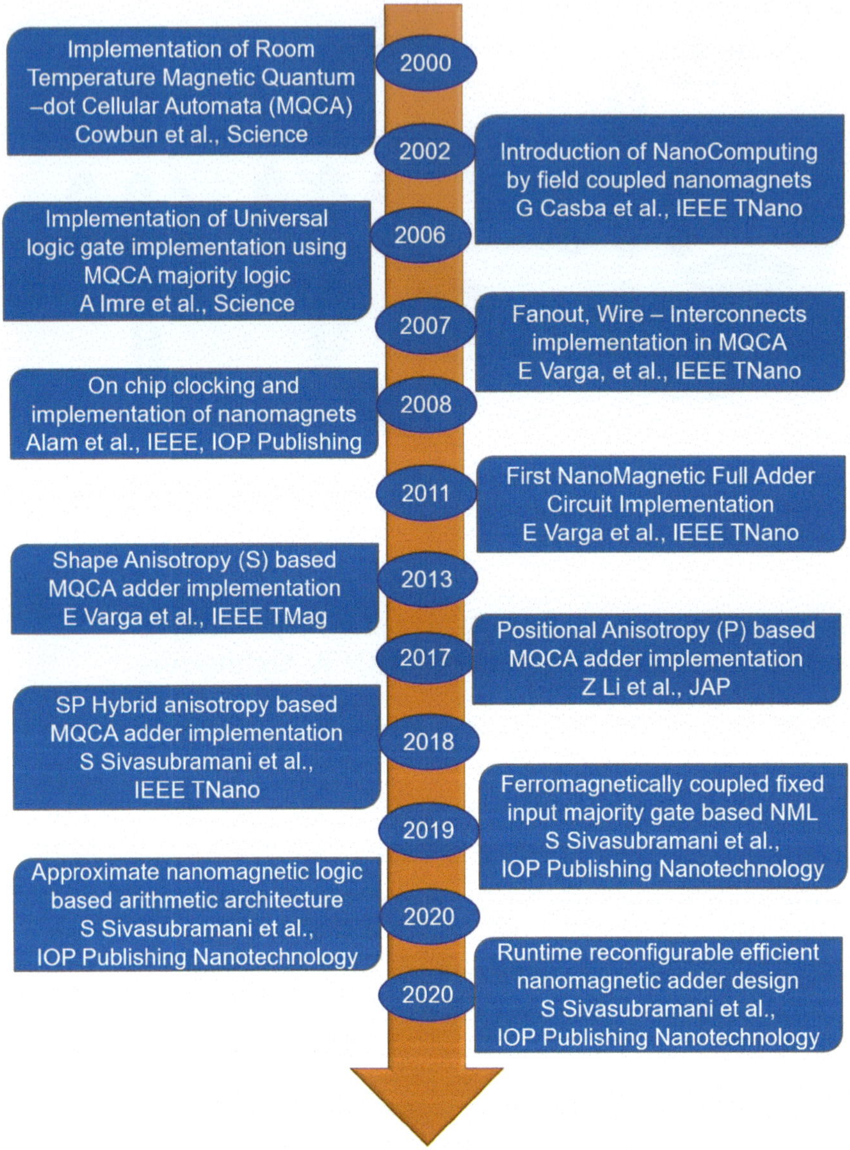

Fig. 13.7 MQCA-based NML architecture design methodology—an chronological order overview

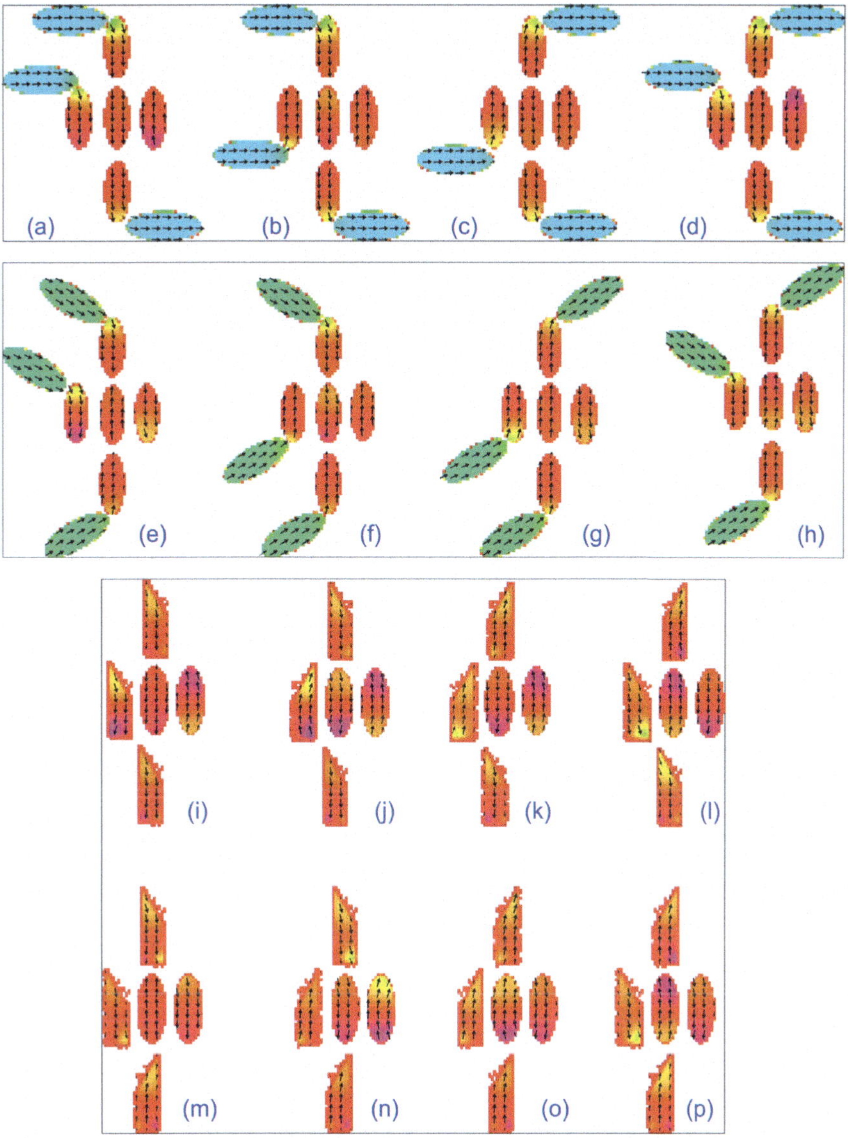

Fig. 13.8 Micromagnetic simulation tool–demonstration of the working principle of MQCA-based NML UMG (**a**–**d**) traditional drivers with C_i set to 0 (**e**–**h**) 45 degree aligned drivers with C_i set to 1 (**i**–**p**) standalone slanted edge input nanomagnets replacing input driver nanomagnets C_i set to 0 and 1

- Shape (S) and positional (P) hybrid anisotropy-based nanomagnetic adder architecture design and
- Ferromagnetically coupled fixed input majority gate-based nanomagnetic adder and subtractor architecture along with its mapping logic and logic optimization

13.4.1.1 SP Hybrid Anisotropy-Based NML Design (Sivasubramani et al. 2018)

This subsubsection briefly introduces the readers to the SP hybrid anisotropy-based NML design to implement carry-out (C_o) and sum (S) outputs of the 1-bit binary full adder. Figure 13.9 illustrates the optmized envisaged model and the individual advantages of both the S and P anisotropy. Combining the advantages of both designs will lead to a better architecture. Figure 13.10 (a–j) depicts the micro-magnetic simulation results of the SP hybrid-based nanomagnetic adder-based design. Two layouts have been proposed (Sivasubramani et al. 2018) for enhancing IC scalability using horizontal and vertical layouts. Figure 13.11 portrays the test loop design in verifying the proposed SP hybrid anisotropy-based design.

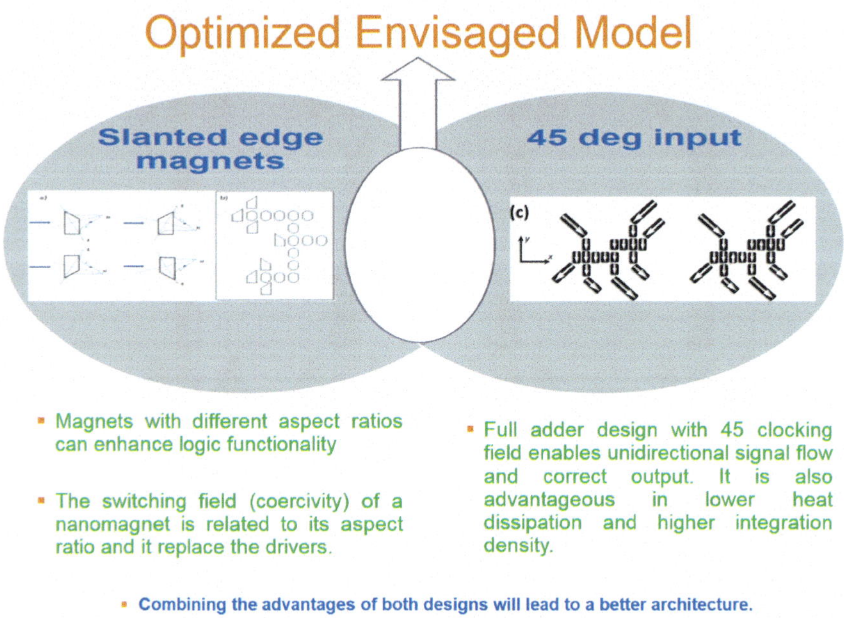

Fig. 13.9 Optimized envisaged model to perform nanomagnetic adder implementation along with the advantages of both shape and positional anisotropy of the nanomagnets

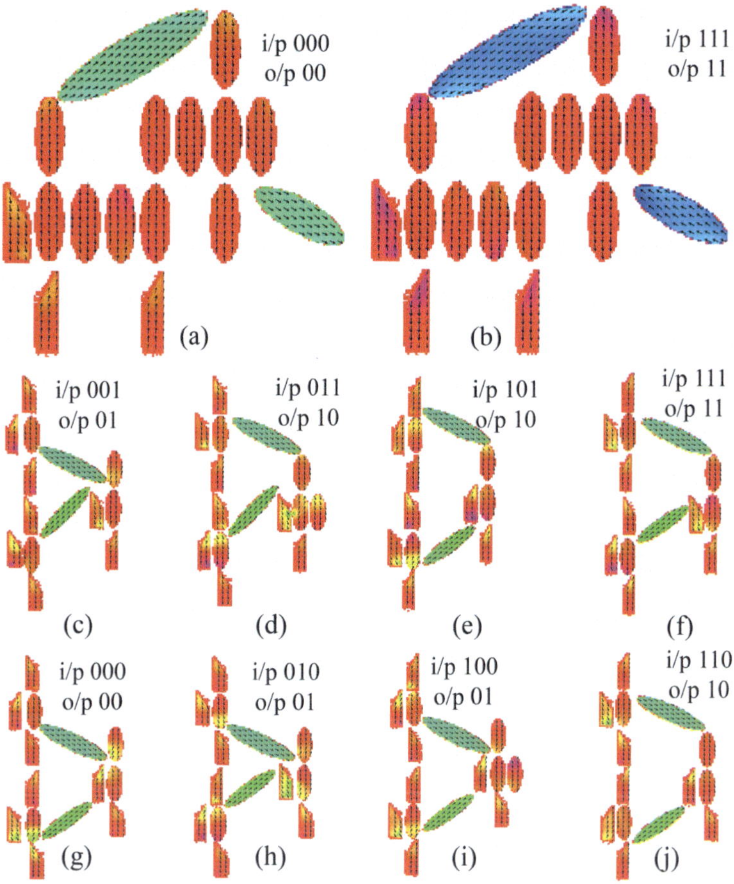

Fig. 13.10 OOMMF (Porter et al. 1999)-based micromagnetic simulation results (Sivasubramani et al. 2018)

13.4.1.2 FMG-Based NML Design Sivasubramani et al. 2019

Figure 13.12a, b illustrates the ferromagnetically coupled fixed input majority gate with two variations of the fixed input. The corresponding architecture representation of the introduced FMG module is depicted in Fig. 13.12c–f. Four different physical configurations are designed to achieve all the eight different input logic variations. Figure 13.13 depicts the performance comparison metrics of this FMG-based design with the comparative % reduction to the state of the art. 36%–69% reduction in the number of nanomagnets, 50%–75% reduction in the number of clock cycles and 33%–50% reduction in the number of MG operations is achieved using the proposed designs. Figure 13.14 portrays the micromagnetic simulation results of the (a–d) FMG-based module using traditional oval-shaped 45° aligned elongated

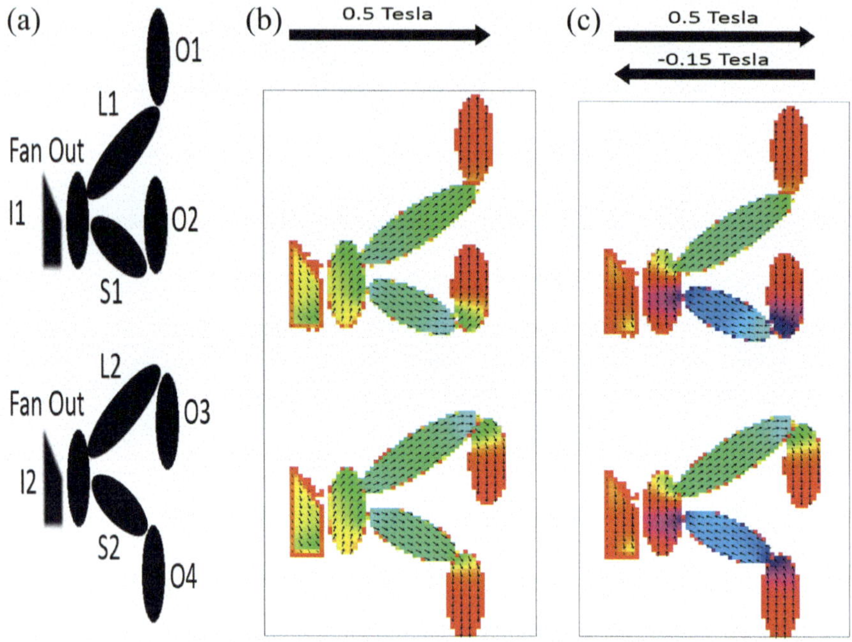

Fig. 13.11 OOMMF-based verification of the test loop design (Sivasubramani et al. 2018)

driver nanomagnet; (e–h) FMG-based module using standalone input nanomagnets; (i–l) 1 bit full adder using FMG module and (m-p) adder design built using module with enhanced structural optimization using slanted edge nanomagnets. Figure 13.15 illustrates the Karnaugh map illustration, and the researchers proposed mapping logic of the introduced FMG-based module to the binary 1-bit nanomagnetic full adder design. Similarly as aforementioned, the proposed ferromagnetically coupled fixed input majority gate design and mapping logic is not only limited to adder design, but it is generic and hence as a proof-of-concept demonstration here we introduce the 1 bit binary full subtractor architecture design using SMG and/or FMG (cf. Figure 13.16). All other design methodology and explanations and discussions hold similar to the adder design detailed above. Figure 13.16 details the K-map representation, its corresponding mapping logic of the module to subtractor and derivations. The proofs are similar as aforementioned. Nanomagnets are replaced in the subtractor architecture design using traditional oval-shaped input driver nanomagnets where B_i set to 0, 1 and using standalone input nanomagnets where B_i set to 0 and 1. Thus, it is evident that the proposed module and mapping logic of the ferromagnetically coupled fixed input majority gate design is generic and thus can be used to design varying arithmetic architectures (adder, subtractor) as shown here. In consequence, recently, researchers have reported approximate nanomagnetic logic-based arithmetic computing architecture designs using dipole-coupled single domain nanomagnets by exploiting inplane magnetization and reversal dynamic magnetization exhibiting runtime reconfigura-

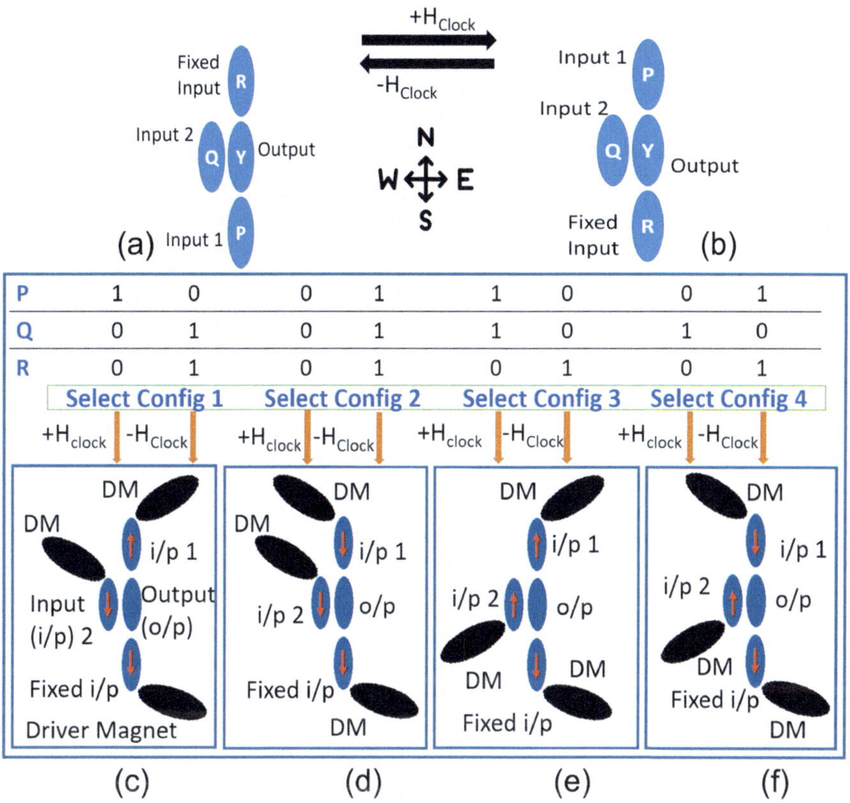

Fig. 13.12 **a–f** Ferromagnetically Coupled fixed input majority gate design Sivasubramani et al. 2019

bility Sivasubramani et al. (2019, 2020). Majorities of the simulations performed are on the sub 50 nm nanomagnetic implementation as a proof-of-concept (POC) demonstration. However, it is to noted that the design methodologies proposed for the nanomagnetic architecture design and development for rebooting computing and its application on resource-constrained AI in this chapter are generic and are independent of the design nodes sub-50 nm (POC), sub-180 nm and sub-250 nm. Hence, these architectures can be implemented using varying parameters for the dimension of nanomagnets adhering to the design rules and methods as proposed.

This subsection summarizes the design methodologies to design an efficient binary adder architecture using dipole-coupled inplane nanomagnets. The proposed design methodology of using SP hybrid anisotropy yields 28 % reduction in the number of nanomagnets leading to 0.032175 μm^2. The proposed design methodology of using FMG based module and mapping logic yields $\sim$ 36–69 % , $\sim$ 50–75 % and $\sim$ 33–50 % reduction in the resources. However the proposed design methodologies yield superior performance in terms of area, speed and error-free operations, it requires

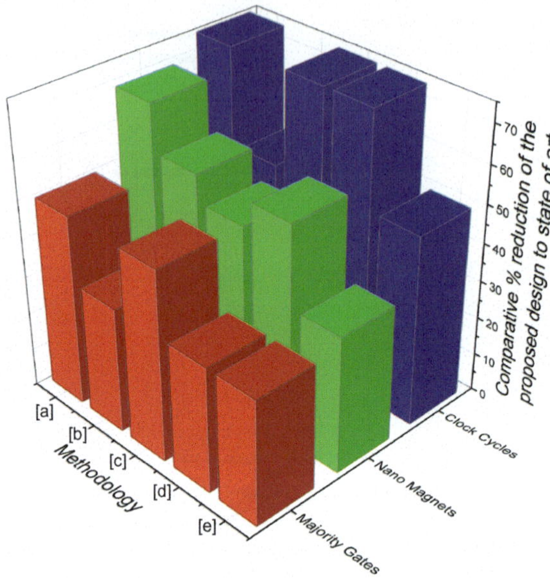

Fig. 13.13 Comparison of the performance metrics **a**—Varga et al. (2013), **b**—Varga et al. (2011), **c**—Varga et al. (2013), **d**—Li and Krishnan (2017), **e**—Sivasubramani et al. (2018), Sivasubramani et al. (2019)

multiple design layouts to perform arithmetic computation. The proposed design should also harness this advantage of CMOS which uses one layout configured at runtime for performing varying input logic combinations (Table 13.4).

13.4.2 PNML-IEC-Based Arithmetic Design

13.4.2.1 Theoretical Background of IEC-Based Coupling:

Today, the primary concern for magnetic logic and storage technology is that scaling these devices is limited by the superparamagnetic limit imposed by thermal variations. As the dimensions of the nanomagnets are reduced the energy barriers between the nanomagnetic states also gets reduced, that leads to increased susceptibility towards soft error. Soft error indicates temperature undulations that can arbitrarily flip the nanomagnetic state and erase the information saved in it. The frequency of flipping for a single nanomagnet at zero applied external field has a exponential dependence on the height of the energy barrier. This is the major challenge faced by the magnetic HDD industry at the present times. The areal density of a HDD is limited by the bit size (Wu et al. 2013), and as the bit size gets reduced beyond the superparamagnetic limit, the bits can randomly flip, destroying the control over them. This is an extremely serious problem in devices that are made of an assembly of small nanomagnets interacting with each other by magnetic dipole of the assembly. Moreover, it can be also noted that the logic functionality of the nanomagnetic

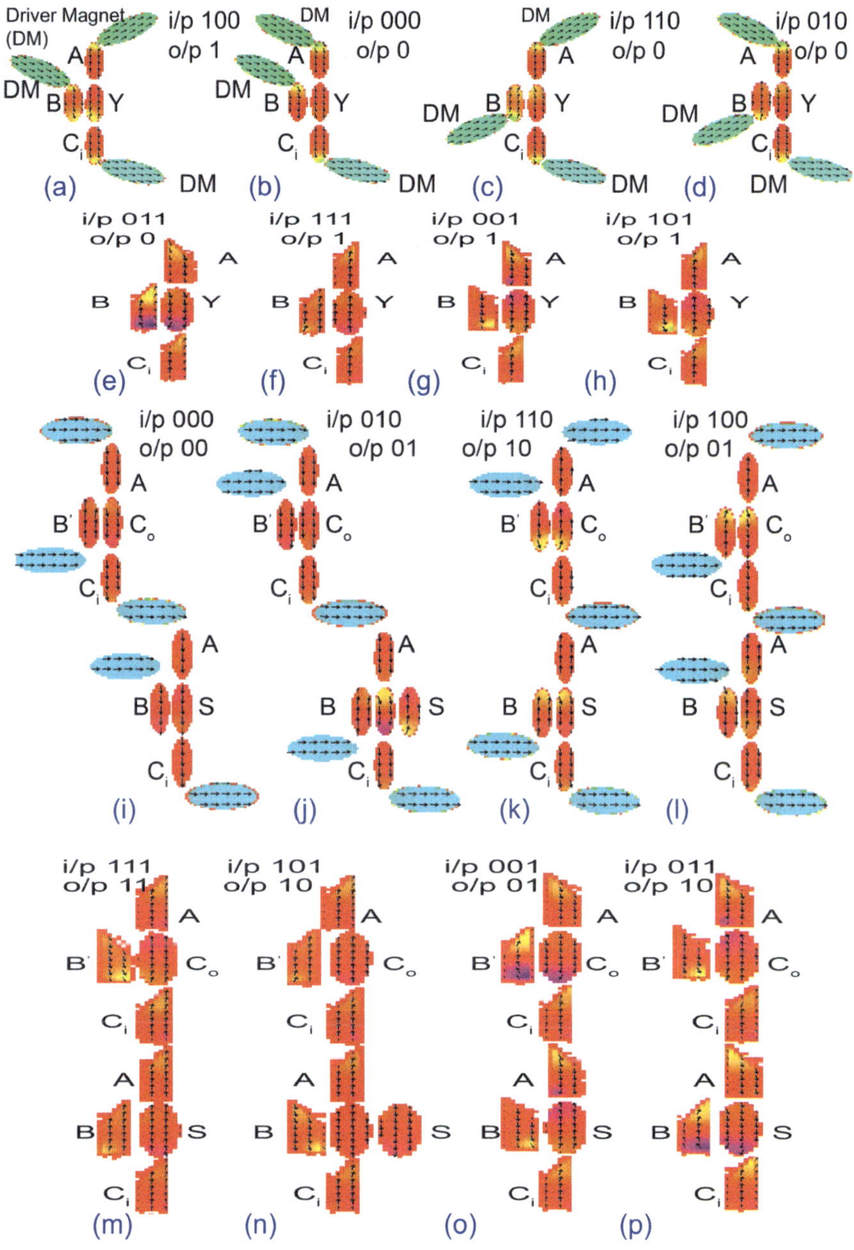

Fig. 13.14 **a–p** Micromagnetic simulation results of the FMG and FMG-based adder design (Sivasubramani et al. 2019)

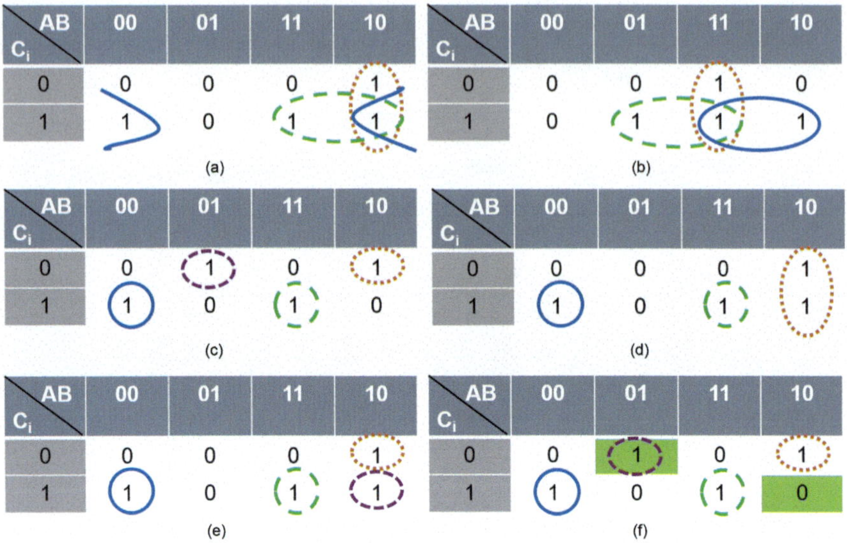

Fig. 13.15 **a–f** Boolean Optimization and the corresponding Karnaugh map (K-Map) depiction of the proposed (Sivasubramani et al. 2019).

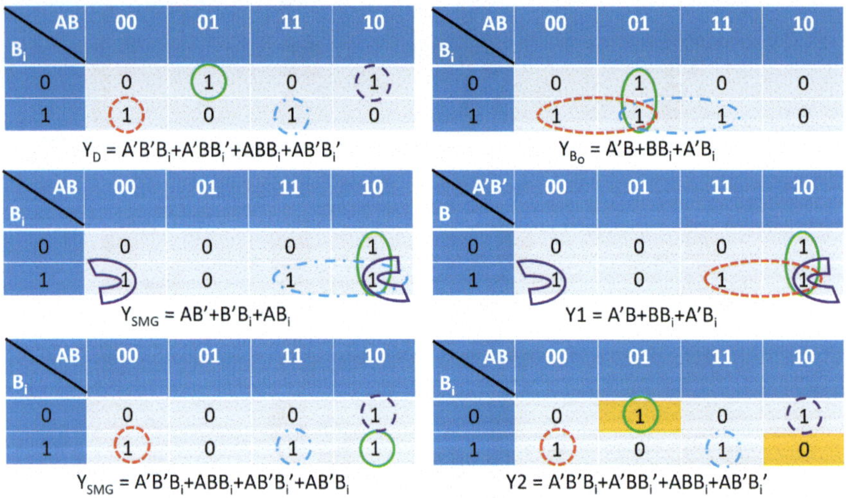

Fig. 13.16 Karnaugh map (K-map) representation of *Y* to design a binary full subtractor architecture using SMG similar to Fig. 13.15. The module's corresponding mapping logic to the borrow out and difference of the subtractor is shown subsequently

Table 13.4 Simulation parameters

S.No	Parameter	Value
1	Exchange stiffness constant	13×10^{-12} J/m
2	Material	Permalloy
3	| m x h |	10^{-5} A/m
4	Damping coefficient	0.25
5	Saturation magnetization	800×10^{3} A/m

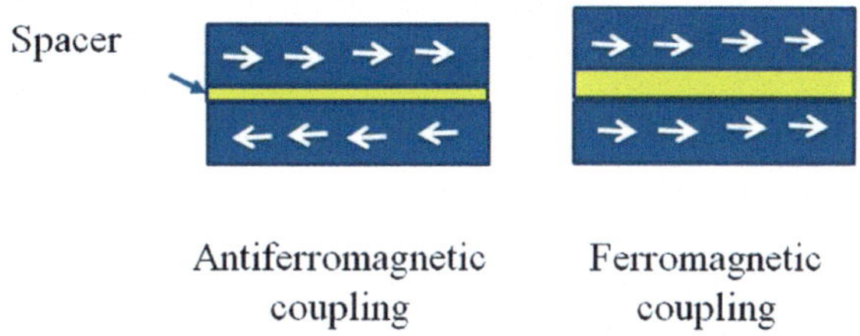

Fig. 13.17 Interlayer exchange-coupled (IEC) magnetic multilayers; **a** antiferromagnetic ground state, **b** ferromagnetic ground state

circuits is determined by the interrelation between the nanomagnets, and thus for the sake of successful data transfer between the nanomagnetic dots, the coupling energy in the nanomagnets should be much greater than the thermal noise (Csaba and Porod 2010). In order to address this issue, researchers, from University of Notre Dame, had been working on the development of a novel coupling scheme where a system based on interlayer exchange coupling (IEC) has been proposed. Interlayer exchange-coupled system comprises a non-magnetic metal layer (known as spacer) sandwiched between two ferromagnetic layer (Liu et al. 2014). The two ferromagnetic layers can be coupled ferromagnetically (FM) or antiferromagnetically (AFM) reliant on the depth of the spacer layer (Parkin et al. 1991). Figure 13.17a, b shows the two orientations of such a system. In Fig. 13.17a, the two magnetic layers (blue) are in antiferromagnetic directions and in Fig. 13.17b, in ferromagnetic directions. It can be noted that the difference between Fig. 13.17a, b is in the spacer thickness, as small as a few angstroms, with, rest being the same. A schematic representation of the coupling strength between the two layers is shown in Fig. 13.18, where the magnitude of the IEC strength is represented by J, and the nature of the coupling is determined by the sign of J. When the sign of J is negative, antiferromagnetic coupling occurs, and when J is positive, ferromagnetic coupling occurs. Thus, interlayer exchange-coupled (IEC) scheme has been demonstrated in Dey et al. (2016, 2015), and has been shown to offer stronger interaction between thin nanomagnets, resulting in greater scalability and better data retention at the deep sub-micron level, hence allowing magnetic interaction to be manipulated both in the vertical and lateral directions at the same time.

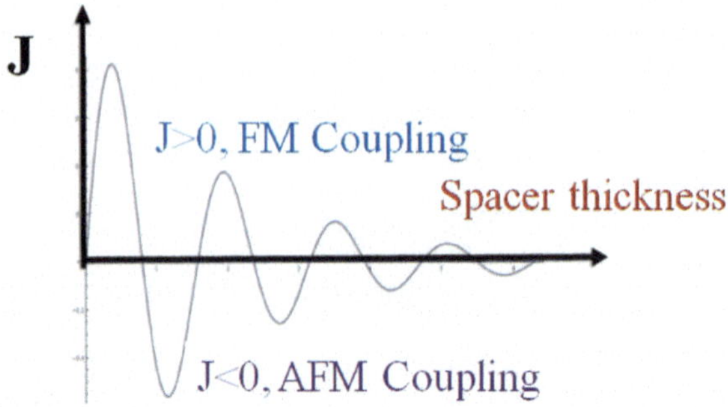

Fig. 13.18 Oscillatory variation of interlayer exchange coupling with spacer thickness

13.4.2.2 Research Development in the Field of IEC-Based Coupled Systems

The interlayer exchange coupling (IEC) has led to substantial interests owing to its potential applications in magnetic memory (Yu et al. 2008; Grünberg et al. 1986; Zhang et al. 1994; Parkin and Mauri 1991; Bruno and Chappert 1991; Gonzalez-Chavez et al. 2013; Pham et al. 2009; Cherepov et al. 2011). However, logic devices based on interlayer-coupled multilayers coupling vertically with each other was demonstrated for the first time in Dey et al. (2016). The authors have demonstrated that if two nanomagnets are laterally placed close to each other above a bottom magnet separated by a spacer, the top magnets will always be ferromagnetically coupled as in contrast to the state-of-the-art dipole coupling, where two nanomagnets placed laterally will get antiferromagnetically coupled. This ferromagnetic coupling happens because of the strong interlayer coupling between the bottom and top magnets mediated by the spacer layer with spacer layer thickness such that ($J<1$). Though exchange-coupled nanomagnets leads to stronger interaction between nanomagnets, but when placed laterally, due to their inherent nature, they will always ferromagnetically couple for data propagation. However, this is a ongoing research domain for digital circuits developments. On the other hand, researchers have also exploited the out-of-plane/perpendicular magnetization leading to three-dimensional arithmetic architecture exploration (Fig. 13.19)

13.4.2.3 3D Magnetic Computing

Nanomagnets with perpendicular magnetic anisotropy (PMA) was proposed as logic computing devices in the year 2002 by the authors of Csaba et al. (2002). Signal flow in one direction was accomplished using partial focused ion beam (FIB) irradiation on the nanomagnets by the authors of Breitkreutz (2001). Data propagation in a chain of nanomagnets followed by majority gate implementation was demonstrated

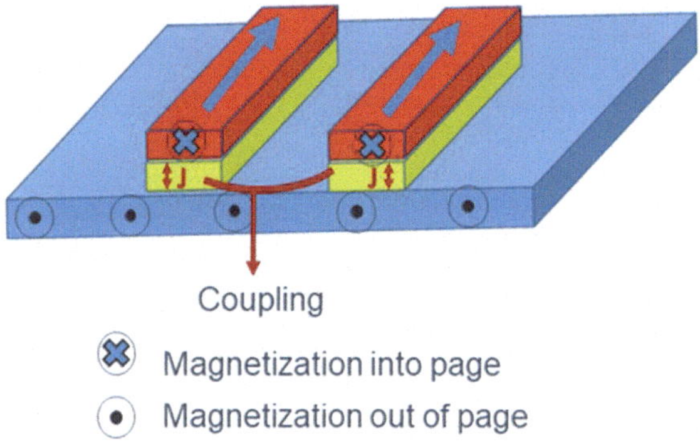

Fig. 13.19 Lateral configuartion of the top two nanomagnets and the bottom layer showing strong exchange coupling

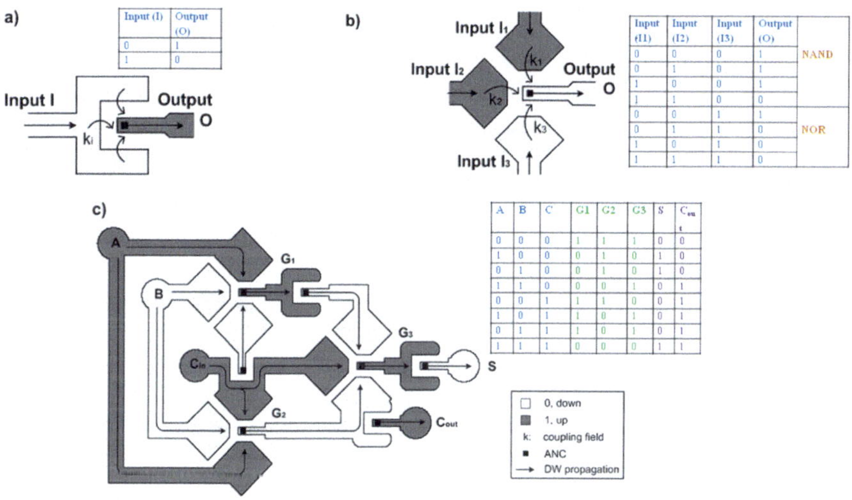

Fig. 13.20 **a** Inverter, **b** majority gate **c** a 1-bit full adder implemenation using input A, input B and input Cin and output can be observed from the sum S and carry-out Cout which is realized by implementing three majority and four inverter structures connected through wires. Stephan Breitkreutz et al. (2013)

in Eichwald (2012); Breitkreutz (2012). Simulations carried out using micromagnetic framework demonstrate the probable use of perpendicular Nanomagnetic Logic (pNML)-based devices for the imminent demands on non-volatile logic applications (Ju et al. 2013). Stephan Breitkreutz et al. (2013) have implemented an inverter structure using pNML as shown in Fig. 13.20a, where the artificial nucleation canter of the output magnet denoted by O is bounded by an input magnet I. Majority logic gate has been implemented in Fig. 13.20b, whose state is decided by the superposition of the couplings fields generated by the three input nanomagnets. Figure 13.20c shows

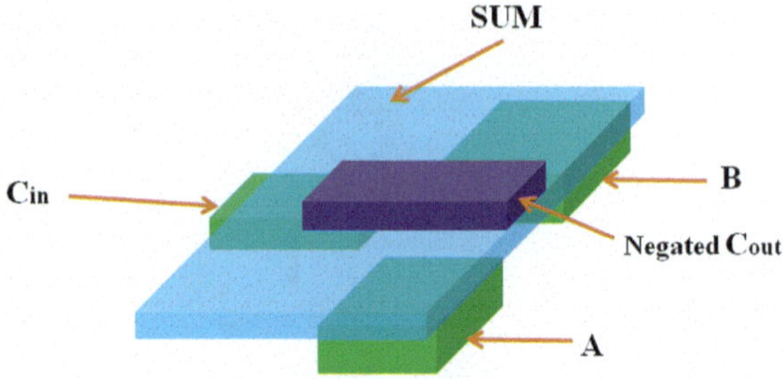

Fig. 13.21 Initial 3D pNML threshold full adder design. SUM is placed above the three input magnets as well as the Cout magnet. The Cout magnet is twice as large as the three input magnets (Robert Perricone et al. 2014)

the of a 1-bit full adder design in pNML that consists of 3 inputs (input A, input B and input c_{in}). For implementing the 1-bit adder design, three majority gates and four inverter gates have been used. The results indicate that complex logic circuitry can be implemented using pNML. Perricone et.al have presented a methodology in Robert Perricone et al. (2014) that has been used to design 3D pNML-based full adder using just five magnets for realizing the entire structure (including the three input magnets.) The 3D full adder shown here has been proved to reduce critical path, by almost 9X smaller than the 2D designs (Fig. 13.21).

Data transmission in 3D Integrated circuit by introducing electron spin: Figures 13.22, 13.24, 13.23 depict and illustrates the interlayer signal propagation in three-dimensional integrated circuit by introducing magnetic quantum cellular automata (MQCA). Authors of Reference Debroy et al. (2017) have shown electron spin for interlayer data propagation instead of charge ensuing area efficiency (Debroy et al. 2017). Through Silicon Via (TSV) diameter ranges in between 5 and 70 μm, thus resulting in occupying larger area than the logic components in the three-dimensional IC and thus needs to be scaled down so as to achieve higher packaging densities. As discussed by Kim et al. Kim et al. (2012) that a TSV with an area footprint of $10 \times 10\,\mu m^2$ in 45 nm technology will occupy 50 gates. Thus, it can be understood that if such 1 million TSVs are used in a IC, it will occupy the area equivalent to 50 million gates, which are exorbitant. Authors of Reference Debroy et al. (2017) have given a solution to this problem by introducing a new methodology that consumes significantly less area allowing more logic on each layer. There has been a lot of research for the replacement of copper-based TSVs (Ghosh et al. 2013; Rhett Davis et al. 2005; Ching-Hua Wang et al. 2013) for better compatibility, higher yield and decreased cost. One such is carbon nanotube-filled TSV with diameter less than 5 μm for less area occupancy (Ghosh et al. 2013). Authors of Reference Debroy et al. (2017) have shown that if electron spin-based interconnects are used in 3D IC as compared to copper and other state-of-the-art solution to TSV as provided in Ghosh et al. (2013), Rhett Davis et al. (2005), Ching-Hua Wang et al. (2013), area saving of 90% and above can be obtained in each layer. Authors of Reference Debroy et al. (2017) have used MQCA for data transmission instead of copper-based TSVs. The

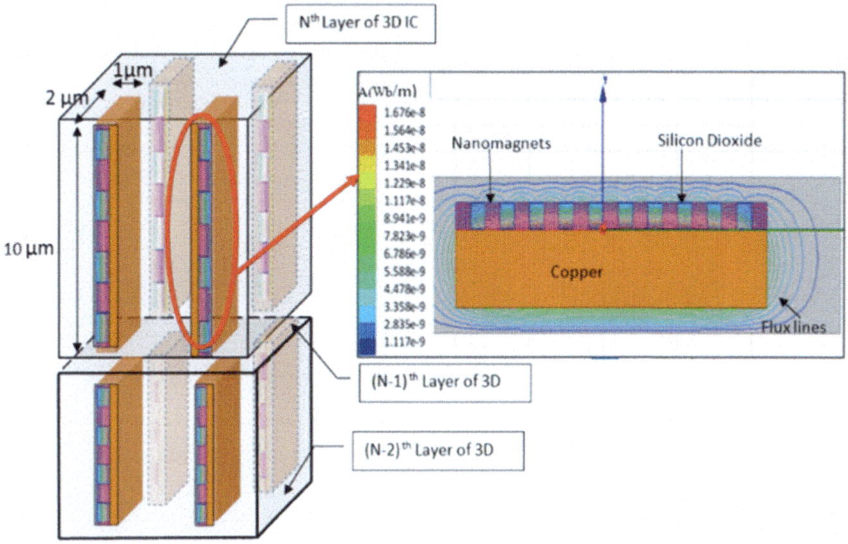

Fig. 13.22 Pictorial representation of the proposed structure (Debroy et al. 2017)

concept of the proposal has been pictorially shown in Fig. 13.22. It can be seen in Fig. 13.22 that a copper wire has been used as a current carrying wire. The current passed through copper wires creates a magnetic field perpendicular to the electric field which in turns drives the driver magnet. A logic signal of 1 or 0 gets propagated through the nanomagnets depending on the direction of the current signal. Figure 13.23 shows the fabrication flowchart. Silicon dioxide was grown on silicon wafer through wet oxidation followed by spinning electron beam resist PMMA for optical lithography patterning of the copper interconnect patterns. Copper was deposited in the patterns using electron beam evaporator and metal lift-off. In the next step, nanomagnets were patterned of dimension 100 nm × 70 nm × 30 nm using RAITH 150 two. The entire wafer was spin coated with electron beam resist (PMMA (2%) and EL9) and exposed to electron beam lithography followed by e-beam evaporation of permalloy. Figure 13.24 shows the results carried out using electromagnetic simulator ANSYS Maxwell 3D. It can be observed that copper wire with a dimension of 2 μm(width) and 1 μm (thickness) was able to produce a magnetic field of 45 mT (Fig. 13.24 a) which is the required field for data transmission through the nanomagnets (Fig. 13.24 c). Figure 13.24b displays the external magnetic field due to different dimensions of the copper interconnect and Fig. 13.24d displays the magnetic field due to the different dimensions of the nanomagnets and copper interconnect.

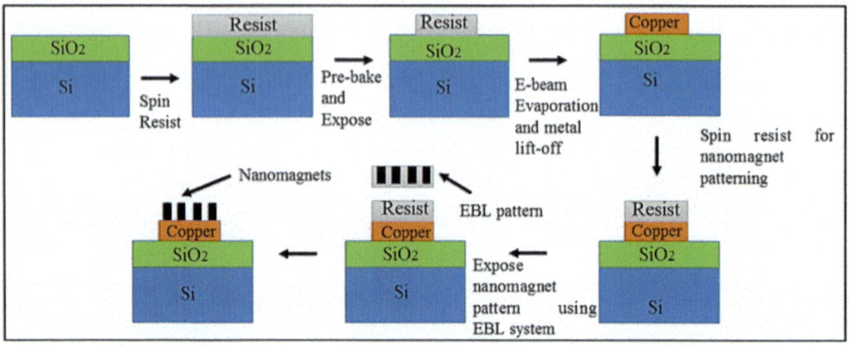

Fig. 13.23 Process flow for fabricating nanomagnets above copper interconnect (Debroy et al. 2017)

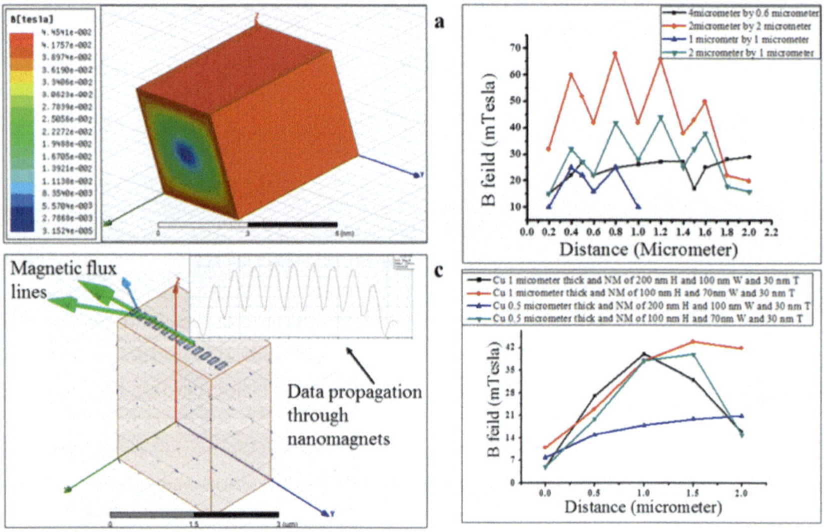

Fig. 13.24 **a** Displays the magnetic field produced due to the copper interconnect using Ansys Maxwell 3D simulator, **b** shows the magnetic feiilds generated due to different dimensions, **c** shows the signal transmission in the nanomagnets that are placed above the copper interconnect, **d** shows the field generated due to different dimension of the nanomagnets and copper interconnect (Debroy et al. 2017)

13.5 Conclusion and Future Research Scope

NML devices offer non-volatility, as well as no energy requirement to maintain data states when not performing computation. In correspondence to the adverse effects of More than Moore and with the emerging demands of computing on the edge device necessitates a significant improvement in the energy and area-efficient rebooting computing architecture design. To address this, the holistic approach from the architectural and system perspective to explore various design methodologies by exploiting those aforementioned inherent nature of nanomagnets for performing arithmetic computation is introduced in this chapter for the first time to the best of the author's knowledge. This chapter presents the theoretical modelling and micromagnetic simulation analysis of such nanomagnetic logic-based arithmetic architectures and their corresponding implementation considering the constraint of resources in terms of the number of NMs, MGs and CCs leading to energy and area efficiency. Overview of the proposed design methodology along with the demonstration of the hybrid approach of using slant edged input and 45 degree aligned nanomagnets for optimized binary full adder design. Asymmetric shape anisotropy nanomagnets pave the way for standalone inputs, whereas positional anisotropy reduces the signal loss in transmission of data and enables lossless information propagation. Consequently, RRN adder and add/sub has been proposed. Additionally, APN subtractor, adder architectures have been introduced Thus, this chapter provides an summary of the proposed architecture designs and its implementation yielding superior performance.

- In resource-constraint application that requires ultra-low-power and low area consuming systems. In engine controllers, nuclear plants and space mission that demand electronic circuits to withstand very high temperature and radiation hard devices for a certain period of their lifetime. NML can also be useful in circuit implementation of medical devices, chemical and biological sensors, etc.
- As a magnetic co-processor and a micro-controller replacing the traditional co-processors
- quadratic optimization, image and signal processing, pattern recognition
- Robotics system for low-power real-time sensing, control and decision-making.

The authors envisage the electronic transport properties, and magnetic properties (Sivasubramani et al. 2018) reported can be further explored, and its possible applications and connection in the nanomagnetic computing designs presented in this chapter can open up a challenging research venue. Nanomagnetic chip-based laptops would not overheat as the magnetic systems are unique in that they have no moving parts unlike moving electrons which are the source of heat generation in traditional computers. "Performing AI computing on edge with approximate nanomagnetic logic deployed on the magnetic ICs is an attempt towards the futuristic computations". The authors believe the work presented in this chapter paves the way towards achieving such a vision. With the designed architectures becoming successful, researchers now aim for a bigger goal by porting some pocwer-hungry AI applications on such indigenously developed ultra-low-power computing platform.

References

Berto F, Tagliabue J (2017) Cellular automata. In: Zalta EN (ed) The Stanford encyclopedia of philosophy (Fall 2017 Edition). https://plato.stanford.edu/archives/fall2017/entries/cellular-automata/

Bhanja S, Karunaratne DK, Panchumarthy R, Rajaram S, Sarkar S (2016) Non-Boolean computing with nanomagnets for computer vision applications. Nature Nanotechnol 11(2):177

Blank Steve (2018) What the global foundries' retreat really means. IEEE Spectrum 10
Breitkreutz S et al (2011) Nanomagnetic logic: demonstration of directed signal flow for field-coupled computing devices. In: Proceedings of the IEEE 41st Europe solid-state device research conference ESSDERC, pp 323–326
Breitkreutz S et al (2012) Majority gate for nanomagnetic logic with perpendicular magnetic anisotropy. IEEE Trans Magn 48(11):4336–4339
Breitkreutz S, Kiermaier J, Eichwald I, Hildbrand C, Csaba G, Schmitt-Landsiedel D, Becherer M (2013) Experimental demonstration of a 1-bit full adder in perpendicular nanomagnetic logic. IEEE Trans Magn 49(7), July 2013
Bruno P, Chappert C (1991) Oscillatory coupling between ferromagnetic layers separated by a nonmagnetic metal spacer. Phys Rev Lett 67:1602
Cherepov SS, Koop BC, Dzhezherya YI, Worledge DC, Korenivski V (2011) Resonant activation of a synthetic antiferromagnet. Phys Rev Lett 107:077202
Cowburn RP (2002) Probing antiferromagnetic coupling between nanomagnets. Phys Rev B 65(9):092409
Cowburn RP, Welland ME (2000) Room temperature magnetic quantum cellular automata. Science 287(5457):1466–1468
Csaba G, Imre A, Bernstein GH, Porod W, Metlushko V (2002) Nanocomputing by field-coupled nanomagnets. IEEE Trans Nanotechnol 1(4):209–213
Csaba G, Imre A, Bernstein GH, Porod W, Metlushko V (2002) Nanocomputing by field-coupled nanomagnets. IEEE Trans Nanotechnol 1(4):209–213
Csaba G, Lugli P, Porod W (2004) Power dissipation in nanomagnetic logic devices. In: 4th IEEE conference on nanotechnology. IEEE, pp 346–348
Csaba G, Porod W (2010) Behavior of nanomagnet logic in the presence of thermal noise. In: 2010 14th Int Work Comput Electron, pp 1–4, Oct 2010
Debroy S, Acharyya A, Singh SG, Acharyya SG (2017) Area-efficient interlayer signal propagation in 3D IC by introducing electron spin. In: 2017 European Conference on Circuit Theory and Design (ECCTD). IEEE, pp 1–4
Debroy S, Sivasubramani S, Acharyya SG, Acharyya A (2019) Nanomagnetic computing for next generation interconnects and logic design. In: Dhiman R, Chandel R (eds) VLSI and Post-CMOS electronics. Volume 2: devices, circuits and interconnects. The Institution of Engineering and Technology publishing, September 2019. https://doi.org/10.1049/PBCS073G_ch8. e-ISBN: 9781839530548
Dey H, Csaba G, Bernstein GH, Porod W (2016) Experimental demonstration of exchange coupling between laterally adjacent nanomagnets. Nanotechnology 27(39)
Dey H, Csaba G, Hu XS, Niemier M, Bernstein GH, Porod W (2013) Switching behavior of sharply pointed nanomagnets for logic applications. IEEE Trans Magnetics 49(7):3549–3552
Dey H, Csaba G, Shah F, Bernstein G, Porod W (2015) Shape-dependent switching behavior of exchange-coupled nanomagnet stacks. IEEE Trans Mag 52(4), Oct 2015
Eichwald I et al (2012) Nanomagnetic logic: Error-free, directed signal transmission by an inverter chain. IEEE Trans Magn 48(11):4332–4335
Ghosh K, Yap CC, Tay BK, Tan CS (2013) Integration of CNT in TSV for 3D IC application and its process challenges. In: 3D Systems Integration Conference (3DIC), IEEE International
Gonzalez-Chavez DE, Dutra R, Rosa WO, Marcondes TL, Mello A, Sommer RL (2013) Interlayer coupling in spin valves studied by broadband ferromagnetic resonance. Phys Rev B 88:104431
Grünberg P, Schreiber R, Pang Y, Brodsky MB, Sowers H (1986) Layered magnetic structures: evidence for antiferromagnetic coupling of Fe layers across Cr interlayers. Phys Rev Lett 57:2442
Gu Z, Nowakowski ME, Carlton DB, Storz R, Im M-Y, Hong J, Chao W et al (2015) Sub-nanosecond signal propagation in anisotropy-engineered nanomagnetic logic chains. Nature Commun 6:6466
Hesjedal T, Phung T (2010) Magnetic logic element based on an S-shaped Permalloy structure. Appl Phys Lett 96(7):072501
IEEE (2016) Rebooting computing. International roadmap for devices and systems. https://rebootingcomputing.ieee.org

IEEE Rebooting Computing. https://en.wikipedia.org/w/index.php?curid=51002633
Imre A, Csaba G, Ji L, Orlov A, Bernstein GH, Porod W (2006) Majority logic gate for magnetic quantum-dot cellular automata. Science 311(5758):205–208
Ju X, Niemier MT, Becherer M, Porod W, Lugli P, Csaba G (2013) Systolic pattern matching hardware with out-of-plane nanomagnet logic devices. IEEE Trans Nanotechnol 12(3)
Kim JS, Oh CS, Lee H et al (2012) A 1.2 V 12.8 GB/s 2 Gb mobile wide-I/O DRAM with 4 X 128 I/Os using TSV based stacking. IEEE J Solid-state Circ 47:107
Lapadeus M (2018) Big trouble at 3nm. Semiconductor engineering, 1 June 2018
Li Z, Krishnan Kannan M (2017) A 3-input all magnetic full adder with misalignment-free clocking mechanism. J Appl Phys 121(2):023908
Li Z, Kwon BS, Krishnan KM (2014) Misalignment-free signal propagation in nanomagnet arrays and logic gates with 45-clocking field. J Appl Phys 115(17):17E502
Liu XM, Nguyen HT, Ding J, Cottam MG, Adeyeye AO (2014) Interlayer coupling in Ni80Fe20/Ru/Ni80Fe20 multilayer films: ferromagnetic resonance experiments and theory. Phys Rev B 90:064428. https://doi.org/10.1103/PhysRevB.90.064428
Madami M, Gubbiotti G, Tacchi S, Carlotti G (2017) Magnetization dynamics of single-domain nanodots and minimum energy dissipation during either irreversible or reversible switching. J Phys D Appl Phys 50(45):453002
Moore Samuel K (2018) The good, the bad, and the weird: 3 directions for Moore's law. IEEE Spectrum 26
Moore Samuel K (2019) Another step toward the end of Moore's law. IEEE Spectrum 31
Niemier MT, Varga E, Bernstein GH, Porod W, Alam MT, Dingler A, Orlov A, Hu XS (2010) Shape engineering for controlled switching with nanomagnet logic. IEEE Trans Nanotechnol 11(2):220–230
Orlov A, Imre A, Csaba G, Ji L, Porod W, Bernstein GH (2008) Magnetic quantum-dot cellular automata: recent developments and prospects. J Nanoelectronics Optoelectronics 3(1):55–68
Parkin SSP, Mauri D (1991) Spin engineering: direct determination of the Ruderman-Kittel-Kasuya-Yosida far-field range function in ruthenium, Phys Rev B 44:7131
Parkin S, Bhadra R, Roche K (1991) Oscillatory magnetic exchange coupling through thin copper layers. Phys Rev Lett 66(16):2152–2155
Perricone R, Hu XS, Nahas J, Niemier M (2014) Design of 3D nanomagnetic logic circuits: a full-adder case study. In: Design Automation & Test in Europe Conference & Exhibition (DATE). https://doi.org/10.7873/DATE.2014.132
Pham H, Cimpoesu D, Plamadăa A-V, Stancu A, Spinu L (2009) Dynamic critical curve of a synthetic antiferromagnet. Appl Phys Lett 95:222513
Porter MJ et al (1999) OOMMF User's Guide, Version 1.0; National Institute of Standards and Technology, Gaithersburg, MD, Interagency Report NISTIR 6376
Pulecio Javier F, Pendru Pruthvi K, Anita K, Sanjukta B (2011) Magnetic cellular automata wire architectures. IEEE Trans Nanotechnol 10(6):1243–1248
Rhett Davis W, Wilson J, Mick S et al (2005) Demystifying 3D ICs: the pros and cons of going vertical. IEEE Des Test Comput 22:498
Sivasubramani S, Acharyya A (2018) Investigation on electronic transport and magnetic properties of graphene for its applications in nanomagnetic computing, Master's thesis. Indian Institute of Technology Hyderabad, India
Sivasubramani S, Debroy S, Acharyya SG, Acharyya A (2018) Tunable intrinsic magnetic phase transition in pristine single-layer graphene nanoribbons. Nanotechnology 29(45):455701. https://doi.org/10.1088/1361-6528/aadcd8
Sivasubramani S, Mattela V, Pal C, Acharyya A (2019) Dipole coupled magnetic quantum-dot cellular automata-based efficient approximate nanomagnetic subtractor and adder design approach. Nanotechnology 31(2):025202. https://doi.org/10.1088/1361-6528/ab475c
Sivasubramani S, Mattela V, Pal C, Saif Islam M, Acharyya A (2018) Shape and positional anisotropy based area efficient magnetic quantum-dot cellular automata design methodology

for full adder implementation. IEEE Trans Nanotechnol 17(6):1303–1307. https://doi.org/10.1109/TNANO.2018.2874206
Sivasubramani S, Mattela V, Rangesh P, Pal C, Acharyya A (2020) Nanomagnetic logic based runtime reconfigurable area efficient and high speed adder design methodology. Nanotechnology 31(18):18LT02. https://doi.org/10.1088/1361-6528/ab704b
Sivasubramani S, Mattella V, Pal C, Acharyya A (2019) Nanomagnetic logic design approach for area and speed efficient adder using ferromagnetically coupled fixed-input majority gate. Nanotechnology 30(37):37LT02. https://doi.org/10.1088/1361-6528/ab295a
The International Roadmap for Devices and Systems (2016) More Moore. https://www.irds.ieee.org
The International Roadmap for Devices and Systems (2017) Beyond CMOS
The IRDS Roadmap: Emerging Research Devices (ERD) (2018) https://www.irds.ieee.org
Varga E (2013) Chapter-10, Experimental study of novel nanomagnet logic devices. Doctoral Dissertation, University of Notre Dame, USA, April 2013
Varga E, Csaba G, Bernstein GH, Porod W (2011) Implementation of a nanomagnetic full adder circuit. In: 2011 11th IEEE international conference on nanotechnology. IEEE, pp 1244–1247
Varga E, Niemier MT, Csaba G, Bernstein GH, Porod W (2013) Experimental realization of a nanomagnet full adder using slanted-edge magnets. IEEE Trans Magnetics 49(7):4452–4455
Varga E, Orlov A, Niemier MT, Sharon Hu X, Bernstein GH, Porod W (2010) Experimental demonstration of fanout for nanomagnetic logic. IEEE Trans Nanotechnol 9(6):668–670
Wang C-H, Dai K-Y, Shen K-H et al (2013) Magnetic wireless interlayer transmission through perpendicular MTJ for 3D-IC applications. IEEE IEDM, 13-613
Wolfgang Porod, Michael Niemier (2015) Better computing with magnets-The simple bar magnet, shrunk down to the nanoscale, could be a powerful logic device. IEEE Spectrum 52(9):44–60
Wolfgang P, Craigs L, Bernstein Gary H, Orlov Alexei O, Islamsha H, Snider Gregory L, Merz James L (1999) Quantum-dot cellular automata: computing with coupled quantum dots. Int J Electronics 86(5):549–590
Wu AQ, Kubota Y, Klemmer T, Rausch T, Peng C, Peng Y, Karns D, Zhu X, Ding Y, Chang EKC, Zhao Y, Zhou H, Gao K, Thiele JU, Seigler M, Ju G, Gage E (2013) HAMR areal density demonstration of 1+ tbpsi on spinstand. IEEE Trans Magn 49(2):779–782
Yu C, Javorek B, Pechan MJ, Maat S (2008) Magnetic coupling of pinned, asymmetric CoPt / Ru / CoFeCoPt / Ru / CoFe trilayers. J Appl Phys 103:063914
Zhang Z, Zhou L, Wigen PE, Ounadjela K (1994) Angular dependence of ferromagnetic resonance in exchange-coupled Co/Ru/Co trilayer structures. Phys Rev B 50:6094

Chapter 14
Design Space Exploration of DSP Hardware Using Adaptive PSO and Bacterial Foraging for Power/Area-Delay Trade-Off

Anirban Sengupta, Mahendra Rathor, and Pallabi Sarkar

Abstract Digital signal processing (DSP) hardware is ubiquitous in the current generation of consumer electronics systems including camera, camcorders, set-top boxes, smartphones, etc. The very large-scale integration (VLSI) design process of DSP hardware is entirely dependent on high-level synthesis framework that comprises design space exploration (DSE) of power/area-delay trade-off. Since DSP hardware is application specific by nature, and thus, exploration of its low power, high performance architectural solution is crucial. However, the exploration process is intricate and involves a number of convoluted factors such as modelling of the objective function (such as power and delay), accuracy/efficiency of the optimization framework, loop dependency, data pipelining, seed encoding, scheduling algorithm, resource binding, ability to escape local minima and terminating criteria. In this chapter, we present a number of emerging evolutionary design space exploration techniques based on bacterial foraging and particle swarm optimization algorithm that is capable to consider the aforesaid complex factors while performing power/area—delay trade-off of DSP hardware. The chapter also discusses the analysis on case studies for each DSE technique with respect to power–delay trade-off.

Keywords Exploration · DSP hardware · PSO · BFOA · Trade-off · Optima

---

A. Sengupta (✉) · M. Rathor
Discipline of Computer Science and Engineering, Indian Institute of Technology, Indore, Madhya Pradesh 453552, India
e-mail: asengupt@iiti.ac.in

M. Rathor
e-mail: mrathor@iiti.ac.in

P. Sarkar
School of Electrical and Electronics Engineering, VIT Bhopal, Bhopal, Madhya Pradesh 466114, India
e-mail: pallabi.sarkar@vitbhopal.ac.in

R. Dhiman and R. Chandel (eds.), *Nanoscale VLSI*, Energy Systems in Electrical Engineering,
https://doi.org/10.1007/978-981-15-7937-0_14

14.1 Introduction

In the current era of technology, consumer electronics (CE) products such as television, laptop, wearable tech gadgets, digital cameras have taken over the consumer market. Today's human life is unimaginable without CE products. A system-on-chip (SOC) employed inside a CE device integrates digital signal processing (DSP) cores as key components. DSP cores in modern CE devices are of paramount importance as they serve several major applications such as real-time processing, video encoding/decoding, image compression/decompression, de-noising and signal attenuation. (Schneiderman 2010; Sengupta 2016, 2017).

In the design process of an electronic device, a designer has to consider design objectives such as area, power and delay. This is because, electronic systems are required to be operated under specified power, delay and area constraints. However, to manage trade-off between these design objectives has been a great challenge for designers. Among the entire design space, there are several design solutions possible which impact design objectives differently. Therefore, the need of design space exploration (DSE) arises to attain an optimal design solution that satisfies power, area and delay constraints (Ascia et al. 2007; Mishra and Sengupta 2014). For example, some data/computation intensive applications such as multimedia and communication process a huge amount of data. However, this processing is required to be done at the expense of minimal power. Therefore, the hardware such as DSP core which handles such applications needs to be explored to obtain an optimal design which would result into higher performance or smaller package size at minimal power consumption. DSE helps in suggesting trade-off between design objectives and providing an optimal design solution. The DSE process is integrated with the high-level synthesis (HLS) process (Sengupta et al. 2010) for exploring datapath of highly complex circuits such as DSPs (Sengupta and Mishra 2014). The importance of HLS for DSP hardware is discussed in the next section.

14.2 Why High-Level Synthesis for DSP Hardware

HLS offers automatic synthesis of the design at higher abstraction level and thus shortens the design cycle time. Further, HLS is capable to generate several design solutions for the same specification. The DSE integrated with HLS helps in exploring different design solutions and yielding an optimal solution satisfying design objectives (such as area, power and delay.). The generic process of generating optimal datapath during HLS using DSE is shown in Fig. 14.1. In the context of DSP hardware, the importance of HLS is highlighted as follows:

1. *Ease of implementation and handling capacity*: The DSPs are highly complex designs comprising thousands of gates. Moreover, the RTL/gate level structures

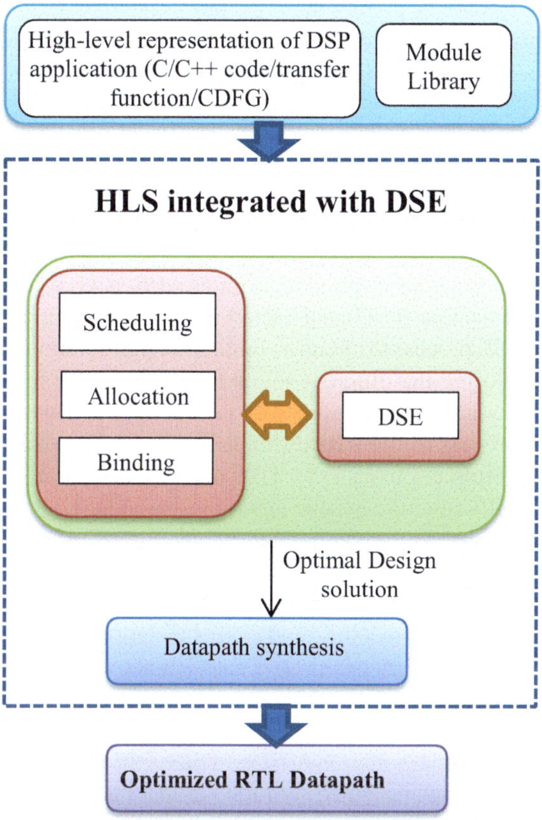

Fig. 14.1 Generic process of generating optimized datapath during HLS using DSE

of DSP circuits are not readily available. The DSP applications are either available in high-level description such as algorithmic/mathematical/C/C++ or intermediate representation such as control data flow graph (CDFG). In this scenario, HLS plays an important role in transforming the high-level description of a DSP application to the equivalent register transfer level (RTL) design. Thus, HLS handles the complex designs and makes their implementation easier.

2. *Possibility of controlling DSP datapath architecture using DSE*: Several possible DSP datapaths can be generated through HLS for the same DSP application. Using DSE during HLS, trade-off between different design objectives such as security-power, area-power, area-latency and latency-power can be explored for various possible datapaths. This exploration helps to obtain the datapath satisfying the user-specified power, area and delay constraints.
3. *Parametric Modelling*: To determine the design parameters such as power, delay and area of resulting RTL datapath in advance, the parametric modelling is crucial. The HLS integrated with DSE exploits parametric modelling of design

parameter to estimate or predict the area, power and delay of various possible datapaths. Based on this prediction, DSE process becomes able to produce an optimal datapath of DSP design through HLS.

14.3 Discussion on Selected DSE Approaches for DSP Hardware

There are evolutionary/nature inspired algorithms such as genetic algorithm (GA), particle swarm optimization (PSO) and bacterial foraging algorithm (BFOA) which can be mapped to DSE process to obtain an optimal design solution of DSP hardware by intelligently exploring the different solutions in the design space. This chapter will discuss PSO (Sengupta and Mishra 2014) and BFOA (Bhadauria and Sengupta 2015)-based DSE process for DSP hardware in details in the subsequent sections.

However, this section discusses other DSE approaches in brief. (Sengupta et al. 2012) performed multi-structure genetic algorithm-based DSE which helps in determining Pareto fronts among various design points. However, this approach takes relatively larger convergence time due to intrinsic nature of the algorithm and gets stuck at local optimal points in most of the situations. Moreover, this approach does not perform accurate power modelling and causes violation to power constraints for some applications. Further, GA-based DSE has also been performed in (Gallagher et al. 2004; Dhodhi et al. 1995; Heijlingers et al. 1995). These approaches do not essentially produce superior design solution always and further result into higher implementation run-time. Another GA-based DSE approach proposed by Harish Ram et al. (2012) integrates weighted sum PSO to find the optimal design solution. However, this approach fails to consider the actual velocity function during updating the particles position. Moreover, this approach does not consider the power and delay constraints (user specified) during fitness/cost evaluation, therefore fails to explore delay–power trade-off for various possible design solutions. (Krishnan and Katkoori 2006) employed a node-priority mechanism for exploring various possible datapaths during HLS. This DSE approach is able to balance the trade-off between area and delay. However, power constraint has not been handled in this approach. Additionally, this mechanism is computationally expensive. Additionally, there is another approach which performs binary encoding of the chromosomes for exploring design space (Torbey and knight 1998a, b) during HLS. This approach is also computationally expensive and does not perform delay–power trade-off. Besides, (Mishra and Sengupta 2014) leveraged PSO algorithm to explore the design space for synthesizing an optimal datapath. However, this approach lacks capability of exploring any high-level transformation parameter [such as loop unrolling factor (UF)] along with resource configuration. Hence, this approach is not efficient for exploring datapaths for CDFGs representing loop-based applications. The more efficient PSO-driven DSE is discussed in Sect. 14.4 followed by discussion on BFOA-DSE in Sect. 14.5.

14.4 Adaptive PSO-DSE for Exploration of Power–Delay Trade-Off of DSP Cores and Its Applications

14.4.1 Overview of PSO-DSE (Sengupta and Mishra 2014)

The PSO-DSE framework is used to solve multi-objective DSE problem. This section discusses the PSO-based DSE process which explores power–delay trade-off for DSP designs and produces an optimal design solution. The PSO-DSE framework (Sengupta and Mishra 2014) has significant features such as (i) it simultaneously explores the design solution and loop unrolling factor (UF) for DSP cores through multidimensional PSO (ii) it uses an evaluation model for delay estimation of a loop unrolled CDFG (representing a DSP application) (iii) the trade-off between power performance and execution latency can be balanced through PSO-DSE framework (iv) this framework offers sensitivity analysis of swarm size. Moreover, the impact of swarm size on exploration time and quality of result (QoR) of DSE can be assessed. The overview of the adaptive PSO-DSE framework is shown in Fig. 14.2. As shown in the figure, the inputs to the PSO-DSE framework are (i) control data flow graph (CDFG/DFG) representing DSP application to be optimized (ii) module library (iii)

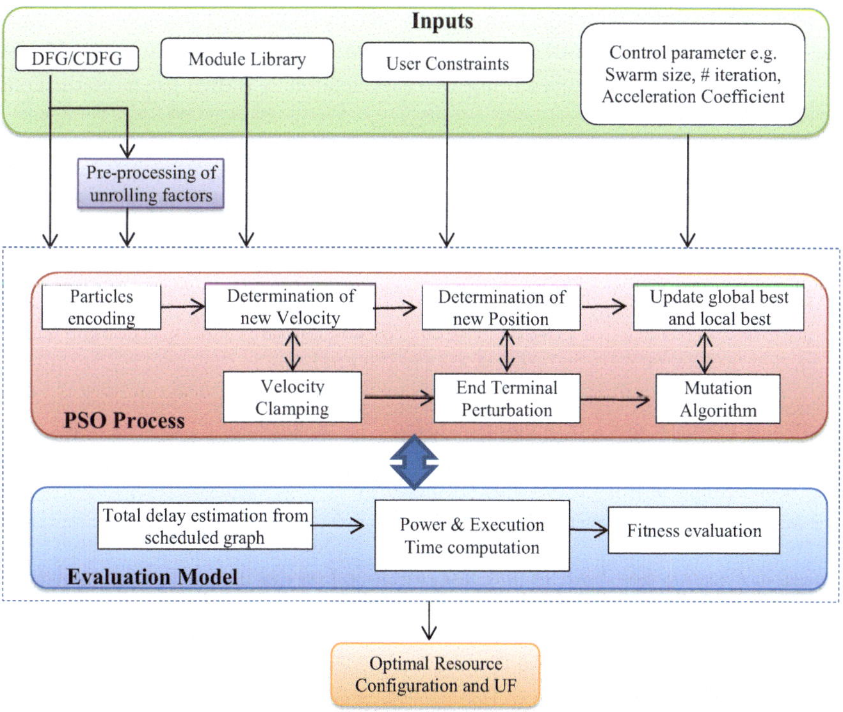

Fig. 14.2 Block diagram of PSO-DSE process (Sengupta and Mishra 2014)

user constraints (iv) PSO control parameters such as swarm size, number of iterations and acceleration coefficient. The PSO process execution along with the evaluation model produces an optimal resource configuration and unrolling factor. As shown in Fig. 14.2, the PSO-DSE process starts with swarm particles encoding which maps PSO algorithm to the DSE process. Further, following steps are performed in the PSO-DSE process: (i) determination of new velocity of each particle which is controlled through velocity clamping (ii) determination of new position of each particle which is controlled through end terminal perturbation (iii) updation of global best and local best which is controlled through a mutation algorithm.

The fitness value of each particle is evaluated using an evaluation model which is based on total delay estimated from the scheduled graph and power consumption. The evaluation model used in the PSO-DSE framework is illustrated in the next subsection.

14.4.2 Evaluation Model Used in PSO-DSE Framework

The fitness of a particle is evaluated using following models:

(a) **Execution time evaluation model**: The execution time is modelled separately for loop-based DSP applications represented by CDFG and non-loop DSP applications represented by DFG. The formulation of execution time evaluation model for a CDFG representing DSP application is shown for three different cases as follows:

Case 1 UF = 1: In this case, the loop body of the CDFG is not unrolled. In order to calculate the execution time (T^{e}), operations in the CDFG are scheduled in control steps (CS) based on a chosen scheduling algorithm such as soon as possible (ASAP) scheduling and as late as possible (ALAP) scheduling and LIST scheduling. Once the CDFG is scheduled, the execution time (T^{e}) is evaluated as follows:

$$T^{\mathrm{e}} = \lambda * \left(C^{\mathrm{body}} * \mu\right) \tag{14.1}$$

where λ is delay of one CS in nanoseconds, C^{body} is the number of CSs required to execute loop body once, and $\mu = \frac{I}{\mathrm{UF}}$. Here, I is the maximum iteration count of the loop body. Since in this case, UF = 1, therefore $\mu = I$. Hence, execution time can be evaluated as $T^{\mathrm{e}} = \lambda * \left(C^{\mathrm{body}} * I\right)$.

Case 2 *UF evenly divides I*: In this case, the loop body of the CDFG is unrolled UF times. An example of unrolling loop body (maximum iteration count $I = 36$) of a sample CDFG with UF = 2 is shown in Fig. 14.3. In general, the number of control steps required to execute once the unrolled portion of the CDFG is given as follows:

$$C^{\mathrm{body}} = \left(C^{\mathrm{first}} + (\mathrm{UF} - 1)\right) * C^{\mathrm{II}} \tag{14.2}$$

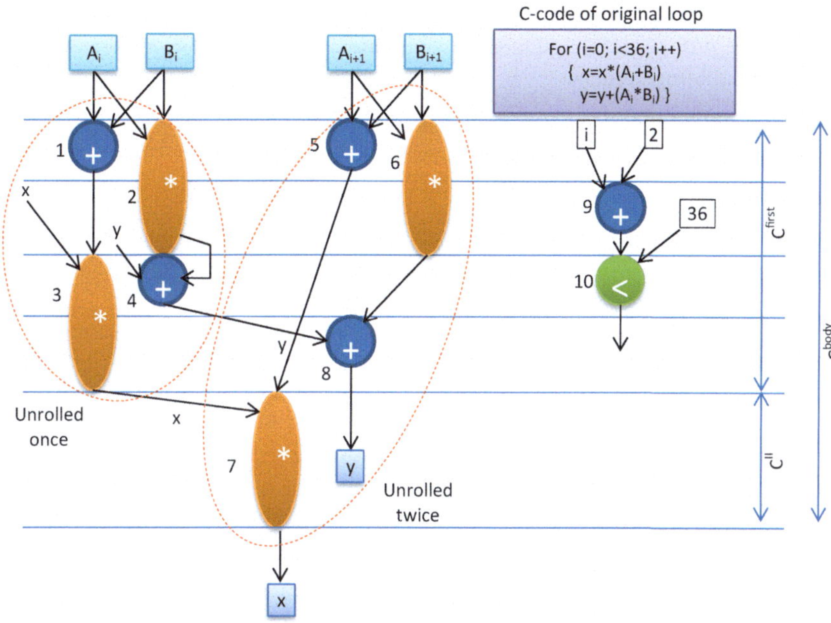

Fig. 14.3 Loop unrolling with UF = 2, complying with resource constraints 2 (*), 2(+), 1(<) (Sengupta and Mishra 2014)

where C^{first} is the number of CSs required to execute first iteration, C^{II} is the number of CSs required between initiations of consecutive iterations. Since, C^{body} executes $\mu = \frac{I}{\text{UF}}$ times, therefore the execution time is evaluated as follows:

$$T^{\text{e}} = \lambda * \left(\left(C^{\text{first}} + (\text{UF} - 1) * C^{\text{II}} \right) * \mu \right) \quad (14.3)$$

Case 3 *When UF unevenly divides I*: In this case, total control steps required for unrolled loop are = $((C^{\text{first}} + (\text{UF} - 1) * C^{\text{II}}) * \mu)$, and total control steps required for sequential loop are = $(I \bmod \text{UF}) * C^{\text{first}}$.

Therefore, the execution time is evaluated as follows:

$$T^{\text{e}} = \lambda * \left(\left(C^{\text{first}} + (\text{UF} - 1) * C^{\text{II}} \right) * \mu + (\text{I mod UF}) * C^{\text{first}} \right) \quad (14.4)$$

Additionally, the estimated execution time (Mishra and Sengupta 2014; Sengupta et al. 2012) for DFGs is given as follows:

$$T^{\text{e}} = L + (\varphi - 1) * T^{\text{c}} \quad (14.5)$$

where L indicates latency of a scheduling solution, φ indicates the number of input samples to be processed by a functionally pipelined datapath and T^{c} indicates cycle time of a scheduling solution.

(b) **Power model:** The dynamic power (P^{dy}) and static power (P^{st}) together result into total power consumption given as follows:

$$P^{T} = P^{dy} + P^{st} \tag{14.6}$$

where average P^{dy} consumption is given by the following formula (Sengupta and Mishra 2014):

$$P^{dy} = \frac{\text{Total energy consumption}}{\text{Total execution time}}$$

For a CDFG:

$$P^{dy} = \frac{\mu * \left(E^{FU} + E^{M/D}\right)}{\lambda * \left(\left(C^{first} + (UF - 1) * C^{II}\right) * \mu + (I \bmod UF) * C^{first}\right)} \tag{14.7}$$

For a DFG:

$$P^{dy} = \frac{\varphi * \left(E^{FU} + E^{M/D}\right)}{L + (\varphi - 1) * T^{c}} \tag{14.8}$$

where E^{FU} indicates energy consumption of the FU resources and $E^{M/D}$ indicates energy consumed by a multiplexer/de-multiplexer. C^{first}, C^{II}, $\underline{I}$, UF, μ, L, φ and T^{c} have already been defined.

The static power consumption is given by the following formula:

$$P^{st} = \left[\sum_{j=1}^{v}\left(N^{Fj} * K^{Fj}\right) + \left(N^{M/D} * K^{M/D}\right)\right] * P^{a} \tag{14.9}$$

where N^{Fj} is number of instance of functional unit (FU) resource Fj, K^{Fj} is area occupied by FU resource Fj, $N^{M/D}$ is number of the multiplexers or de-multiplexers, $K^{M/D}$ is area occupied by a multiplexer or de-multiplexer and P^{a} is power dissipated per area unit.

(c) **Fitness Function Model**: The fitness function based on execution time and power consumption of a design solution is formulated as follows (Sengupta and Mishra 2014):

$$C_f^{Z_i} = \sigma_1 \frac{T^{\mathrm{e}} - T^{\mathrm{con}}}{T^{\max}} + \sigma_2 \frac{P^T - P^{\mathrm{con}}}{P^{\max}} \tag{14.10}$$

where $C_f^{Z_i}$ indicates fitness of particle Z_i, T^{con} indicates execution time constraint specified by the user, P^{con} indicates power constraint specified by the user, $T^{\max}$ and $P^{\max}$ indicate maximum execution time and power consumption, respectively. Further, σ_1 and σ_2 indicate the user-defined weightage to execution time and power, respectively.

14.4.3 Role of Unrolling Factor

The loop body of the CDFG is unrolled based on the UF value chosen by the designer. However, some UFs may result into increase in design cost rather than optimizing execution time, area and power. Therefore, such FUs are considered to be unfit. These unfit UFs are filtered out through a pre-processing algorithm (Sengupta and Mishra 2014) which ensures the inclusion of only good/fit candidates in the swarm particle. Input to the pre-processing algorithm is the value of I (maximum loop iteration count), and output is filtered set of UFs. Steps of the algorithm are as follows (Sengupta and Mishra 2014):

1. Initialize UF = 2.
2. If I mod UF is less than UF/2 and UF <= I/2, then accept UF as a good candidate.
3. Add UF value to the list of fit UFs.
4. Increment the UF value by 1.
5. Repeat steps 2, 3 and 4 unless UF = I.

To add in the list such fit FUs that may have been filtered out through the above process, the following more steps are executed:

6. Again, initialize UF with 2.
7. If I mod UF is less than UF/2, then accept UF as a good candidate.
8. Add UF value to the list of fit UFs.
9. Increment the UF value by 1.
10. Repeat step 7, 8 and 9 till the following condition is true: (I mod UF) < (UF/2).
11. Terminate the algorithm.

14.4.4 Particle Encoding, Local and Global Best

To map the PSO algorithm to the DSE process, the swarm particles are encoded as follows:

1. Position of a particle (Z_i) is encoded as (resource configuration, UF).
2. Total dimensions (D) for each particle are (number of resource types +1).

3. Velocity of ith particle in dth dimension (V_{di}) is encoded as exploration drift.

The position of ith particle is given as follows (Sengupta and Mishra 2014):

$$Z_i = \left(N^{F1}, N^{F2}, \ldots, N^{Fy}, \ldots, N^{F(D-1)}, \mathrm{UF}\right)$$

where F_y indicates yth FU resource type and N^{Fy} indicates number of instances of resource type F_y. Here, UF value indicates the Dth dimension of a position.

The positions of swarm particles are initialized in order to evenly cover the entire design space as follows (Sengupta and Mishra 2014):

$$\begin{aligned}
Z_1 &= (\min(F_1), \min(F_2), \ldots, \min(F_{D-1}), \min(\mathrm{UF})) \\
Z_2 &= (\max(F_1), \max(F_2), \ldots, \max(F_{D-1}), \max(\mathrm{UF})) \\
Z_3 &= ((\min(F_1) + \max(F_1))/2, (\min(F_2) + \max(F_2)) \\
&\quad /2, \ldots, (\min(F_{D-1}) + \max(F_{D-1}))/2, \max(\mathrm{UF})/2)
\end{aligned}$$

If swarm size is 'W', then rest of the positions for particles (Z_4, Z_5, …, Z_W) are randomly initialized complying with UF and resource constraints. Additionally, all particles velocities are initialized to zero, and acceleration coefficients ($c1$ and $c2$) are initialized to values in the range given in (Engelbrecht 2005; Kennedy and Eberhart 1995; Trelea 2003). The value of inertia weight 'θ' is linearly decreased between 0.9 and 0.4 to achieve faster convergence (Eberhart and Shi 2000).

Determination of local best: The local best position for a particle is determined by evaluating its fitness function value. The local best is calculated for each particle. Initially, each particle position is local best position. In any iteration, the local best of a particle is updated when the fitness value of the particle position is evaluated to be higher (or cost is computed to be lower) than the previous position (Mishra and Sengupta 2014).

Determination of global best: That particle position is considered to be global best whose fitness is highest or cost is lowest among all the particles. The global best in the swarm population is determined using the following function (Sengupta and Mishra 2014):

$$Z_{\mathrm{gb}} = Z_i\left[\min\left(C^{Z_1}_{f\mathrm{lb}1}, C^{Z_2}_{f\mathrm{lb}2}, C^{Z_3}_{f\mathrm{lb}3}, \ldots, C^{Z_w}_{f\mathrm{lb}w}\right)\right] \tag{14.11}$$

where Z_{gb} is global best position of the population, $C^{Z_i}_{f\mathrm{lb}i}$ is local best cost of particle Z_i.

14.4.5 Velocity Clamping and Terminal Perturbation

A particle position is updated during iterations in the PSO-DSE algorithm. In order to update the particle position, the exploration drift parameter is added in the current position as shown below (Sengupta and Mishra 2014):

$$F_{di}^{+} = F_{di} + V_{di}^{+} \tag{14.12}$$

where F_{di}^{+} is new resource value or UF value of particle Z_i in dth dimension, F_{di} is resource value or UF value of particle Z_i in dth dimension and V_{di}^{+} is new velocity of ith particle in dth dimension. The value of exploration drift or velocity is computed as follows (Sengupta and Mishra 2014):

$$V_{di}^{+} = \theta V_{di} + c1r1\left[F_{d_{\mathrm{lb}i}} - F_{di}\right] + c2r2\left[F_{d_{\mathrm{gb}}} - F_{di}\right] \tag{14.13}$$

where V_{di} is velocity of ith particle in dth dimension, θ is inertia weight, $c1$ and $c2$ are acceleration coefficients, $r1$and $r2$ are random numbers between 0 and 1, $F_{d_{\mathrm{lb}i}}$ is resource value of $Z_{\mathrm{lb}i}$ in dth dimension and $F_{d_{\mathrm{gb}}}$ is resource value of Z_{gb} in dth dimension. However, the exploration drift is only possible between minimum and maximum value of resources /UF. Therefore, velocity clamping (Mishra and Sengupta 2014) is performed to get the drift within the valid range. When a particle's exploration drift outreaches the $\pm V_{di}^{m}$, then velocity clamping is applied as follows (Sengupta and Mishra 2014):

$$V_{di}^{+} = \begin{cases} +V_{di}^{m} \text{ if } V_{di}^{+} > +V_{di}^{m} \\ -V_{di}^{m} \text{ if } V_{di}^{+} < -V_{di}^{m} \\ V_{di}^{+} \quad \text{else} \end{cases}$$

The value of $\perp V_{di}^{m}$ is given as follows:

$$V_{di}^{m} = \pm\frac{\max(F_d) - \min(F_d)}{2}$$

Additionally, when the exploration drift is added in the dth dimension to upgrade the current position, the particle may violate the boundary of the design space. In this case, the end terminal perturbation algorithm is adopted using the following steps (Sengupta and Mishra 2014):

1. Check if $((F_{di} < \min(F_d)) \| F_{di} > \max(F_d))$ is true, then go to step 2; otherwise go to step 4.
2. If $(F_{di} < \min(F_d))$, then $F_{di} = F_{di} + \emptyset$.
3. Else if $F_{di} > \max(F_d)$, then $F_{di} = F_{di} - \emptyset$.
4. Terminate the algorithm.

Where Ø is chosen randomly under the range of $\min(F_d)$ to $\max(F_d)$.

14.4.6 Mutation and Stoping Criteria

The mutation operation (Sengupta and Mishra 2014) helps in improving the DSE convergence by avoiding premature convergence. Performing mutation on local best positions of each particle ensures the exploration of design space in every corner. The mutation is performed with probability 1.0 (i.e. after each iteration). An adaptive rotation mutation algorithm (Sengupta and Mishra 2014) is used in the PSO-DSE process. İnput to the algorithm is local best resource configuration $Z_{\text{lb}i}$ for each particle, and output is new mutated local best resource configuration. Steps of the adaptive rotation mutation algorithm are as follows (Sengupta and Mishra 2014):

1. Initial variable $i = 1$.
2. If (i%2 ==0), then go to next step, otherwise jump to step 9.
3. Initialize d = 1.
4. Store value of F_{di} into a temporary variable (t).
5. Store the value of $F_{(d+1)i}$ into F_{di}.
6. Store the value of 't' into $F_{(d+1)i}$.
7. İncrement 'd' by 1.
8. Repeat steps 4, 5, 6 and 7 until $d = D$.
9. If (i%2 ==1), then go to next step, otherwise jump to step 9.
10. Initialize $d = 1$.
11. Perform $F_{di} = F_{di} \pm G$. Here, G is a random number in the range 1–3.
12. İncrement 'd' by 1.
13. Repeat steps 11 and 12 until $d = D$.
14. İncrement 'i' by 1.
15. Repeat steps 2–15 until $i = W$ (population size).
16. End algorithm.

As the stopping criteria of the algorithm reach, the position of the best fit particle in the swarm represents the optimal design solution (resource configuration) for synthesizing datapath of the corresponding DSP application.

Stopping criteria: The PSO-DSE process terminates when either of the conditions satisfies (Sengupta and Mishra 2014).

1. Maximum number of iterations exceeds 100.
2. No improvement is observed in Z_{gb} over q = 10 number of iterations.

14.5 BFOA-DSE for Exploration of Power–Delay Trade-Off of DSP Cores and Its Applications

14.5.1 Overview of BFOA-DSE (Bhadauria and Sengupta 2015)

A regular bacterial foraging optimization algorithm (BFOA) is mapped to the DSE process to explore the power–delay trade-off in the design space of a DSP core (Bhadauria and Sengupta 2015). The BFOA mimics the biological behaviour of an Escherichia coli (*E. coli*) bacterium to obtain an optimal design point satisfying the user constraints (power and delay). The BFOA-DSE framework has significant features such as (Bhadauria and Sengupta 2015): (a) exploration drift driven by chemotaxis algorithm (b) multidimensional bacterium encoding to evenly cover the design space (c) bacterium position manipulation using customized replication algorithm (d) introducing diversity during DSE using elimination-dispersal (ED)algorithm (e) handling boundary outreach problem during DSE using adaptive schemes such as step size clamping and resource clamping achieving reduction in runtime of >4% and improvement in QoR of >35% with respect to contemporary DSE approaches.

Figure 14.3 depicts the flow chart of the BFOA-DSE framework (Bhadauria and Sengupta 2015). Inputs to the BFOA-DSE framework are (i) data flow graph (DFG) representing DSP application (ii) module library (iii) user constraints (iv) control parameters such as bacterium population size (p), maximum number of chemotactic steps (N^{c}), maximum number of replication steps (N^{rp}), maximum number of elimination-dispersal steps (N^{ed}), maximum number of times elimination-dispersal has to be done (E) and maximum number of times replication has to be done (R). As evident from the flow chart, the BFOA-DSE process operates within an effective temperature range [t_{min}, t_{max}]. Since an *E. coli* bacterium can survive within motility range [25 °C, 45 °C] and is eliminated beyond 40 °C, therefore the BFOA-DSE process also adapts this fact during exploration. Further, the DSE process progresses while imitating the following three basic mechanisms of bacterial foraging: (a) chemotaxis (b) replication (c) elimination-dispersal (ED).

The chemotaxis algorithm is run for a designer specified number of chemotactic steps (N^{c}). Further, the replication and ED mechanism occur in corresponding periodic interval specified by the designer. At pth chemotactic step, the occurrence of replication mechanism is determined by the following relation (Bhadauria and Sengupta 2015):

$$p = n.\left(\frac{N^{c}}{N^{rp}}\right), \quad \text{where } 1 \leq \text{n} \leq N^{rp} \tag{14.14}$$

Here, N^{rp} is the maximum number of replication steps. A variable 'R' is used in the algorithm to keep track the occurrence of replication steps as shown in Fig. 14.4.

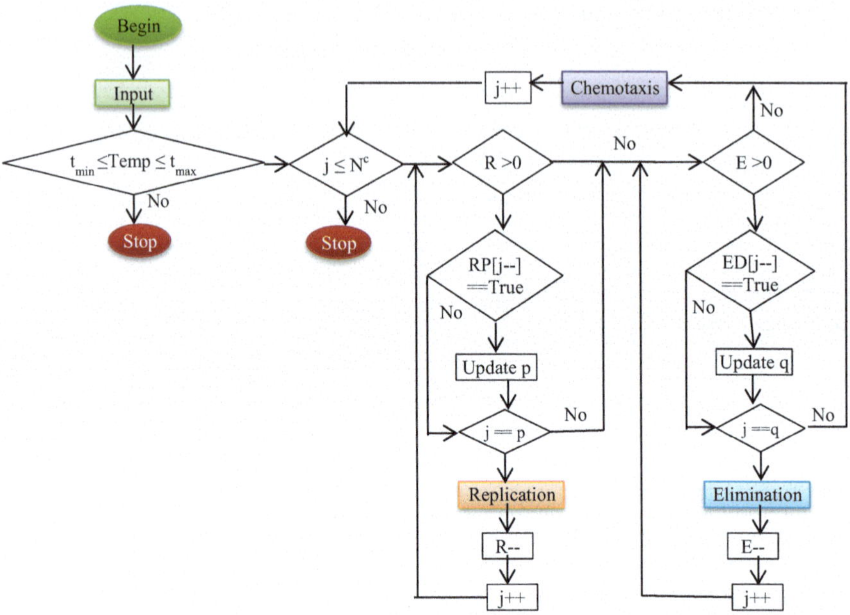

Fig. 14.4 Flow chart of BFOA-DSE framework (Bhadauria and Sengupta 2015)

Further, array RP [j–] in the replication algorithm is used to check whether in the last iterative step, the replication has been performed or not. İf it is true, then p is updated to its next value determined from the relation shown in Eq. (14.14). However, if RP [*j*–] is not true and the iteration count (*j*) of chemotactic step matches with *p*, then replication is performed.

Likewise, at *q*th chemotactic step, the occurrence of ED mechanism is determined by the following relation (Bhadauria and Sengupta 2015):

$$q = n.\left(\frac{N^{\mathrm{c}}}{N^{\mathrm{ed}}}\right), \quad \text{where } 1 \leq n \leq N^{\mathrm{ed}} \tag{14.15}$$

Here, N^{ed} is the maximum number of ED steps. A variable '*E*' is used in the algorithm to keep track the occurrence of replication steps as shown in Fig. 14.4. Further, array ED [*j*–] in the ED algorithm is used to check whether in the last iterative step, the elimination-dispersal has been performed or not. İf it is true, then q is updated to its next value determined from the relation shown in Eq. (14.15). However, if ED [*j*–] is not true and the iteration count (*j*) of chemotactic step matches with *y*, then elimination-dispersal is performed.

14.5.2 Evaluation Models Used in BFOA-DSE Framework

(a) **Execution time evaluation model**: For a scheduled DFG representing a DSP application, the execution time is evaluated using the following model (Bhadauria and Sengupta 2015):

$$T^{\mathrm{e}} = L + (\varphi - 1) * T^{\mathrm{c}} \tag{14.16}$$

where L, φ and T^{c} have already been defined in the previous section.

(b) **Power evaluation model**: For a specific resource configuration, the total power consumption (P^{T}) is evaluated as a summation of static power (P^{st}) and dynamic power (P^{dy}) (Bhadauria and Sengupta 2015):

$$P^{\mathrm{T}} = P^{\mathrm{dy}} + P^{\mathrm{st}} \tag{14.17}$$

Here, average P^{dy} is represented in terms of power consumption in dynamic activity of resources and is given as follows:

$$P^{\mathrm{dy}} = \frac{\varphi * E^{\mathrm{FU}}}{L + (\varphi - 1) * T^{\mathrm{c}}} \tag{14.18}$$

where E^{FU} indicates total energy consumption of the resources.

Further, P^{st} is represented as follows:

$$P^{\mathrm{st}} = \left[\sum_{j=1}^{v} \left(N^{Fj} * K^{Fj} \right) + \left(N^{\mathrm{M/D}} * K^{\mathrm{M/D}} \right) \right] * P^{\mathrm{a}} \tag{14.19}$$

where N^{Fj}, K^{Fj}, $N^{\mathrm{M/D}}$, $K^{\mathrm{M/D}}$ and P^{a} have already been defined in the previous section.

(c) **Fitness Function evaluation model**: The fitness of bacterial is evaluated using the following cost function (Bhadauria and Sengupta 2015):

$$C_{\mathrm{f}}^{B_i} = \sigma_1 \frac{T^{\mathrm{e}} - T^{\mathrm{con}}}{T^{\mathrm{max}}} + \sigma_2 \frac{P^{\mathrm{T}} - P^{\mathrm{con}}}{P^{\mathrm{max}}} \tag{14.20}$$

where $C_{\mathrm{f}}^{B_i}$ indicates fitness of bacterium B_i, T^{con} indicates execution time constraint specified by the user, P^{con} indicates power constraint specified by the user, T^{max} and P^{max} indicate maximum execution time and power consumption, respectively. Further, σ_1 and σ_2 indicate the user defined weightage to execution time and power, respectively.

14.5.3 Bacterial Encoding and Terminating Criteria

To map the BFOA to the DSE process, the positions of bacteria are encoded in terms of resource configuration. As the terminating criteria of the algorithm reach, the position of the best fit bacterium represents the optimal design solution (resource configuration) for synthesizing datapath using HLS.

For ith bacterium, the position (B_i) is represented as follows (Bhadauria and Sengupta 2015):

$$B_i = \left(N(F_1), N(F_2), \ldots, N\left(F_y\right), \ldots, N(F_D)\right)$$

where F_y indicates yth FU resource type and $N(F_y)$ indicates number of instances of resource type F_y. Further, D indicates total FU resource types.

The positions of bacteria are initialized in order to evenly cover the entire design space as follows:

$$\begin{aligned}
B_1 &= (\min(F_1), \min(F_2), \ldots \min(F_D)) \\
B_2 &= (\max(F_1), \max(F_2), \ldots \max(F_D)) \\
B_3 &= ((\min(F_1) + \max(F_1))/2, (\min(F_2) + \max(F_2)) \\
&\quad /2, \ldots (\min(F_D) + \max(F_D))/2)
\end{aligned}$$

If bacteria population size is 'K', then positions for rest of the bacteria (B_4, B_5, …, B_K) are initialized using the following relation (Bhadauria and Sengupta 2015):

$$N\left(F_y\right) = \left(\min\left(F_y\right) + \max\left(F_y\right)\right)/2\omega$$

Here, ω is a random number between the min($N(F_y)$) and max($N(F_y)$).

Terminating criteria: The BFOA-DSE process terminates when one of the following conditions satisfies (Bhadauria and Sengupta 2015):

1. Maximum temperature (45 °C) is reached.
2. Designer specified maximum chemotactic steps (N^c) is reached.
3. The global best position among the bacteria population does not improve over last 10 chemotactic steps.

14.5.4 Role of Chemotaxis, Replication and Elimination-Dispersal

(a) **Chemotaxis Mechanism**:

The function of chemotaxis mechanism is to provide exploration drift in bacteria position. With respect to the last position (B_i^{last}), the new position $\left(B_i^{\text{new}}\right)$ of a bacterium 'i'

is determined using the following function (Passino 2002; Das et al. 2009; Bhadauria and Sengupta 2015):

$$B_i^{\text{new}} = B_i^{\text{last}} + S(i)\frac{\delta(i)}{\sqrt{\delta^t(i) * \delta(i)}} \tag{14.21}$$

where $S(i)$ indicates the step size by which a bacterium moves in a random direction. Here, two basic moves are considered: (i) a bacterium can move in the same direction for certain iterations (ii) a bacterium can tumble in a certain direction. The move of the bacterium using $S(i)$ is determined by the tumble. The tumbling is controlled through a random vector called 'δ' whose value lies within the range [−1, 1] (Bhadauria and Sengupta 2015).

The major steps of the chemotaxis mechanism (Bhadauria and Sengupta 2015) for jth iteration are as follows:

1. Step size is set as $S(i) = S(i) + 2$.
2. In case of violation, step size clamping is performed as follows:

$$\text{If } \left(S(i) > \max\left(N\left(F_y\right)\right)\right), \text{ then } S(i) = S(i)^{\text{new}} - \left(S(i)^{\text{new}} - \left(S(i)^{\text{last}} - 2\right)\right)$$

$$\text{Else if} \left(S(i) < \min\left(N\left(F_y\right)\right)\right), \text{ then } S(i) = S(i)^{\text{new}} - \left(S(i)^{\text{new}} - \left(S(i)^{\text{last}} + 2\right)\right)$$

3. Generate a random vector for tumbling wherein total elements are equal to total resource types and each element is a random number between [−1, 1].
4. Compute cost for each bacterial position.
5. Perform a bacterial move using Eq. (14.21).
6. In case the boundary outreach problem occurs, resource clamping is performed as follows:

$$\text{if}\left(B\left(F_y\right)_i^{\text{new}} < 0\right),$$

$$\text{Then } B\left(F_y\right)_i^{\text{new}} = B\left(F_y\right)_i^{\text{new}} + 2\left|B\left(F_y\right)_i^{\text{new}}\right|$$

$$\text{Else if } \left(B\left(F_y\right)_i^{\text{new}} > \max\left(\text{N}\left(\text{F}_\text{y}\right)\right)\right),$$

$$\text{Then } B\left(F_y\right)_i^{\text{new}} = \text{N}\left(\text{F}_\text{y}\right)^{\text{new}} - \left(\text{N}\left(\text{F}_\text{y}\right)^{\text{new}} - \left(\text{N}\left(\text{F}_\text{y}\right)^{\text{max}} - 1\right)\right)$$

$$\text{Else if } \left(B\left(F_y\right)_i^{\text{new}} < \min\left(N\left(F_y\right)\right)\right),$$

$$\text{Then } B\left(R_y\right)_i^{\text{new}} = N\left(F_y\right)^{\text{new}} - \left(N\left(F_y\right)^{\text{new}} - \left(N\left(F_y\right)^{\text{min}} + 1\right)\right)$$

7. If new position $B(F_y)_i^{\text{new}}$ has not been explored yet, then compute cost/fitness for the new position. Otherwise, new bacterial move is performed using Eq. (14.21) at step 5.
8. If cost of the new position is less than that of last position of the bacteria, then both position and cost of the bacteria are updated with new one. Otherwise, go to steps 5 and perform tumbling using Eq. (14.21) with a random vector 'δ' and repeat steps 5–8 until the cost of the new position is evaluated to be lesser than the last position or terminating criteria reaches.
9. Temperature is increased by δt.

(b) **Replication Mechanism**:

The function of replication mechanism is to manipulate the bacterial position in order to explore the untouched design solutions. In context of DSE, the replication algorithm produces new position using a random variable 'ω'. In the replicated position, the original ordering of resource types in the bacterial position is kept unchanged.

The major steps of replication mechanism (Bhadauria and Sengupta 2015) are as follows:

1. Generate a random number 'ω', such that $\min(N(F_y)) \le \omega \le \max(N(F_y))$.
2. For each dimension of a bacterium B_i, perform $N(F_y)^{\text{new}} = N(F_y) \pm \omega$.
3. In case the boundary outreach problem occurs, resource clamping is performed as follows:

$$\text{If } \left(B(F_y)_i^{\text{new}} > \max(N(F_y))\right),$$

$$\text{Then } B(F_y)_i^{\text{new}} = N(F_y)^{\text{new}} - \left(N(F_y)^{\text{new}} - \left(N(F_y)^{\max} - 1\right)\right)$$

$$\text{Else if } \left(B(F_y)_i^{\text{new}} < \min(N(F_y))\right),$$

$$\text{Then } B(F_y)_i^{\text{new}} = N(F_y)^{\text{new}} - \left(N(F_y)^{\text{new}} - \left(N(F_y)^{\min} + 1\right)\right)$$

4. If new position $B(F_y)_i^{\text{new}}$ obtained due to replication has already been explored, then repeat steps 1–4 performing replication again.
5. The new replicated position is accepted if its cost is lesser than that of its original position.
6. Temperature is increased by δt.
7. The replication is performed N^{rp} times by repeating steps from 1 to 6. Each replication is performed at pth chemotactic step obtained from Eq. (14.14).

(b) **Elimination-Dispersal Mechanism**:

The function of ED mechanism is to eliminate the lesser fit bacteria and disperse the better fit bacteria within the bacteria population. This introduces diversity in the design space. The ED mechanism in BFOA-DSE mimics the elimination phenomenon of *E. coli* bacterium where a rise in temperature beyond a certain limit eliminates a group of bacteria (least fit). In BFOA-DSE, the elimination is performed if temperature reaches at 40 °C. The eliminated bacterium is replaced with a new bacterium whose position is still unexplored and is yet a better fit. The major steps (Bhadauria and Sengupta 2015) are as follows:

1. Elimination is performed only if the temperature is greater than or equal to 40 °C.
2. Among the bacterial population, the least fit and best fit bacterium are determined.
3. The least fit bacterium is eliminated from the population.
4. To perform dispersal, a mid-point configuration ($B_{\rm m}$) between least fit bacterium and best fit bacterium is obtained as follows:

$$B_{\rm m} = \left(B_{\rm lf}\left(N\left(F_y\right)\right) + B_{\rm bf}\left(N\left(F_y\right)\right)\right)/2$$

where $B_{\rm lf}$ and $B_{\rm bf}$ are least and best fit bacteria, respectively.

5. Dispersal is performed as follows:

$$B_{\rm rp}\left(N\left(F_y\right)\right) = B_{\rm m} - \eta.$$

Where $B_{\rm rp}$ indicates new replacement and η is a random configuration whose range is $1 \le \eta \le B_{\rm bf}(N(F_y))$.

6. If the obtained new position (replacement) has already been explored, then go to step 5 to perform dispersal again.
7. Cost of the new bacterial position is evaluated. If this cost is lesser than the cost of eliminated bacterium (least fit), then the least fit bacterium ($B_{\rm lf}$)is replaced by the new bacterium ($B_{\rm rp}$) as follows: $B_{\rm lf}(N(F_y)) = B_{\rm rp}(N(F_y))$. Otherwise, dispersal is performed again.
8. Temperature is increased by δt.
9. The ED algorithm is performed $N^{\rm ed}$ times. Each ED is performed at qth chemotactic step obtained from Eq. (14.15).

14.6 Analysis on Case Studies

The PSO-DSE (Sengupta and Mishra 2014) and BFOA-DSE (Bhadauria and Sengupta 2015) approaches have been analysed in terms of quality of results (QoR) and exploration run-time for different DSP benchmarks. Following subsections discuss the performance of PSO-DSE and BFOA-DSE approaches.

$$B_{rp}(N(F_y)) = B_m - \eta.$$

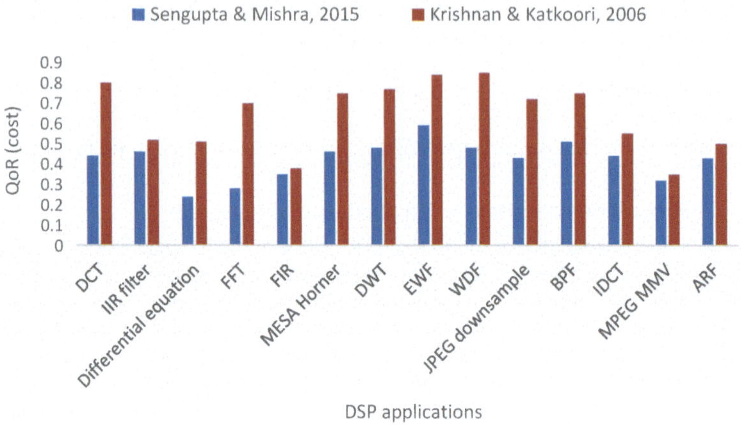

Fig. 14.5 QoR comparison between (Sengupta and Mishra 2014) and (Krishnan and Katkoori 2006)

14.6.1 Comparative Study Between (Sengupta and Mishra 2014) and (Krishnan and Katkoori 2006)

(a) QoR (cost) comparison
As evident from Fig. 14.5, (Sengupta and Mishra 2014) result into better QoR (i.e. lower cost) in comparison with (Krishnan and Katkoori 2006). This is because, (Krishnan and Katkoori 2006) do not consider loop unrolling factor during exploration of design space. Therefore, (Krishnan and Katkoori 2006) fail to choose the optimal UF value for loop-based applications (CDFGs) and result into higher cost than (Sengupta and Mishra 2014) as shown in Fig. 14.5. (Sengupta and Mishra 2014) achieve average improvement in QoR of ~28%.

(b) Exploration time comparison
As evident from Fig. 14.6, PSO-DSE (Sengupta and Mishra 2014) results into highly lesser exploration run-time in comparison with (Krishnan and Katkoori 2006). The PSO-DSE (Sengupta and Mishra 2014) approach leads to reduction in exploration run-time ~92% with respect to (Krishnan and Katkoori 2006).

14.6.2 Comparative Study Between (Bhadauria and Sengupta 2015) and (Sengupta et al. 2012)

The BFO-DSE (Bhadauria and Sengupta 2015) approach has been compared with (Sengupta et al. 2012) in terms of QoR and exploration run-time. The results of

(a) QoR (cost) comparison

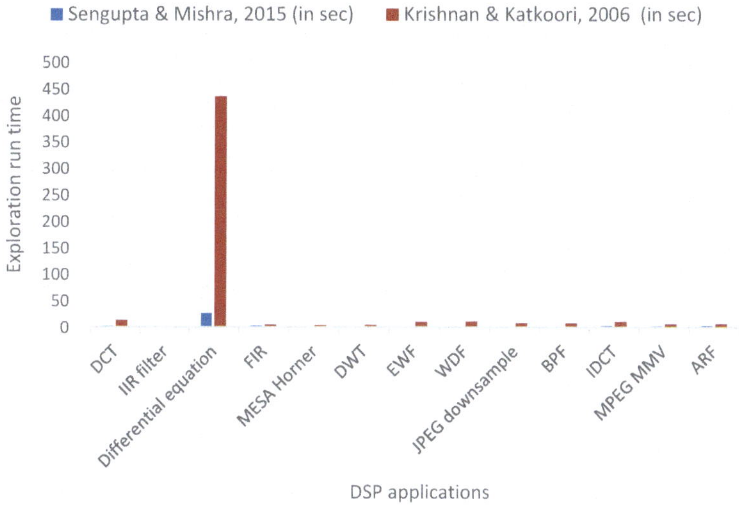

Fig. 14.6 Exploration run-time comparison between (Sengupta and Mishra 2014) and (Krishnan and Katkoori 2006)

(Bhadauria and Sengupta 2015) have been obtained based on the following values: $N^{c} = 120$, $N^{rp} = 5$, $N^{ed} = 4$, $K = 3$, $\sigma_1. = \sigma_2 = 0.5$.

(a) QoR (cost) comparison
As evident from Fig. 14.7, BFOA-DSE (Bhadauria and Sengupta 2015) results into better QoR (i.e. lower cost) in comparison with (Sengupta et al. 2012). The BFOA-DSE (Bhadauria and Sengupta 2015) achieves ~48% improvement in QoR.

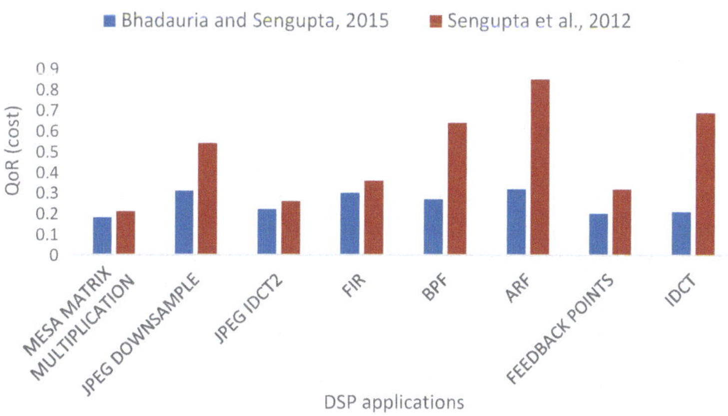

Fig. 14.7 QoR comparison between (Bhadauria and Sengupta 2015) and (Sengupta et al. 2012)

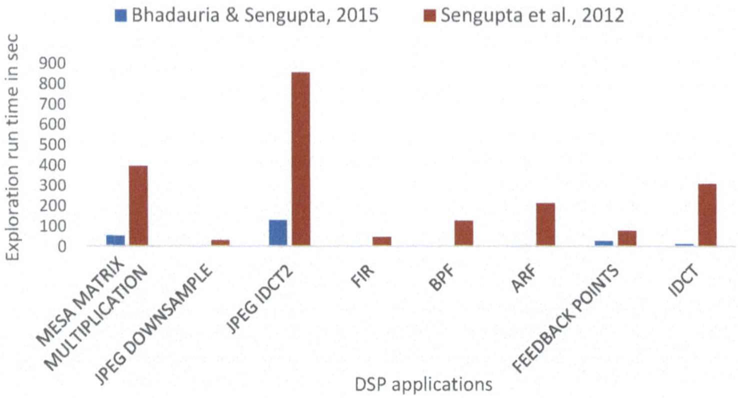

Fig. 14.8 Exploration run-time comparison between (Bhadauria and Sengupta 2014) and (Sengupta et al. 2012)

(b) Exploration time comparison
As evident from Fig. 14.8, BFOA-DSE (Bhadauria and Sengupta 2015) is capable to explore the design space with highly lesser exploration run-time in comparison with (Sengupta et al. 2012). The BFOA-DSE (Bhadauria and Sengupta 2015) approach leads to reduction in exploration run-time ~90% with respect to (Sengupta et al. 2012).

14.7 Conclusion

This chapter highlighted two efficient approaches, viz. PSO-DSE (Sengupta and Mishra 2014) and BFOA-DSE (Bhadauria and Sengupta 2015) for design space exploration of DSP hardware. These approaches are efficient in terms of exploration run-time and achieves better QoR with respect to similar approaches (Krishnan and Katkoori 2006; Sengupta et al. 2012).

At the end of this chapter, the reader is able to understand the following concepts:

- Importance of HLS for DSP hardware.
- Mapping of PSO algorithm to the DSE process.
- Details of PSO algorithm to perform DSE for DSP hardware.
- Mapping of BFO algorithm to the DSE process.
- Details of BFO algorithm to perform DSE for DSP hardware.
- Performance evaluation of PSO-DSE and BFOA-DSE approaches in terms of QoR and exploration run-time.
- Comparison of PSO-DSE and BFOA-DSE approaches with the contemporary approaches.

References

Ascia G, Catania V, Di Nuovo AG, Palesi M, Patti D (2007) Efficient design space exploration for application specific systems-on-a-chip. Elsevier J Syst Arch 53:733–750
Bhadauria S, Sengupta A (2015) Adaptive bacterial foraging driven datapath optimization: exploring power-performance tradeoff in high level synthesis. Appl Math Comput 269:265–278
Das S, Biswas A, Dasgupta S, Abraham A (2009) Bacterial foraging optimization algorithm: theoretical foundations, analysis, and applications. Stud Comput Intell 203:23–55
Dhodhi MK, Hielscher FH, Storer RH, Bhasker J (1995) Datapath synthesis using a problem-space genetic algorithm. IEEE Trans Comput-Aided Des 14:934–944
Eberhart RC, Shi Y (2000) Comparing inertia weights and constriction factors in particle swarm optimization. In: Proceedings of IEEE congress evolutionary computation, San Diego, CA, pp 84–88
Engelbrecht AP (2005) Fundamental of computational swarm intelligence. Wiley, England
Gallagher JC, Vigraham S, Kramer G (2004) A family of compact genetic algorithms for intrinsic evolvable hardware. IEEE Trans Evol Comput 8(2):111–126
Harish Ram DS, Bhuvaneswari MC, Prabhu SS (2012) A novel framework for applying multiobjective GA and PSO based approaches for simultaneous area, delay, and power optimization in high level synthesis of datapaths. VLSI design. Hindawi. Article ID 273276, 12 p
Heijlingers MJM, Cluitmans LJM, Jess JAG (1995) High-level synthesis scheduling and allocation using genetic algorithms. In: Proceedings of Asia South Pacific design automation conference, pp 61–66
Kennedy J, Eberhart RC (1995) Particle swarm optimization. In: Proceedings of the 1995 IEEE international conference on neural networks, pp 1942–1948
Krishnan V, Katkoori S (2006) A genetic algorithm for the design space exploration of datapaths during high-level synthesis. IEEE Trans Evol Comput 10(3):213–229
Mishra VK, Sengupta A (2014) MO-PSE: adaptive multi-objective particle swarm optimization based design space exploration in architectural synthesis for application specific processor design. Elsevier J Adv Eng Softw 67:111–124
Passino KM (2002) Biomimicry of bacterial foraging for distributed optimization and control. IEEE Control Syst Mag 22(3):52–67
Schneiderman R (2010) DSPs evolving in consumer electronics applications. IEEE Signal Process Mag 27(3):6–10
Sengupta A (2016) Intellectual property cores: protection designs for CE products. IEEE Consum Electron Mag 5(1):83–88
Sengupta A (2017) Hardware security of CE devices. IEEE Consum Electron Mag 6(1):130–133
Sengupta A, Sedaghat R, Zeng Z (2010) A high level synthesis design flow with a novel approach for efficient design space exploration in case of multi-parametric optimization objective. Microelectron Reliab 50(3):424–437
Sengupta A, Sedaghat R, Sarkar P (2012) A multi structure genetic algorithm for integrated design space exploration of scheduling and allocation in high level synthesis for DSP kernels. Elsevier J Swarm Evolut Comput 7:35–46
Sengupta A, Mishra VK (2014) Automated exploration of datapath and unrolling factor during power–performance tradeoff in architectural synthesis using multi-dimensional PSO algorithm. Expert Syst Appl 41(10):4691–4703
Torbey E, Knight J (1998a) High-level synthesis of digital circuits using genetic algorithms. In: Proceedings of the international conference on evolutionary computation, pp 224–229
Torbey E, Knight J (1998b) Performing scheduling and storage optimization simultaneously using genetic algorithms. In: Proceedings of the IEEE Midwest symposium on circuits systems, pp 284–287
Trelea IC (2003) The particle swarm optimization algorithm: convergence analysis and parameter selection. Inf Process Lett 85(6):317–325

Chapter 15
Register-Transfer-Level Design for Application-Specific Integrated Circuits

Dilip Singh and Rajeevan Chandel

Abstract Over the years, a rapid growth has been witnessed in electronics semiconductor industry because of the huge demand for system-level designs. System-level designs are prominently used for various applications such as high-performance computing, controls, telecommunications, image and video processing, consumer electronics and others. Hence to accomplish such applications using very large-scale integration (VLSI) design, it is recommended to have an efficient register-transfer-level (RTL) design abstraction, as it can provide a low power and high-performance outcome (Wu and Liu in IEEE Trans Very Large Scale Integr (VLSI) Syst 6:707–718, Wu and Liu 1998). In digital integrated circuit (IC) design, RTL models a synchronous digital circuit in terms of the flow of digital signals or data between hardware registers and the logical operations performed on these signals. RTL abstraction is used in hardware description languages (HDLs) to create high-level representations of a circuit (Chinedu et al. in 3rd IEEE international conference on adaptive science and technology (ICAST 2011). IEEE, pp 262–267, Chinedu et al. 2011). From these lower-level representations, ultimately actual circuitry can be derived. Design at the RTL level is a typical practice in modern digital system designs. This chapter mainly focuses on design of RTLs for application-specific integrated circuits (ASICs) and how it differs for field-programmable gate arrays (FPGAs). The examples and modules discussed in this chapter are written in HDL, viz. Verilog language.

Keywords Register transfer level (RTL) · ASIC · VLSI design styles · RTL guidelines · Synthesis · FPGA

D. Singh (✉) · R. Chandel
Department of Electronics and Communication Engineering, National Institute of Technology Hamirpur, Hamirpur, Himachal Pradesh 177005, India
e-mail: dilipgovindsingh@gmail.com

R. Chandel
e-mail: rchandel@nith.ac.in

R. Dhiman and R. Chandel (eds.), *Nanoscale VLSI*,
Energy Systems in Electrical Engineering,
https://doi.org/10.1007/978-981-15-7937-0_15

15.1 VLSI Design Styles

A VLSI system designer is required to figure out the best suitable VLSI design style for digital designs into silicon. This in turn reduces the tape-out time and design complexity. During the 1970s, it was a very time exhausting process to design a circuit on silicon. For designing a complex circuit with more than 100,000 transistors, years of effort of more than 10 members were required to debug, synthesize, floor plan, and layout the entire design (Rosenberg 1980). In recent years, computer-aided design (CAD) tools are being used to achieve complex designs with reduced man efforts, resulting in different types of VLSI design styles. Figure 15.1 shows the commonly used different VLSI design styles. Selecting a suitable design style depends on the specific application of the design (Sherwani 1999).

To reduce the complexity of a design, the following points need to be considered:

- Proceeding with hierarchal approach is best suited. Hierarchal approach breaks a design into different numbers of levels. Each level is derived from the previous level, thus providing more details to each stage. This is also known as top-down approach which prevents the designer from losing the sight of the details of the whole design.
- Using the basic structures such as RAMs, ROMs and PLAs reduces the design time.

15.1.1 Choice of Design Style

Design styles ought to be chosen in a way such that the designer can extract all the benefits of silicon area, with reduced computational time and complexity. Following are the number of factors regarding the choice of the design style:

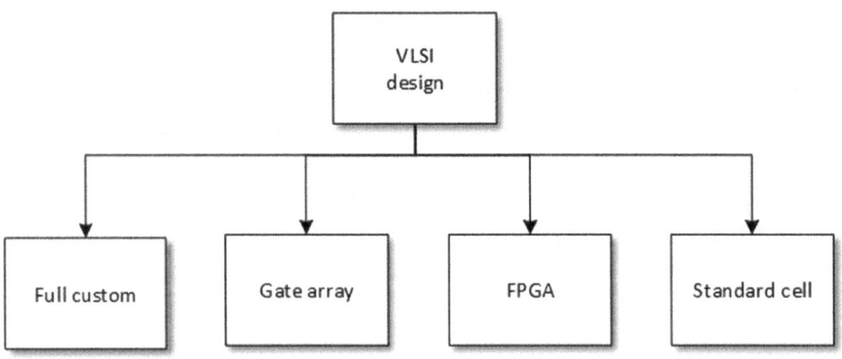

Fig. 15.1 Various VLSI design styles

- If the design is having a large volume, full-custom approach is the best option as it provides optimized performance and conservation of chip area.
- If performance is not that important aspect and production, i.e., time to market, is the main factor, then standard cell design technique is best suited because it utilizes macros which are already designed and stored in the library.
- For modest quantities where fast turnaround is essential, the gate array approach is appropriate. Currently, manufacturers may themselves offer a placement and routing service, so that the risk involved in the gate arrays is no greater than in production of printed circuit boards.
- For designs carried out only for functional verification, field-programmable gate arrays (FPGAs) are best suited.

15.1.2 Full-Custom Design

Full-custom design style carries out the design from scratch to utilize full silicon area. For very complex circuits such as a microprocessor, which is a mass produced product (more than 100,000 units per annum), there is a need to use every square micron of silicon area efficiently to achieve maximum yield and hence arrive at a minimum cost. Such a design style is referred to as full-custom design.

In full-custom approach, a design is divided into several sub-designs where each sub-design contains some portion of the information of the overall design, hence dividing the system into a hierarchal format. A VLSI chip is made using cluster of units, where each unit contains a functional block. In full-custom design style, designer has the freedom to change the height and width of the block to best utilize the silicon area. Hence, blocks can be placed anywhere on the silicon chip and much more compact designs can be made. If only aspect of a designer is area, then full custom is the best approach. However, when it comes for routing the sub-blocks full-custom approach introduces more complexity than any other design style. Consequently, it is used when requirement is only of reduced area and high efficiency. Several design steps involved in this design style are carried out manually. Much more of the optimization in full-custom design is done by layout compaction. Layouts generated using CAD tools are not much area efficient; hence to well utilize the Si area, layouts are manually designed.

Figure 15.2 depicts a typical full-custom design structure. To interconnect different chips and routing different blocks, the input/output (I/O) pads are placed at the corners of the overall design. I/O pads are rectangular shaped blocks made up of metal. Several metal layers are used for routing the functional block in a design. Higher layers have via of larger size, than the lower-level vias. For example, as shown in Fig. 15.2, size of the metal layer M2 is more than that of M1. As the number of routing layers increases, the routing area for interconnection of blocks reduces. Usually, chip area is determined by the area of the transistors inside a design. Hence, that is the reason most of the routing takes place on the top of transistors in the additional metal layers. However, at times the complexity of circuit is in such

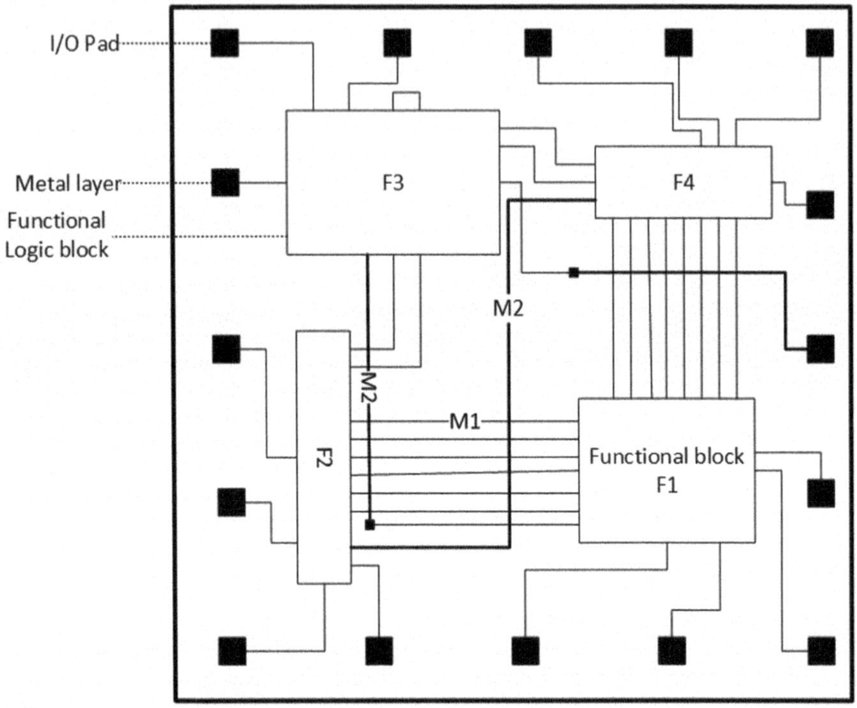

Fig. 15.2 A typical full-custom structure

a way that a greater number of routing layers are needed to interconnect the sub-blocks and functional blocks. In such a case, the die size is determined by the area occupied by interconnects and the transistor area serves as lower bound on the die size of the chip. Full-custom design requires a large amount of time hence not suitable for fast turnarounds. This approach is mostly used for processor design and high-performance circuits.

15.1.3 Standard Cell

Standard cell is similar to full-custom approach in a way that design is divided into sub-blocks and then routed to minimize the area. But the difference between the two is that the height of the sub-blocks is fixed and some of the sub-blocks are predefined which are also known as cells or standard cells. Each predefined cell is analyzed, has its own properties and is tested extensively to perform the given operation. All predefined cells, which are ready to be used in a design, are stored in the cell library. Usually, there are more than 1200 cells stored in a cell library. Cells are then instantiated and connected, making a complete design. This approach is more expensive than any other approach as design requires a complete mask for target technology.

Cells are placed in a row in such a way that the space between two different cells in a same row is minimal. The space between two rows is called a channel, whereas if interconnection is to be made between two cells lying on different rows or non-adjacent rows then interconnection wire is passed through empty space between cells. The empty space between two cells is called feedthrough. Initially, interconnection is made for cells on non-adjacent rows; thereafter, feedthrough is carried out amongst cells inside the same row. Generally, only one metal layer is used for interconnection between cells. However, if complexity increases more numbers of layers are introduced. It is impossible to achieve a channel less design; hence, more than two metal layers are used for feedthrough cells. Figure 15.3 shows the structure of the standard cell design.

Standard cell design has advantage over full custom in terms of simplicity. As most of the cells are predefined, time required for making a design becomes less and modern tools can analyze and synthesize the complete design with greater speed.

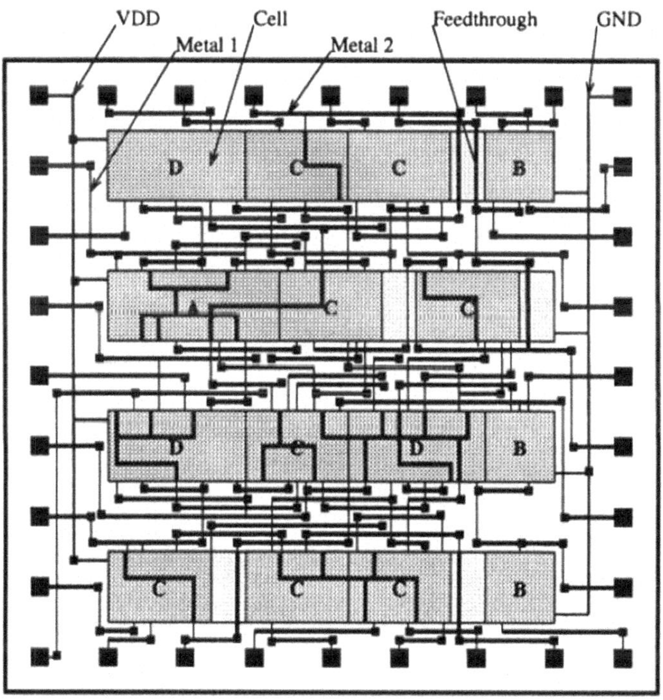

Fig. 15.3 Standard cell architecture [see Fig. 15.6 in Sherwani (1999)]

Standard cell designs are used for design of control logic used in full-custom design. When area is taken into consideration, standard cell designs are not area efficient.

15.1.4 Gate Arrays

This approach is known as 'gate arrays' because a cell may be simply a logic gate such as a three-input NAND gate. Gate arrays consist of cells as in standard-cell-based designs, but cells are identical. Each gate array-based designed chip is thus made up of identical gates or cells. Unlike standard cell, there are horizontal as well as vertical channels. In simple terms, the cells are placed in fixed space, i.e., horizontally and vertically. Gate arrays do not follow hierarchy-based design like full custom and standard cell. The design in this approach is divided in such a way that each sub-block or cell is identical to each of the other cells. During the placement phase, each sub-block is mapped onto a prefabricated cell on the chip. The number of sub-blocks placed or partitioned in a chip must be less than or equal to the total number of cells on the chip. Once partition of each cell is done, these must be interconnected horizontally and vertically. Figure 15.4 shows an uncommitted gate array, also known as a prefabricated cell.

Routing between the cells is done using metal layers on vertical and horizontal channels. It should be noted that only fixed number of layers can be used for routing in channels. Two layers of interconnections are widely used for less complex circuits.

Gate arrays have advantage over full custom and standard cell in terms of cost and simplicity to fabricate. All gate array designs start with prefabricated chip. Hence, the initial step required for gate array is the same for any design. However, it differs only at the last stage of layout depending on the type of application of the design.

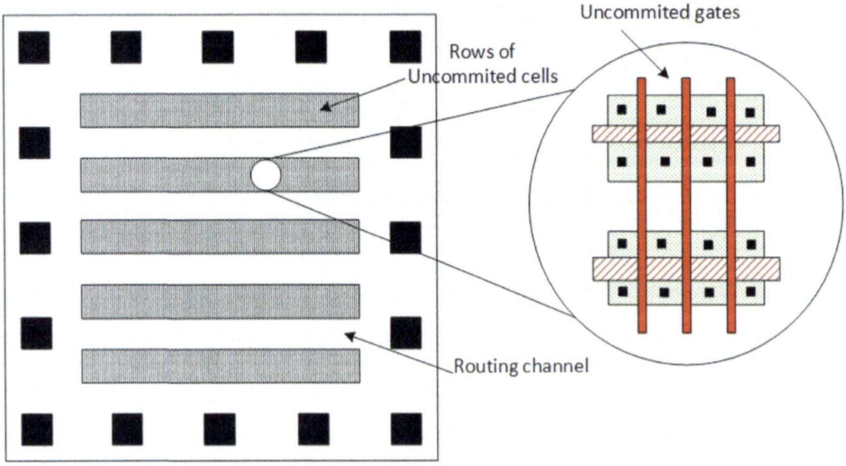

Fig. 15.4 Uncommitted gate array structure

15.1.5 Field-Programmable Gate Arrays (FPGAs)

The FPGA is the fastest design style amongst all the other design styles in terms of turnaround time. When production requirement is low, FPGAs are at much advantage. FPGAs are easy to program because cells and interconnects are prefabricated. In FPGAs, programmable logic blocks are placed horizontally, which are connected by routing metal layers. Figure 15.5 shows a generalized structure of FPGA. All cells of an FPGA are identical to each other and have the same layout. It may be considered that cells are more like memory blocks which store the values in the form of truth table, hence making a lookup table (LUT). Each lookup table is assigned to store a function of a design. Whenever a program is simulated, lookup tables are searched for their respected outputs. This makes an FPGA a fast and high-performance design. Thus, for different functions each logic block can be programmed accordingly. To represent k-bit input and 1-bit combinational output, 2^k-bits are required in a logic block. There are two types of interconnection used in programming the logic blocks. The first one is through anti-fuses, and second is cross-fuse. The empty space between the horizontal logic blocks is equipped with routing wires.

Anti-fuse is used to provide connection between the horizontal segments. Cross-fuse provides connection between the vertical segments. The limitation when using fuse-based programmable FPGAs is that these cannot be reprogrammed. For reprogrammable FPGAs, pass gates are used.

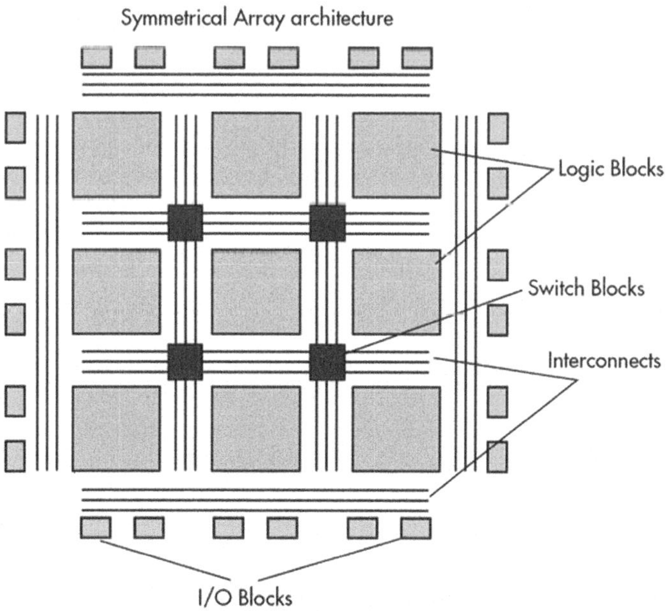

Fig. 15.5 A generic FPGA structure

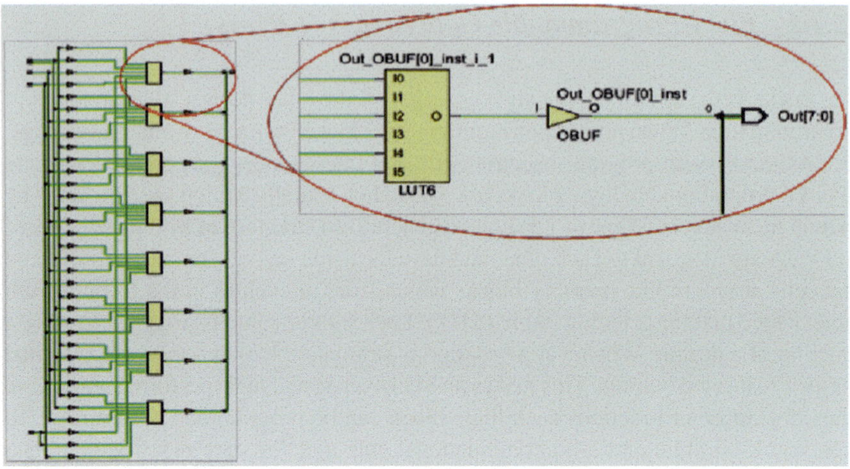

Fig. 15.6 Synthesized design of Verilog code 1.1 on FPGA using Xilinx Vivado tool

For better understanding, Fig. 15.6 shows a synthesized design of *Verilog code 1.1* on FPGA using Xilinx Vivado tool (Xilinx Vivado 2017). An encirclement is used to show the magnified view of the 6-input LUTs used. LUTs are retaining the output values for all possible inputs making synthesis and implementation easier and faster.

15.2 ASIC Design Flow

An application-specific integrated circuit (ASIC) is a chip designed or customized for a special use, for example, a particular kind of transmission protocol or a palmtop computer. It may be differentiated from general-purpose ICs, such as the microprocessor and the random access memory chips in computers and for mixed signal ICs (Barr 2007). For example, a chip designed to run in a voice coder and decoder (vocoder) is an ASIC.

Developing an ASIC chip as per the given specifications takes a lot of time. To design a chip, there are a number of steps which should be followed for a complete fully functional hardware. There are some automated tools which generate RTL based on the behavior description of the design. One of the tools is MATLAB HLS. It boosts up the design flow by generating the RTL automatically based on reference model and performs logic synthesis. Figure 15.7 provides the detailed ASIC design flow of HLS and the manual flow (Cong et al. 2011).

Before building of the design, it is recommended to draw a top-level architecture. Choosing the best architecture and optimized algorithms improves the latency and performance of an ASIC. Designs which are developed considering specific specifications reduce the workload, time and complexity.

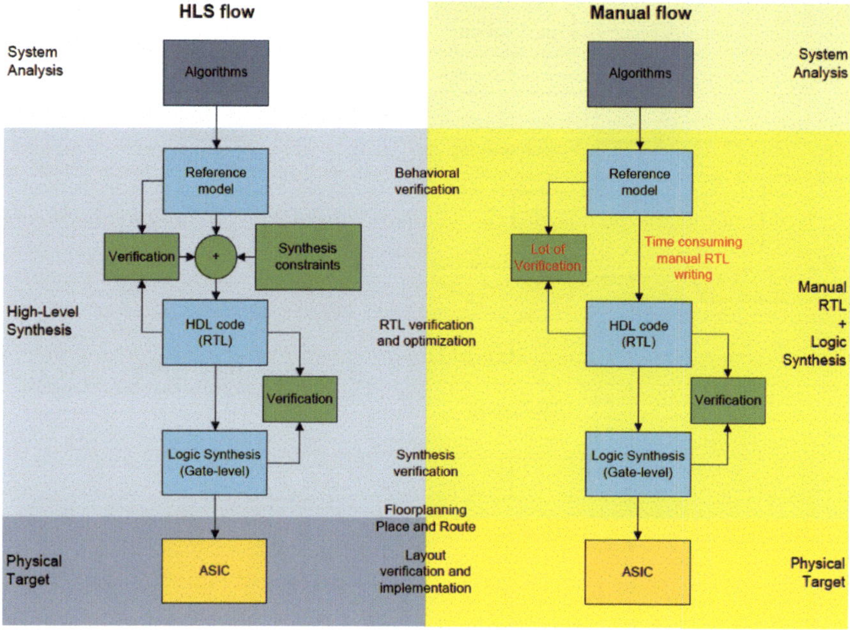

Fig. 15.7 General design flow of ASIC development

Hardware description language codes are designed in the EDA tools like Xilinx Vivado and ModelSim and simulated to check functionality. This model works as reference for observing the behavior of the design. By looking at the reference code, the RTL code is written and then verified whether it is meeting the desired functionality or not. RTL describes the hardware in terms of registers and the combinatorial logic. The generated RTL is then synthesized using synthesis tools and then sent to backend flow.

Logic synthesis is done by RTL for performing gate-level synthesis. Floor plan is performed to identify where the functional cells should be placed in limited silicon area. Place and route are done after floor planning is over, and cells are connected by wires in an optimized manner. After this, layout verses schematic is performed. All the design tests are carried out, and if the designer confirms that specifications are met then the chip is ready to be fabricated. Figure 15.8 shows a typical digital ASIC design flow using various EDA tools (Smith 2008).

15.2.1 Steps Involved in an ASIC Design

1. **Specification**: This is the first step of ASIC design flow where the specifications such as area, functionality, timings, cost and applications of the design are

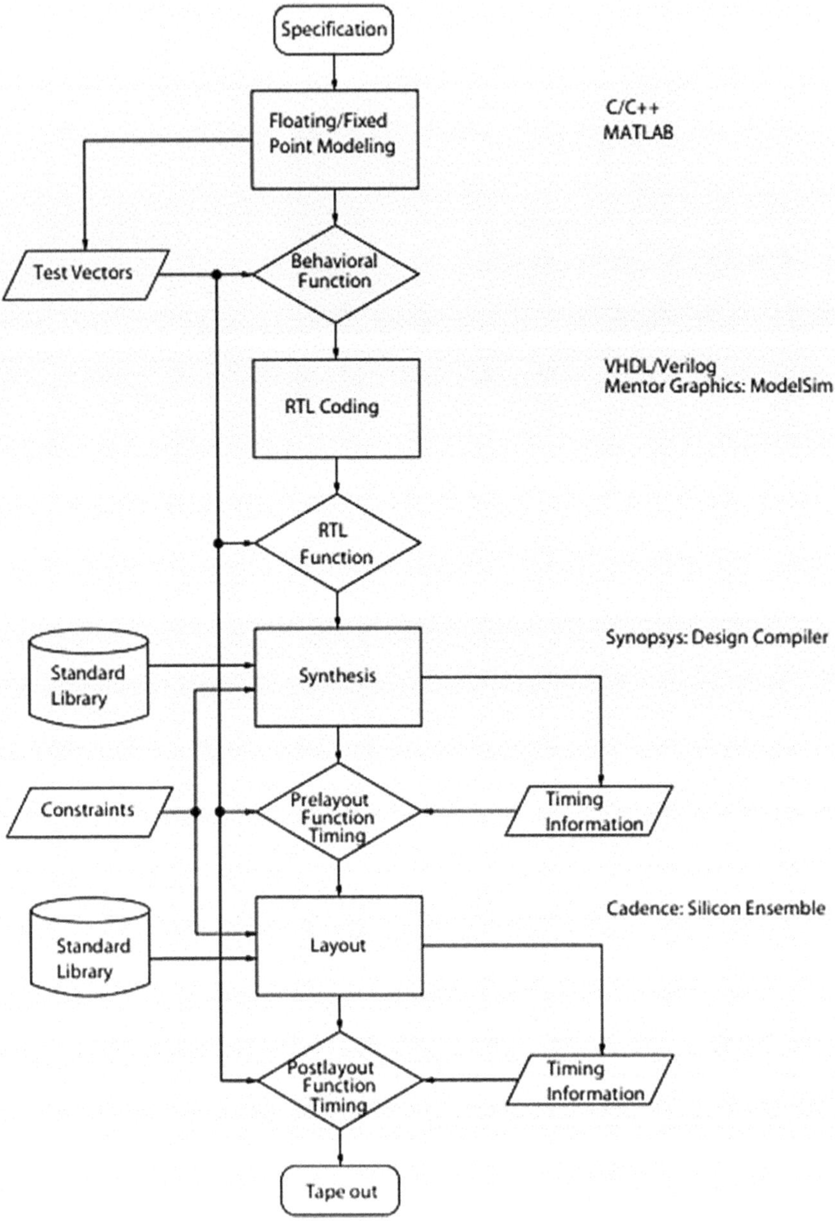

Fig. 15.8 Typical digital ASIC design flow

identified. Depending upon those factors, the designer moves to the next step. Specifications are generally provided by the clients. For instance, a client wants a chip which can process speech and reduce its bit rate. So according to the given design, the specifications will be its operating frequency, the algorithm to be used, area or speed-centric design.

2. **Floating/fixed-point modeling**: Data representation is an important aspect when designing digital systems. Sometimes, data to be processed is in real or complex format. Hence, data representations are used for calculation of real and complex numbers. There are two types of data representation techniques, i.e., fixed point and floating point. A floating-point number is typically expressed as

$$F \times r^e \tag{15.1}$$

where 'F' is fraction, 'r' is the radix and 'e' is the exponent.
For example, the number 76.55 can be represented as 7.655×10^1, 0.7655×10^2, 0.07655×10^3 and so on. The fractional part is normalized in such a way that there is only one nonzero digit left before the radix point. For example, decimal number 12.34567 can be normalized as 1.234567×10^1 and similarly to represent binary number 1110.1011 it can be normalized as 1.1101011×2^3.
Fixed-point data representation is mostly used for high-performance designs. In fixed point, data is represented on limited range that is integer and fraction parts are fixed.
The range is decided by designer according to the required precision. MSB bit is fixed for sign representation. Therefore, if data is negative then sign bit becomes '1' whereas if data is positive then sign bit becomes '0'.
For example, the number -77.66 can be represented as 1100101.101010001111010111000010, if considered 7-bit integer width and 24-bit fraction width.

3. **Behavioral function**: When all the specifications are received and data representation method is selected, next step is to design the Verilog code or VHDL code in behavioral format to check the functionality of the design. The design under test is supported with a test bench to check the code coverage of the designed behavioral function. Simulations are performed on electronics design automation (EDA) tools such as Xilinx Vivado and ModelSim (Mentor Graphics 2016).
For example, behavioral verilog code of 4:1 multiplexer will be written as:

Verilog_code_1.1

```
\\\\\\ Code for 4x1 MULTIPLEXER\\\\\\\\\\\\\\
 MUX_4x1 (Out, In0, In1, In2, In3, Sel);
Input [7:0] In0, In1, In2, In3;
Input [1:0] Sel;
Output reg [7:0] Out;
Always @(*) begin
  Case(Sel)
    2'b00:  Out=In0;
    2'b01:  Out=In1;
    2'b10:  Out=In2;
    2'b11:  Out=In3;
  Endcase
End
Endmodule
\\\\\\\\\\\\\\\\\\\\\\\\\\\\\\\\\\\\\\\\\\\\\\\\\\
```

4. **Register-transfer-level (RTL) coding**: RTL is the most important step in ASIC design. This is because bad RTL leads to bad outputs and synthesis failure. RTL is based on synchronous logic and contains three primary pieces, namely registers which hold state information, combinatorial logic which defines the next state inputs and an input clock that controls the change of states. There are two widely used RTL design approaches:

   - Algorithmic state machine (ASM) chart
   - Datapath and controlpath design.

   Figure 15.9 illustrates the RTL generated for the Verilog code 1.1 using Xilinx Vivado tool. It can be clearly seen that there is a 4 × 1 mux introduced for 'case' statement. Multipliers also fall under the category of combinational logic block; hence, they are essential for efficient RTL designs.

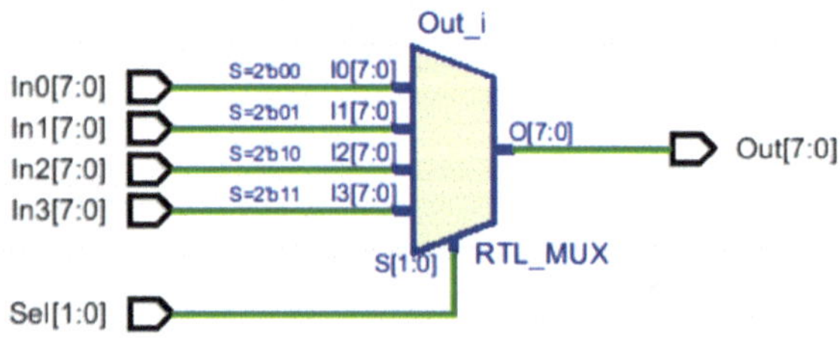

Fig. 15.9 RTL generated using Xilinx Vivado for *Verilog code 1.1*

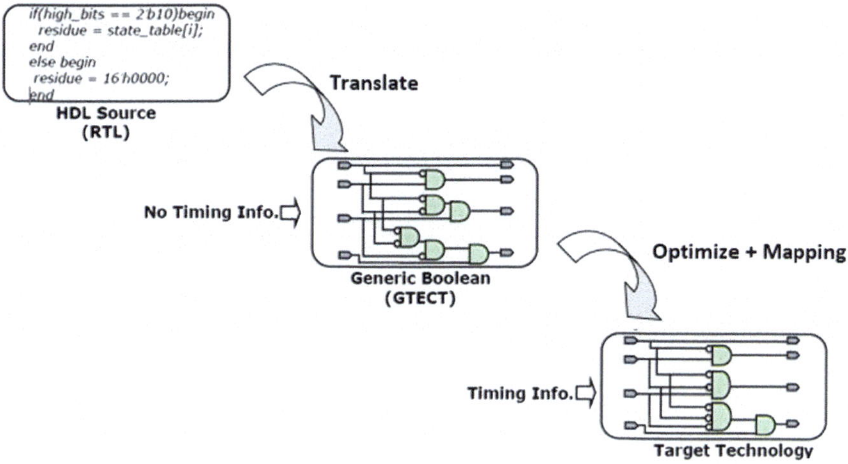

Fig. 15.10 Generalized logic synthesis flow

5. **Synthesis**: Synthesis is performed on an RTL to convert the design into gate-level netlist. A gate-level netlist can be explained as interconnection of basic logic gates such as NAND, NOR, AND, XOR, XNOR, OR and NOT in a structural flow. Logic synthesis uses standard cell library which has simple cells, such as basic logic gates like AND, OR and NOR, or macro-cells, such as adder, multiplexers, memory and specific flip-flops. Figure 15.10 shows the synthesis flow in ASIC design.

$$\text{Synthesis} = \text{translation} + \text{optimization} + \text{mapping}$$

6. **Physical design (layout)**: Physical design means converting gate-level netlist into geometric representation. Geometric representation is achieved by using multiple layers of metals, poly-silicon, diffusion for making transistors, gates and cells. This approach is also known as layout design. To design a layout, there are some design rules which need to be followed to avoid design rule check (DRC) errors. Design rules are guidelines based on the limitations of the fabrication process and the electrical properties of the fabrication materials. Layout is accepted only after it undergoes several validation and verification measures. There are commercial EDA tools available which can convert a netlist into a verified layout. To achieve area-efficient design, layouts are designed manually. Such a design involves complex circuits like microprocessors. For smaller circuits, automatically generated layouts are well suited as this can speed up the time to market process. However, manual design can be a disadvantage for a layout for very large and complex circuit. To overcome such hurdles, global optimization techniques are used. Figure 15.11 shows the layout of 8-bit counter designed using Cadence Innovus tools for RTL to GDS flow. The layout involves

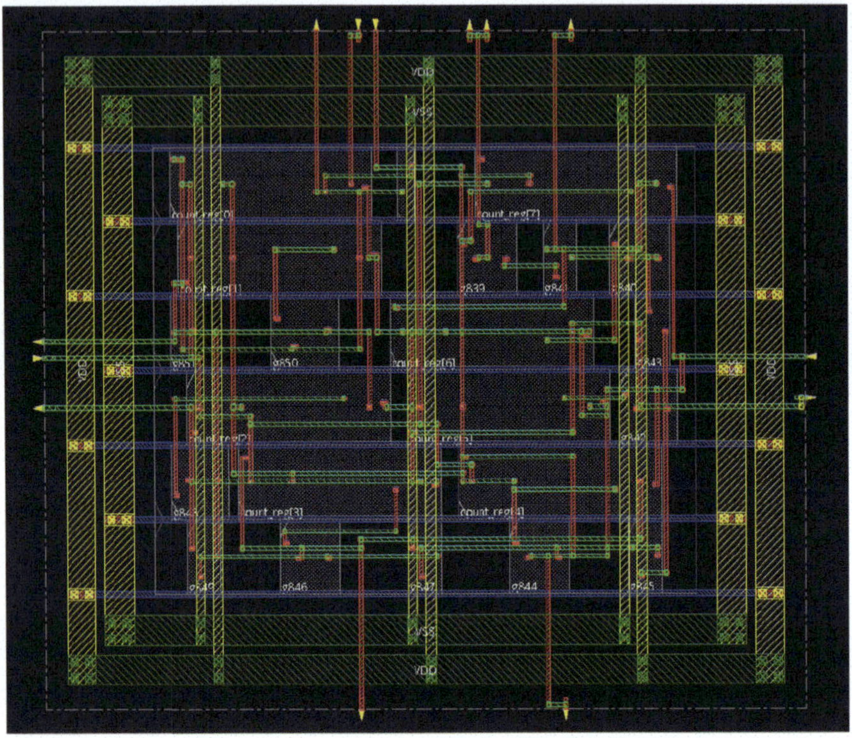

Fig. 15.11 Physical layout of a 8-bit counter generated using Cadence Innovus

all the cells and macros with power grids to avoid voltage drop inside the chip (Cadence 2017).

7. **Tape out**: During physical design, if the layout passes all the verification and validation steps then it is converted into a mask. Mask is the final product of the physical design. It is also known as graphic data system (GDS-II). GDS file contains all the data required for final chip preparation. GDS-II file format is accepted by all the foundries for chip design. The layout data inside GDS-II file is used to create the photolithographic masks of the circuit going to be fabricated. Masks are used to identify the spaces on the wafer, to identify where materials should be deposited, implanted or diffused. There are several steps involved in fabrication process such as diffusion, etching and ion implantation, and for each step one mask is required. For a large design, the number of masks increases. This results in increase in the cost of the chip. To reduce the cost per chip, several numbers of chips are produced on one single silicon wafer. Currently, 300-mm (12 in.) diameter silicon wafers are used to produce chips with die per wafer of 640 mm (ITRS 2013).
   Furthermore, in a VLSI design flow, there are many more steps than discussed above. Each step includes sub-steps. If those criterions are met, then only data

proceeds to the next step. In each design step, a new representation of the system is created and analyzed. Considering the layout design step, the layout is refined until all DRC errors are removed and area utilization is achieved.

15.3 RTL

A digital design mainly consists of combinational as well as sequential circuits. Whenever an RTL is to be designed, both combinational and sequential circuits are used in such a way that the designer can map the overall system on an FPGA or implement it as an ASIC. The term RTL generally means register transfer level. It is clear from the name that in a digital system data transfers take place using registers. To maintain the continuous flow of data without clock mismatch, it is necessary to keep asynchronous and sequential circuits separated. A combinational circuit can be referred as asynchronous, whereas sequential circuits involve feedback and clock triggering along with storage element (Mangassarian et al. 2014; Ramachandran 2007).

15.3.1 RTL Coding Strategies

Given below are some important guidelines for designing an optimized and efficient RTL.

1. **Introducing pipeline registers to improve system speed**

In a large system, there are several numbers of registers and combinational logic blocks for proper functioning of the system. However, due to complex circuitry sometimes system performance degrades due to slow speed. Hence to improve the speed, some of the larger combinational blocks are broken down into small logic blocks. Between each sub-block, a register is introduced which is often termed as pipeline register. Figure 15.12a, b shows the insertion of pipeline registers at output stages of the divided sub-blocks. It should be noted that the divided blocks should have propagation delay similar to each other. Introducing such pipeline registers also prevents setup and hold time violations.

2. **Avoid feedback paths in combinational logic**

Whenever a feedback path is introduced to combinational logic, its behavior becomes unpredictable or '*x*', i.e. a don't care condition. Feedbacks should be synchronous; otherwise, racing problem is seen in the circuit output. Consequently, the expected output is not generated. To avoid such a problem, a D flip-flop is introduced at the feedback path creating a delay of one clock. The functionality of the design remains the same, but the issue of unpredictable output is removed. Figure 15.13a shows

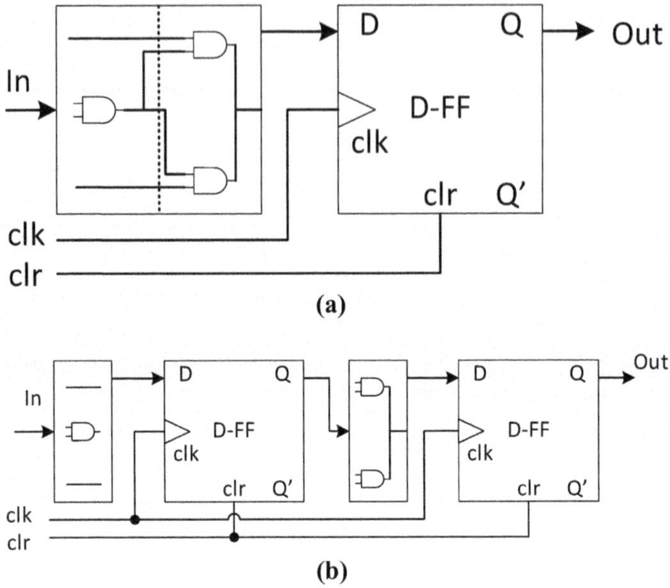

Fig. 15.12 Insertion of pipeline registers between adjacent divided sub-blocks. **a** Combinational logic block directly connected to register and **b** combinational logic block subdivided into sub-blocks, with registers introduced between sub-blocks

the feedback loop and its elimination using D flip-flop, and Fig. 15.13b shows the waveform.

3. **Removing gated clocks to prevent clock skew**

A complete RTL involves large numbers of registers and combinational logics. Some logics are made in such a way that the output of those blocks drives the clocks for the next register. Such kind of design method introduces clock skew. Clock skew can be defined as the time it would take for a system clock to reach the different components. Clock skew occurs due to propagation delays, interconnect path delays and combinational delays resulting in a setup and hold time violation. Clock gating is beneficial if designer wants to achieve low power design (Friedman 2001). This is because clock gating helps in reducing dynamic power of the system. Figure 15.14 shows the technique used to remove clock gating. As we can see, the combinational logic block output is connected to the clock input of the register. Further, the output of the register is connected as input to the combinational logic block. To remove such clock gating, a multiplexer (mux) is introduced at the input terminal of register and clock is driven by system clock. Whenever select is high, data through the combinational block is stored on the D flip-flop. When select is low, previous data is retained until the next clock pulse. System will work exactly how it was before clock gating, but now clock skew will not occur as registers are driven directly by the system clock.

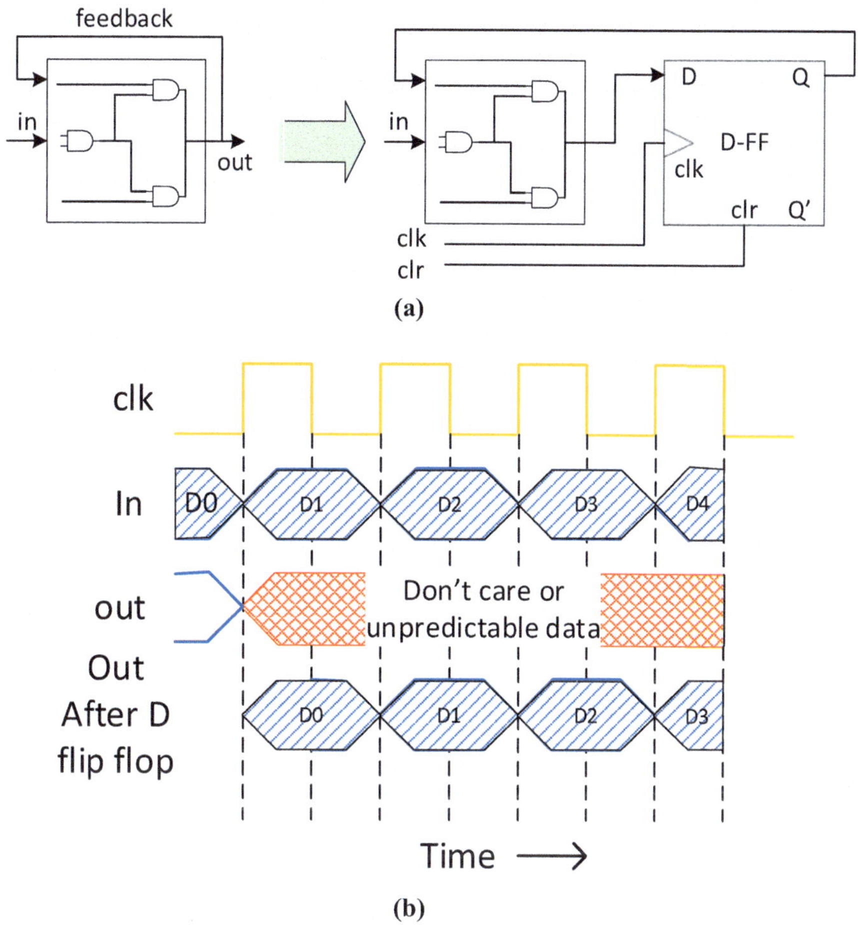

Fig. 15.13 **a** Combinational logic with feedback and after D flip-flop insertion, and **b** waveforms showing output data before and after D flip-flop insertion

4. **Generate single pulse delay using flip-flops and counters**

Single pulse delay is used as an example of push button like reset. To realize such delays, most of the designers use series of buffers. This creates a delay of $N * t_p$, where N is the number of buffers used and t_p is the gate propagation delay. The drawback of this method is that it is technology dependent. Whenever there is a change in technology, the value of t_p will change according to that. Hence, to overcome such a problem, delay is generated using flip-flops. Two D flip-flops are connected in cascade, and then an AND gate is used at the outputs of the D flip-flops. Figure 15.15 illustrates the two flip-flop method for generating single pulse delay.

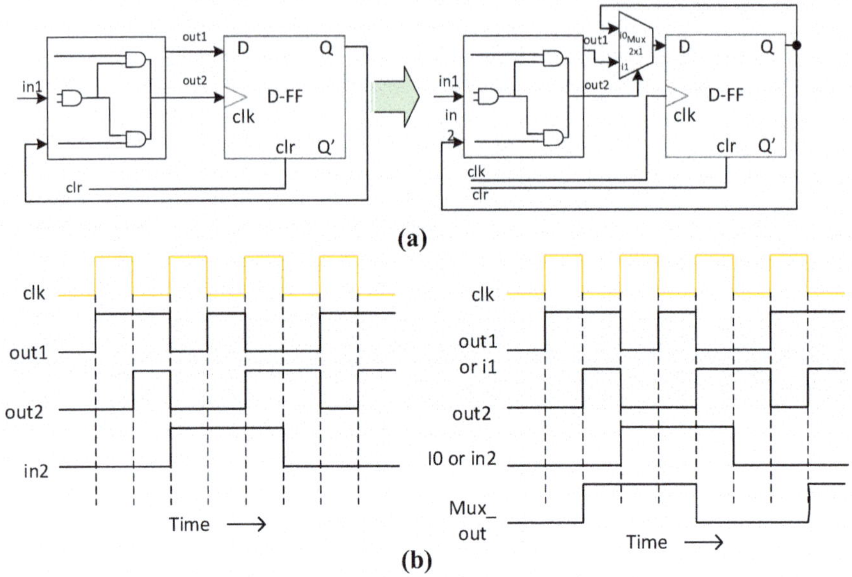

Fig. 15.14 **a** Use of 2 × 1 mux for clock gating removal and **b** timing diagram to check the output

5. **Use flip-flops at output stages of combinational logic to avoid glitches**

Glitches are seen frequently in digital circuits due to delays introduced by components like gates. Glitch can be defined as an unwanted pulse of very narrow width. Glitches can alter the function of the design if transmitted successfully to the next stage. It is recommended to remove the glitch before processing the next stage. To eliminate glitch, a designer should use a D flip-flop at the output stage of the combinational logic. Figure 15.16 illustrates the generation of glitch and its removal using a flip-flop. Considering a random input data 'a', 'b' and 'sel' is going to be processed through the design. Here, 'sel' is select signal. Assuming both inputs 'a' and 'b' as logic high, input 'sel' toggles after some time. When input 'sel' passes through NOT gate, because of propagation delay a glitch is observed at the output 'y'. But if D flip-flop is connected to output port, then only that data will be captured which appeared at the positive or negative edge depending upon the triggering of the flip-flop. Thus, avoiding the glitch and producing the fine expected output.

6. **Avoid setup and hold time violation**

Setup time can be defined the as the time required for a data to reach at the input of flip-flop and remain stable until the clock is edge triggered, whereas hold time is defined as the time for a data to remain stable after the clock is edge triggered. Whenever a flip-flop is used in a design, its setup and hold time are already specified. If input data does not follow the specified time, then it can be considered that either setup or hold time violation has occurred. Figure 15.17 shows an example where

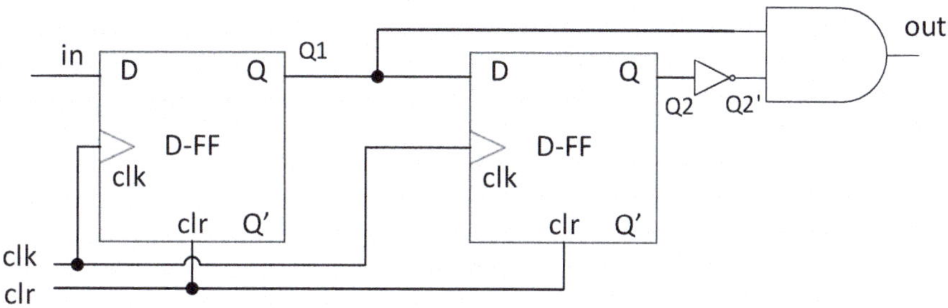

(a)

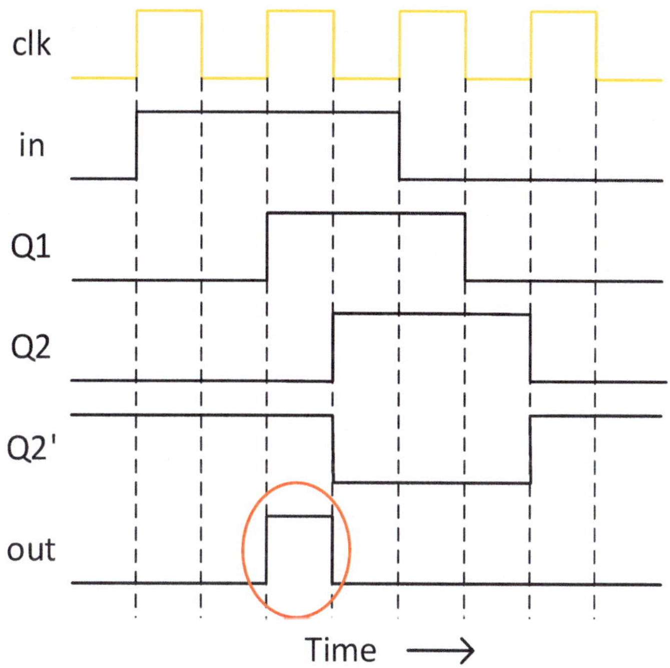

(b)

Fig. 15.15 **a** Single pulse delay generation using two flip-flops and **b** waveforms showing single pulse at 'out' port

output of one flip-flop 'FF1' is connected to clock input of the next flip-flop 'FF2'. Assume 'data2' to be asynchronous input which is initially low. At FF1, 'data1' is fed and assumed logic low. When positive edge of clock arrives at the 'FF1', input 'data1' is stored until the next positive edge clock. The output of 'FF1' is fed to clock input of the 'FF2'. When FF2 gets a positive edge clock input, the 'data2' must be stable according to the timing specifications mentioned for that flip-flop; otherwise, there will be hold time violation. But as shown in Fig. 15.17, 'data2' changes along with the positive edge of clock resulting in hold time violation. This could have been

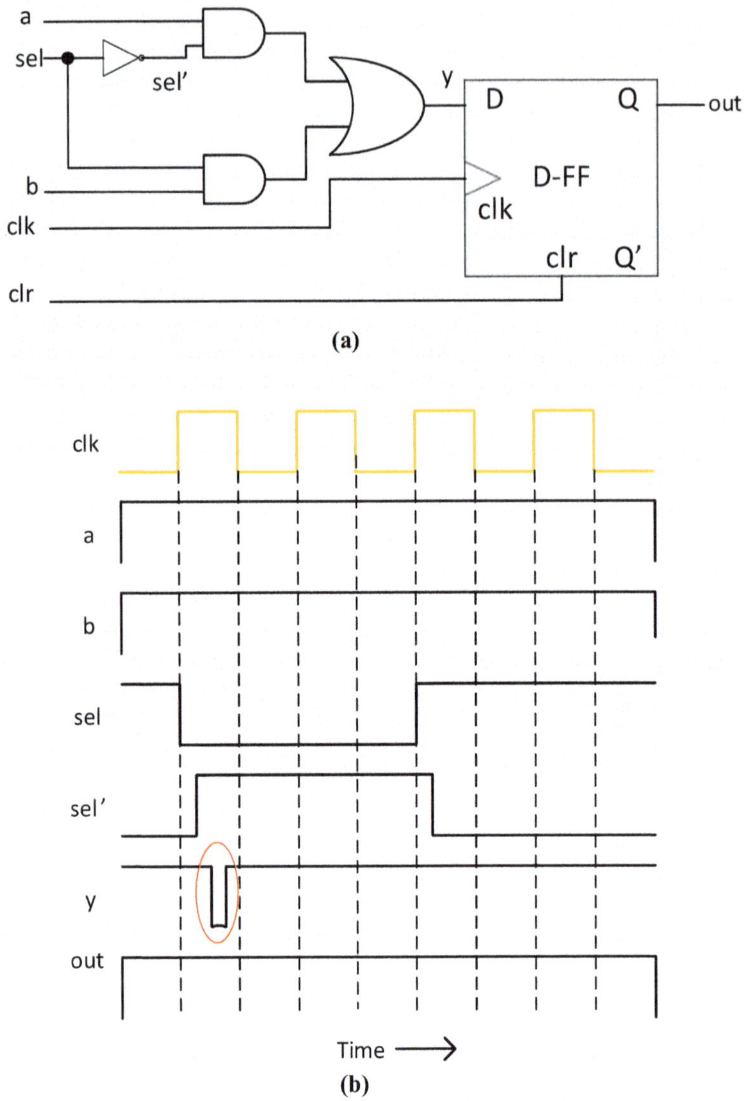

Fig. 15.16 **a** An example showing glitch generation and removal, and **b** waveform showing generated glitch (encircled red)

prevented if 'data2' had remained active low for some duration of time, fulfilling the timing specifications of the flip-flop used in the design (Salman et al. 2007).

7. **Avoid inferring latches in conditional statements**

In Verilog, whenever a logic depends upon another logic, the conditional statements are invoked. Conditional statements such as 'if-else-if' are used which results in

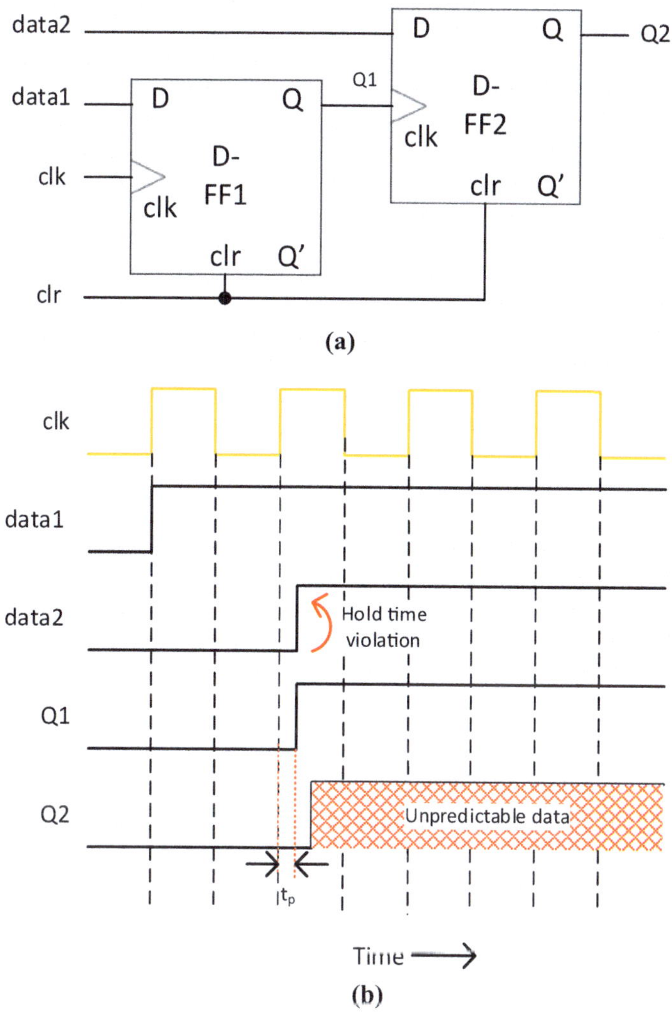

Fig. 15.17 **a** An example showing hold time violation and **b** timing diagram illustrating hold time violation due to change of data2 at rising edge of Q1

synthesis of multiplexers. In most of the cases, designers only mention if statement in the code and ignore the else statements. This is the main reason of latch inferring. The RTL code should contain all the values of conditions to prevent this problem. An example verilog code 1.2 where only if and else if statements are used is considered.

Figure 15.18 shows the RTL generated for the verilog code 1.2 with incomplete conditional statements inferring a latch. A modified code is written later with complete conditions, and RTL is shown in Fig. 15.19 for the same. It can be clearly seen in Fig. 15.19 that latch is removed, and multiplexers are generated. The latch is inferred because the synthesis EDA tool does not know what the output should

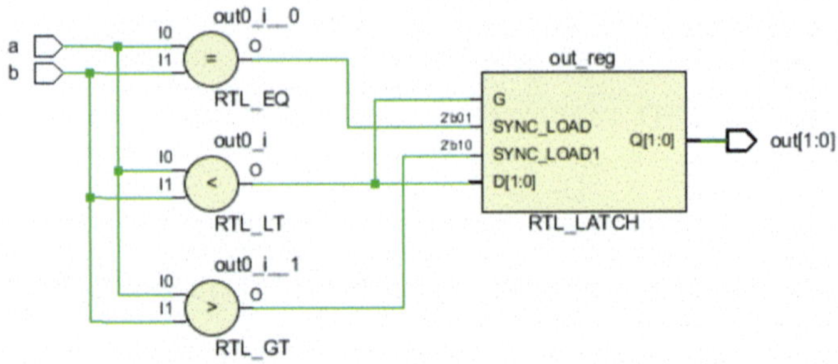

Fig. 15.18 RTL generated in Xilinx Vivado tool for incomplete conditions

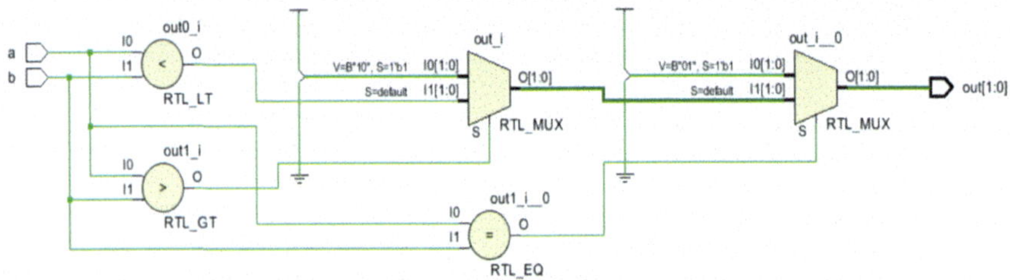

Fig. 15.19 RTL generated in XILINX Vivado for all conditions

be, when a certain condition which is not enlisted is triggered. Therefore, by default it keeps the previous value for that condition which is done only by the memory element, i.e., latch.

Verilog code 1.2

////////////////// Code for conditional block//////////////////

```
Module cond (a,b,out);
Input a,b;
Output reg [1:0] out;
always@(*) begin
  if(a==b)
     out=2'd1;
else if (a>b)
    out=2'd2;
else if (a<b)
    out=2'd3;
end
endmodule
```

///

//////////////////////////Code for modified conditional block/////////////////////////////

```
Module cond (a,b,out);
Input a,b;
Output reg [1:0] out;
always@(*) begin
  if(a==b)
     out=2'd1;
else if (a>b)
    out=2'd2;
else if (a<b)
    out=2'd3;
else
   out=2'd4;
end
endmodule
```

///

Various RTL coding strategies are highlighted to attain efficient VLSI designs. There are some more steps which basically focus on supported and unsupported Verilog syntax for RTL design. Syntax errors are easy to rectify; hence, those steps are not discussed. An ASIC or an FPGA design made using the above guidelines greatly boosts up the front-end flow and reduces post-synthesis errors.

15.4 Summary

This chapter presents the design flow for an FPGA or ASIC chip. Several design style methods have been discussed. FPGAs provide higher performance and time to market advantages if cost factor is not a constraint. In order to achieve higher area optimization and performance, full-custom design is the best candidate. The steps required to choose the design styles make it easier to identify what kind of design style is best suited for a specific application. However, there are a lot of factors considered which change the RTL to GDS-II flow of the design. The stages a designer needs to take care of to design an ASIC are discussed. It can be clearly established that RTL generation takes almost 70–80% of the design time when done manually. Subsequently, some of the do's and don'ts are explained along with their timing diagrams. These greatly help in understanding the ASIC design phenomenon. In each step, most of the problems are rectified using flip-flops. It can be concluded that flip-flops are really helpful to achieve synthesizable designs, as these prevent setup and hold time violations and also keep the design synchronized with clock. The various issues related to RTL design discussed in the present chapter shall be of immense use to researchers and the VLSI and ASIC designers.

Acknowledgements The technical resources and facilities of SMDP-C2SD Project of DeitY, Ministry of Electronics and Information Technology, GoI, New Delhi, India, awarded to NIT Hamirpur (HP) are duly acknowledged.

References

Barr K (2007) ASIC design in the silicon sandbox : a complete guide to building mixed-signal integrated circuits. McGraw-Hill

Cadence (2017) Innovus user guide

Chinedu OK, Genevera EC, Akinyele OO (2011) Hardware description language (HDL): an efficient approach to device independent designs for VLSI market segments. In: 3rd IEEE international conference on adaptive science and technology (ICAST 2011), pp 262–267. IEEE

Cong J, Liu Bin, Neuendorffer S et al (2011) High-level synthesis for FPGAs: from prototyping to deployment. IEEE Trans Comput Aided Des Integr Circuits Syst 30:473–491. https://doi.org/10.1109/TCAD.2011.2110592

Friedman EG (2001) Clock distribution networks in synchronous digital integrated circuits. Proc IEEE 89:665–692. https://doi.org/10.1109/5.929649

ITRS FEP9 (2013) Starting materials technology requirements. http://www.itrs2.net/2013-itrs.html. Accessed 1 Oct 2019

Mangassarian H, Le B, Veneris A (2014) Debugging RTL using structural dominance. IEEE Trans Comput Aided Des Integr Circuits Syst 33:153–166. https://doi.org/10.1109/TCAD.2013.2278491

Mentor Graphics (2016) ModelSim® user's manual. https://www.microsemi.com/document-portal/doc_download/136662-modelsim-me-10-5c-user-u-s-manual-for-libero-soc-v11-8. Accessed 1 Oct 2019

Ramachandran S (2007) RTL coding guidelines. Digital VLSI systems design. Springer, Netherlands, Dordrecht, pp 187–214

Rosenberg LM (1980) The evolution of design automation to meet the challanges of VLSI. In: Proceedings—design automation conference. association for computing machinery, pp 3–11

Salman E, Dasdan A, Taraporevala F et al (2007) Exploiting setup–hold-time interdependence in static timing analysis. IEEE Trans Comput Aided Des Integr Circuits Syst 26:1114–1125. https://doi.org/10.1109/TCAD.2006.885834

Sherwani NA (1993) Design styles. In: Algorithms for VLSI physical design automation. Springer, US, pp 16–20. https://doi.org/10.1007/978-1-4757-2219-2

Smith MJS (2008) Application-specific integrated circuits. Addison-Wesley

Wu A-Y, Liu KJR (1998) Algorithm-based low-power transform coding architectures: the multirate approach. IEEE Trans Very Large Scale Integr (VLSI) Syst 6:707–718. https://doi.org/10.1109/92.736144

Xilinx Vivado (2017) Vivado design suite user guide: hierarchical design. https://www.xilinx.com/support/documentation/sw_manuals/xilinx2017_3/ug893-vivado-ide.pdf. Accessed 1 Oct 2019

The manufacturer's authorised representative in the EU is Springer Nature Customer Service Centre GmbH, Europaplatz 3, 69115 Heidelberg, Germany. If you have any concerns regarding our products, please contact ProductSafety@springernature.com

Printed and bound by CPI Group (UK) Ltd, Croydon, CR0 4YY
30/09/2026
02225328-0006